Water Science & Technology

From Nutrient Remova

Water Science & Technology

From Nutrient Removal to Recovery

From Nutrient Removal to Recovery

Selected Proceedings of the IWA Conference "From Nutrient Removal to Recovery", held at Aquatech 2002, Amsterdam, The Netherlands, 2–4 October 2002

Issue Editor: Mark van Loosdrecht
Kluyverinstitute for Biotechnology, Technical University of Delft, The Netherlands

International Programme Committee
Mark van Loosdrecht (*Chairman*) The Netherlands
Glen Daigger USA
Mogens Henze Denmark
Jurg Keller Australia
Tove Larsen Switzerland
Piet Lens The Netherlands
Takashi Mino Japan
Hallvard Odegaard Norway
Ralf Otterpohl Germany
Cliff Randall USA
Hans Ruedi Siegrist Switzerland
Christopher Thornton France

Organizing Committee
Joost de Jong
Jaap van der Graaf
Roelof Kruize
Piet Lens
Mark van Loosdrecht
Paul Roeleveld
Christopher Thornton
Arie van der Vlies
Jaap Voorhoeve
Jac Wilsenach

British Library Cataloguing in Publication Data
A CIP catalogue record for this book is available from the British Library
ISBN 1 84339 447 2

Contents

vii Foreword

Concepts and opinions

1   **From waste treatment to integrated resource management**   J.A. Wilsenach, M. Maurer, T.A. Larsen and M.C.M. van Loosdrecht

11 **Potential societal and economic impacts of wastewater nutrient removal and recycling**   C.W. Randall

19 **Recycling of wastewater-derived phosphorus in Swedish agriculture – a proposal**   E. Kvarnström, C. Schönning, M. Carlsson-Reich, M. Gustafsson and E. Enocksson

27 **Exergy analysis of nutrient recovery processes**   D. Hellström

37 **Nutrients in urine: energetic aspects of removal and recovery**   M. Maurer, P. Schwegler and T.A. Larsen

47 **How farmers in Switzerland perceive fertilizers from recycled anthropogenic nutrients (urine)**   J. Lienert, M. Haller, A. Berner, M. Stauffacher and T.A. Larsen

57 **Investigating consumer attitudes towards the new technology of urine separation**   C. Pahl-Wostl, A. Schönborn, N. Willi, J. Muncke and T.A. Larsen

Optimisation of biological nutrient removal techniques

67 **The quest for sustainable nitrogen removal technologies**   A. Mulder

77 **A proposed sustainable BNR plant with the emphasis on recovery of COD and phosphate**   X.-D. Hao and M.C.M. van Loosdrecht

87 **Enhanced biological phosphorus removal process implemented in membrane bioreactors to improve phosphorous recovery and recycling**   B. Lesjean, R. Gnirss, C. Adam, M. Kraume and F. Luck

95 **Improved nutrient removal using *in situ* continuous on-line sensors with short response time**   P. Ingildsen and H. Wendelboe

Impact of source separated municipal wastes

103 **Impact of separate urine collection on wastewater treatment systems**   J. Wilsenach and M. van Loosdrecht

111 **Investigation of the effectiveness of source control sanitation concepts including pre-treatment with Rottebehaelter**   D.R. Gajurel, Z. Li and R. Otterpohl

119 **Nitrification and autotrophic denitrification of source-separated urine**   K.M. Udert, C. Fux, M. Münster, T.A. Larsen, H. Siegrist and W. Gujer

Recovery and recycling of minerals from sewage and agriculture

131 **Treatment of domestic sewage in a combined UASB/RBC system. Process optimization for irrigation purposes**   A. Tawfik, G. Zeeman, A. Klapwijk, W. Sanders, F. El-Gohary and G. Lettinga

139 **Optimization of phosphorus precipitation from swine manure slurries to enhance recovery**   R.T. Burns, L.B. Moody, I. Celen and J.R. Buchanan

147 **Selected requirements on a sustainable nutrient management**   C. Lampert

155 **Effect of lime stabilisation of enhanced biological phosphorus removal sludges on the phosphorus availability to plants**   D. Seyhan and A. Erdincler

163 **Development of a high-efficiency phosphorus recovery method using a fluidized-bed crystallized phosphorus removal system** K. Shimamura, T. Tanaka, Y. Miura and H. Ishikawa

171 **Removal and recovery of phosphate and ammonium as struvite from supernatant in anaerobic digestion** M. Yoshino, M. Yao, H. Tsuno and I. Somiya

179 **A new phosphate-selective sorbent for the Rem Nut® process. Laboratory investigation and field experience at a medium size wastewater treatment plant** D. Petruzzelli, L. De Florio, A. Dell'Erba, L. Liberti, M. Notarnicola and A.K. Sengupta

185 **Phosphate recovery from sewage sludge in combination with supercritical water oxidation** K. Stendahl and S. Jäfverström

Industrial waste removal and recovery

191 **High-strength nitrogen removal of opto-electronic industrial wastewater in membrane bioreactor – a pilot study** T.K. Chen, C.H. Ni, J.N. Chen and J. Lin

199 **Conversion of pollutants to fertilisers: ion exchange synthesis of potassium sulphate from acidic mine waters** D. Muraviev

207 **Carbon and nitrogen removal from tannery wastewater with a membrane bioreactor** A. Goltara, J. Martinez and R. Mendez

215 **Nitrogen removal from tannery wastewater by protein recovery** I. Kabdaşlı, T. Ölmez and O. Tünay

Foreword

In the last decades, nutrient removal has become more or less standard practice for wastewater treatment in the developed world. The best available technology was exhibited during the Aquatech Amsterdam conference in October 2002. Most of the exhibitions revealed a focus toward improvements within existing systems. Several conferences were organised to coincide with the exhibition, but one was aimed at presenting innovation and change.

Nitrogen, phosphorus and sulphur removal requires relative large amounts of resources (energy and chemicals). Most techniques currently applied do not allow for proper recovery of finite minerals, which are often disposed of. Wastewater and solid waste treatment can only be really sustainable if concepts and design for wastewater handling are based on an overall minimisation of resource consumption, in which energy consumption and nutrient recovery are taken into account. This implies adaptation of the presently available technologies as well as the complete re-design of wastewater and solid waste collection and treatment scenarios.

This conference "From nutrient removal to recovery" gave an overview of the present situation with relation to improved nutrient removal and/or recovery technologies. Nutrient recovery has applications in both industrial and municipal wastewater. Questions regarding small or large-scale treatment, central or local treatment, etc., which were discussed in a final forum, are not of primary concern. Various oral presentations and posters presented at the conference showed that nutrient recovery becomes feasible when different waste streams are not diluted, but handled and treated separately.

Mark van Loosdrecht
Chairman, Scientific Committee

From waste treatment to integrated resource management

J.A. Wilsenach*, M. Maurer**, T.A. Larsen** and M.C.M. van Loosdrecht*

* Delft University of Technology, Julianalaan 67, 2326 BC Delft, The Netherlands
(E-mail: *J.A.Wilsenach@tnw.tudelft.nl*)
** EAWAG, Überlandstrasse 133, CH-8600 Dübendorf, Switzerland

Abstract Wastewater treatment was primarily implemented to enhance urban hygiene. Treatment methods were improved to ensure environmental protection by nutrient removal processes. In this way, energy is consumed and resources like potentially useful minerals and drinking water are disposed of. An integrated management of assets, including drinking water, surface water, energy and nutrients would be required to make wastewater management more sustainable. Exergy analysis provides a good method to quantify different resources, e.g. utilisable energy and nutrients. Dilution is never a solution for pollution. Waste streams should best be managed to prevent dilution of resources. Wastewater and sanitation are not intrinsically linked. Source separation technology seems to be the most promising concept to realise a major breakthrough in wastewater treatment. Research on unit processes, such as struvite recovery and treatment of ammonium rich streams, also shows promising results. In many cases, nutrient removal and recovery can be combined, with possibilities for a gradual change from one system to another.
Keywords Energy; exergy; minerals; sanitation; sustainability; wastewater

Introduction

Water is used as a medium for waste transportation. The association between sanitation and wastewater results from the historic development of urban hygiene. After the discovery of waterborne diseases, faeces was removed from cities with rain water sewers, which already existed in many cases. This resulted in wastewater treatment to protect both downstream users and surface waters. From this point of view, modern centralised wastewater treatment is very effective. In Europe and North America, water borne diseases are not a significant cause of illness or death any more. Nutrient removal has also become standard technology in wastewater treatment in the last decade.

The responsibility for ensuring safe and good quality wastewater effluent usually rests with an organisation, such as a municipality or water board. The degree of treatment, control and test procedures are agreed upon, standardised and enforced on a national scale. Centralisation of treatment works have until now ensured their relative success. Furthermore, it is generally believed that high-tech biological treatment processes need a reasonable scale. Operation, control and maintenance of wastewater treatment plants are specialised professions. Sludge treatment and incineration also require good control structures. It is still widely understood that the scale of large centralised treatment plants makes them more affordable.

Nevertheless, carbon, nitrogen, phosphorus and sulphur removal requires relatively large amounts of resources (energy and chemicals). Potentially useful minerals are usually disposed of. Removal technologies have to be changed to make wastewater management more affordable and sustainable in terms of nutrient management. This will involve application of presently available technologies as well as completely new concepts in urban water and solid waste management. This has not only technical but also social implications. We report on the conference "From nutrient removal to recovery" where state-of-the-art technology and new concepts were demonstrated.

Integrated resource management

The important resources involved in wastewater are water, energy and minerals. Apart from being used as transport medium, clean water is the main source of animal and plant life, as well as an important habitat. Water is therefore historically the most important resource, with primary emphasis on purification (Larsen and Gujer, 1997). After solving the problems with urban hygiene, the importance once placed thereon gradually shifted towards environmental protection. In some cases nowadays, "sustainable" wastewater treatment seems to be limited to ever-increasing effluent quality standards. Different life cycle assessments show that current investments in wastewater treatment are justified by the improved effluent quality (e.g. Roeleveld *et al*., 1997). This is true in comparison to other polluters in developed countries. However, there is doubt about the validity of the LCA methodology and general statements on environmental impact deduced from such assessments (e.g. Ayres, 1995). "Environment" also means many different things to different people or cultures. Effluent quality can not be the only criterion of sustainable wastewater treatment. Apart from protecting the water resources, future developments must also consider all other resources, including capital, energy and nutrients.

Energy is limited and its use has become more important in the last decade. Furthermore, energy production causes pollution. Waste in the form of COD contains potential energy (e.g. through methane production). Energy is in fact consumed in wastewater treatment to destroy potential energy. Currently, around 5W/p is *consumed* in wastewater treatment, mostly through oxidation. If methane were produced with all available BOD in municipal wastewater, around 4W/p could be *generated* continuously. When this is put in perspective of the total energy consumption in Western Europe of around 5kW/p, it might seem insignificant. However, future scenarios could change the importance of energy consumption in wastewater treatment. Technology involved in all spheres of society, including wastewater management, has to be improved or replaced to realise more sustainable societies as a whole.

Accessibility of energy and sophisticated technology made nutrients widely available for agriculture. In wealthy societies, nutrients from human excretion have therefore lost their value and are now being treated as waste. This leads to high costs for wastewater treatment and causes natural resources to be used faster than their natural recovery rate. A consequence of this approach, is that nutrients might not be available anywhere and at anytime. Evidence for this can already be seen in the problems that developing countries face when adopting the Western approach to sanitation and wastewater treatment (Ujang and Buckley, 2002). An important requirement for sustainable wastewater management is therefore that it should be feasible under poor economic conditions.

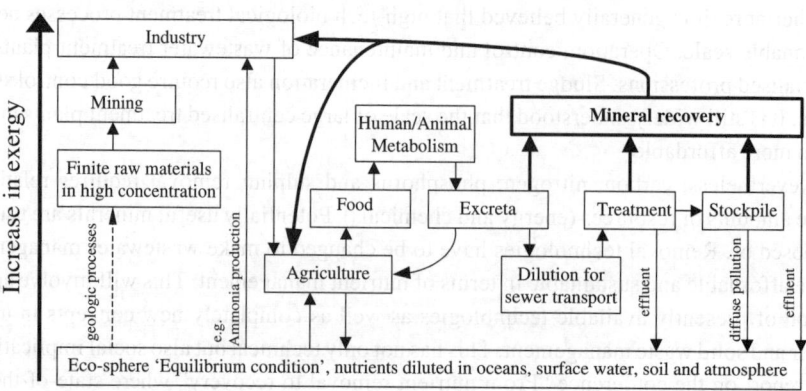

Figure 1 Relation between industry, agriculture, nutrients and sanitation

Although nutrient *removal* is an important aspect of modern wastewater treatment, removal techniques currently applied do not allow for proper *recovery* of these nutrients (minerals). The most important minerals are considered to be nitrogen (N) and phosphorus (P). However, other minerals such as potassium (K) and sulphur (S) should also be taken into account. Recovery techniques are not necessarily limited to end-of-pipe solutions. Complete mineral cycles must be integrated, from mineral production to final use. Figure 1 shows different facets involved in mineral and nutrient cycles: Raw materials are usually mined and processed in industry to produce industrial fertiliser, which is spread out on farmland. Agricultural products (food) contain minerals, which is taken up in the metabolism of animals and humans. Most of these minerals are excreted via faeces and urine. In most urban societies, human excreta are removed through sewers. Minerals are diluted in sewers by a factor of more than 100 and have to be removed in treatment plants to protect surface waters. If global consumption occurs at a higher rate than natural recovery (or anthropogenic removal) processes, dilute minerals accumulate in the eco-sphere (or stockpiles).

Various routes are available for recovering minerals from waste. The first is obvious: prevention of dilution. Urine separation is a means of direct recovery, as 80% of the nitrogen, 50% of the phosphorus and 70% of potassium in municipal wastewater originate from urine. Lienert *et al.* (2003) show that (Swiss) farmers are in general willing to accept a urine-based fertiliser, provided it fulfils the function of industrial fertiliser and does not involve smell, inconvenience or risk. Phosphorus can also be recovered from liquid solutions with various techniques requiring resources in the process.

Fossil fuels and phosphate rock are theoretically renewable resources, but the rates of these natural renewal processes are on a geologic time scale. Recovery of finite minerals is technically possible, even from dilute sources such as sea water, but this would be far too energy intensive. Good quality ore for phosphorus, potassium and sulphur are all limited. Furthermore, production of these minerals co-produces waste, e.g. 1 kg of P produced leads to 2 kg gypsum, contaminated with heavy metals and radioactive elements. To give the phosphate industry a sustainable future, phosphate would have to be recovered and recycled (Driver *et al.*, 1999). Recovery and recycling is currently rather expensive and has to have a political drive to be realised. The Swedish EPA for example proposes that 25% of P in wastewater be recycled to agriculture in 2015 and 40% in 2025 (Kvarnström *et al.*, 2003).

The natural resources for nitrogen are extensive and universally accessible. However, in order to be accessible to plants and most micro-organisms, atmospheric nitrogen has to be converted to ammonia or nitrate; either industrially or naturally by N-fixing organisms. The amount of ammonia technically produced is of the same magnitude as the natural nitrogen fixation, or even greater. Industrial processes require 35 to 50 MJ kg_N^{-1} in the form of fossil fuel for energy supply (Maurer *et al.*, 2002).

In densely populated areas, our aquatic environment has to be protected from excessive nitrogen loads. This is conventionally done with nitrification/denitrification processes, also consuming much energy. The benefit of removal or recovery must not be annulled by the demands of the removal/recovery process. We need to find shortcuts in the mineral cycles, requiring the smallest amount of resources to recycle available resources (bold lines in Figure 1).

Three different "levels" of research can be identified in attempts to find these shortcuts, as shown in Figure 2.

Firstly, there is an integration of different systems involving resources (of which wastewater management is only one part) to consider complete cycles. Then there is work being done on improving the efficiency of processes, within an existing paradigm or bridging dif-

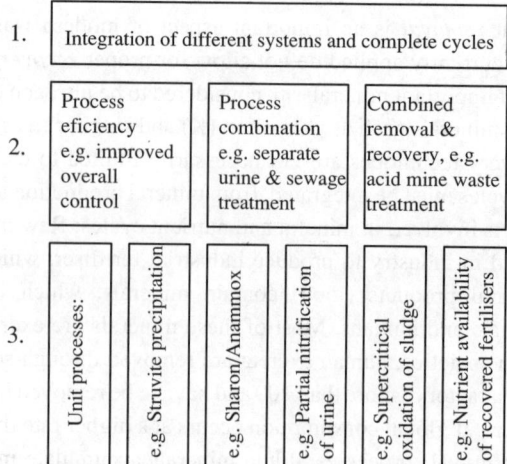

Figure 2 Levels of research activity

ferent paradigms. Lastly there are improvements to unit recovery operations and novel techniques with practical or experimental results. These three different "levels" are discussed in the following sections.

Systems approach

Larsen and Gujer (1997) argue that improvements to urban water management have to fulfil certain boundary conditions (e.g. urban hygiene, flood prevention, recreation), but improvements can be found outside of existing technological paradigms, such as sewers leading to end-of-pipe treatment. In other words, research shouldn't be limited to improving existing systems' efficiency. A real breakthrough in improving urban waste and water management would most likely result from a complete redesign of the system. Separate collection of different waste streams at the source is the most important alternative to the existing system. The two best known examples of source separation are urine separation and vacuum transport of blackwater. Since most of the wastewater nutrients are contained in urine and faeces, source separation of toilet wastewater is an obvious measure for recycling of nutrients (Larsen *et al.*, 1997).

An objective method for comparison of alternatives is based on thermodynamics. The second law of thermodynamics states that the quality of energy can be destroyed. The utilizable energy of a system can be defined in terms of its *exergy* (derived from *ex*tractable en*ergy*). While energy remains constant within systems, the *exergy* decreases during all physical/chemical processes. Minerals also represent energy and can be equated with other forms of energy in terms of their *exergy* content (Szargut *et al.*, 1988). In more sustainable systems, the loss of exergy should be minimised.

The different facets shown in Figure 1 can also be interpreted to represent relative levels of exergy. This means that dilute minerals in the "equilibrium condition" represents the lowest exergy level. On the other hand, concentrated raw materials have higher exergy values. Industrial processes can concentrate minerals available in the "equilibrium condition", but this requires consumption of more exergy. Dilution in sewers destroys some of the exergy. Wastewater treatment again requires consumption of more exergy.

Hellström (2003) demonstrates the overall efficiency of different strategies in wastewater treatment by means of an exergy analysis. Nutrient recovery could be achieved with a lower overall exergy loss in source separation systems, as compared to conventional treatment systems. The increased exergy efficiency is due to at least two aspects: exergy contained within the minerals, as well as a decrease in the exergy demand in treatment.

Hellström assumes that urine can be used directly as fertiliser. Nutrient recovery by other recovery processes (such as reverse osmosis) could prove to be much less efficient, due to a higher overall exergy demand.

Another example of this approach is the evaluation of nitrogen recovery via different routes. Nitrogen is abundantly present in the atmosphere and only energy is required for industrial ammonia fertiliser production in the Haber/Bosch process. Nitrogen removal is therefore an indirect way of ammonia recovery. The Sharon/Anammox process is the most efficient technique of biological nitrogen removal (Van Dongen et al., 2001). Ammonia can also be recovered directly in various techniques. The best available techniques involved in these alternative routes are illustrated in Figure 2.

Figure 3 shows that direct recovery can also be less sustainable (e.g. air stripping), whereas other routes are clearly more viable (thermal volume reduction). Maurer et al. (2003) show that the energy required for the indirect recovery (Sharon/Anammox and production via the Haber/Bosch processes) is 60% higher than energy required for direct recovery via thermal volume reduction of urine, but 60% lower than for air stripping. The notion of "energy" used for this comparison includes the production of chemicals as well as primary energy required for electricity production, so that it in fact approaches an exergy comparison.

In the current debate on nutrient recovery, three different approaches can be distinguished:
1. The "conventional" approach (N: nitrification/denitrification, P: direct application of sludge or extraction of P from treatment plants/sludge).
2. Direct use of urine/faeces in agriculture.
3. Production of a urine-based fertilizer, including volume reduction, removal of micropollutants and attention to specific nutrient demand in agriculture.

Besides exergy, other important issues form part of any discussion on different recovery techniques of nutrients from waste. Theses issues determine the strengths and weaknesses of the different approaches:
- Micropollutants may be a concern in (2), but by definition not in (3). In (1), micropollutants may be a concern for P, but not for N.
- Industrial fertiliser composition is designed on specific plant needs and maximum uptake rates, which could differ considerably from urine, black water, or sewage sludge. Limitation of one nutrient and oversupply of another could again lead to diffuse pollution. This is most significantly a problem for (2).
- Animal manure is a problem in many places with intensive bio-industries and is potentially a greater and more concentrated source of nutrients than municipal wastewater. Resources in wastewater (and its potential recovery) can not be viewed in isolation. For (2), this is a problem due to the limited possibilities of transport. The same is true for the recycling of P via sewage sludge (1). For the other options, transport is possible.

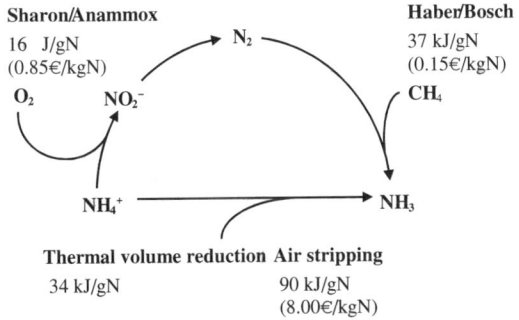

Figure 3 Nitrogen removal is low-cost and indirect ammonia recovery

- Costs of recovery is at the moment much higher than efficient removal (8 times for the case of air stripping of ammonia in Figure 2). Although cost differences and decisions could change in future scenarios, current costs govern water boards, municipalities and agriculture. Furthermore, the cost to pay for any process has to be "earned" with other economic activities. In a consumer driven economy, higher costs can often be associated with a higher exergy consumption.

To summarise, exergy considerations support approach (2), whereas today's economic reality favours approach (1). Experience with new technology, probably based on transition scenarios (see below) would be necessary in order to advance approach (3), which combines a number of advantages.

The strength of the systems approach lies in not only taking one aspect like nutrient recycling into account, but in considering the entire system. Besides nutrient recycling, water pollution control remains of paramount interest. With separate treatment of urine and/or faeces, a number of problems like eutrophication and oxygen depletion could be dramatically improved. Source separating measures can also be adapted to different contexts. Low-tech versions are relevant in rural areas, whereas more high-tech variations can be adapted to urban conditions and even integrated into the existing conventional system (Larsen and Gujer, 1997; Rauch et al., 2003). Source separation is also interesting with regard to household water polluters other than the toilet. If effluent from toilets, washing machines and dishwashers are collected separately, 85% of the COD, N and P in municipal wastewater can be addressed with integrated on-site, or in-pipe, high-tech technologies (Larsen and Gujer, 2001). In non-arid climates, cities will probably still need a drainage system, but simple treatment plants would be sufficient for polishing grey water.

The transition phase from one paradigm to another is also important. Localised treatment applications could first be targeted at hospitals (including treatment of pharmaceutical residues and hormones), public buildings such as airports, shopping areas, sport stadiums (places with high human "strike rate") or office buildings (integrating urban irrigation, landscaping and fertilisation). Innovative use of ideas proposed for new systems can already improve existing systems.

Improvement and optimisation of the conventional treatment system

Sewers may remain an efficient transport method of waste in densely populated urban environments. We expect that in the near future, sewers with centralised treatment plants will still be the most common way of sanitation and waste management. Efforts to improve the system are therefore justified, although this also enforces the system. This is also true for arid regions, provided that wastewater is integrated with irrigation. A process such as the combined upflow anaerobic sludge bed/rotating biological contacter (UASB/RBC) can be used for primary treatment. This process produces an effluent with partial nitrification and *E. coli* removal, suitable for irrigation in parts of the world with less stringent regulations (Tawfik et al., 2003). This is a simple process, requiring little resource input and recycling some nutrients to agriculture.

Some problems prevent direct use of sewage sludge or manure. Where heavy metal and micropollutant content are of no concern, sewage sludge could be directly used as fertiliser even though it is not very efficient. Sewage sludge can only recover a maximum of 30% of the nutrients available in wastewater. Sludge transportation is also a very inefficient (and uneconomical) way to transport nutrients. One should also keep in mind that cities import food from areas outside their own direct agricultural region. Food import/export is a global phenomenon, just as industrial fertiliser import/export is. This obviously limits the direct application of nutrients recovered from waste. Space around cities is to a large extent also not used for agriculture, but rather for industries, transport, recreation, etc.

Over-application of sewage sludge or manure leads to build up of minerals on farmland, which leads to diffuse pollution. The plant availability of nutrients in sewage sludge is also fairly uncertain. Seyhan *et al.* (2003) show that lime stabilised sludge performed well in comparison to triple super phosphate in pot experiments. Sludge application based on phosphorus needs could prevent the over application of any material.

A way of solving the transport problem is to concentrate the nutrients at the treatment plant, which makes truck transport more feasible. Struvite (magnesium ammonium phosphate) is one such product, which can be produced with relatively simple technology. Although it is generally known that struvite is a good slow release fertiliser, it's nutritional value for different crops is unknown. The N:P:K ratio in struvite available to plants still has to be researched (Burns *et al.*, 2003).

Phosphorus accumulating organisms present a low-cost low-energy option for phosphorus concentration from dilute wastewater. Lesjean *et al.* (2003) show how current centralised biological nutrient removal plants could be further optimised with a membrane bio-reactor to recover high amounts of phosphorus with sewage sludge. Up to 99% phosphorus removal was achieved with relatively low sludge production.

Hao and Van Loosdrecht (2003) evaluate a system for optimal use of the resources in wastewater. This process removes a part of the ammonium load via the CANON process (based on the principle of the Sharon/Anammox), uses COD for methane production and recovers phosphorus as struvite. Wilsenach and Van Loosdrecht (2003) show that even partial urine separation could improve nitrogen removal and phosphorus recovery significantly on a centralised scale. This holds for both existing and new treatment concepts, where a main advantage is a net energy production from increased methane production. The immediate benefits of separate urine collection for present wastewater treatment plants provide a bridge between the existing system and possible future systems, such as complete urine separation.

Removal and recovery of minerals are not necessarily fundamentally different. Acid mine drainage is one of the most serious pollutants from mining. Muraviev (2003) demonstrates an ion exchanger to extract and concentrate sulphate from acidic mine wastes to produce K_2SO_4 fertiliser. Addition of KOH to the waste also removed metal ions and increased the pH. This is a perfect example of combining mineral removal and mineral recovery in treatment.

Unit processes for recovery and removal of nutrients

Within existing systems, parts of the system are often a bottleneck for further improvement. Improvements to unit processes are mostly focussed on removal/recovery of nutrients in higher concentrations than municipal wastewater. The capabilities of the phosphate industry to process recovered P is limited because they are set up for utilisation of large quantities of calcium phosphate. Calcium phosphate recovery from liquid waste is technically possible, but might not be the most efficient way. Wastewater treatment boards seem to chose the route of struvite recovery, being a cheap and simple process and requiring less exergy. In Japan, struvite recovery from central treatment plants is becoming more profitable. Shimamura *et al.* (2003) describe a two tank fluidised bed reactor for struvite recovery from anaerobic digester supernatant. Yoshino *et al.* (2003) show that high struvite production is also possible from a similar effluent, but using a continuously stirred tank reactor. Struvite in Japan is used for the fertilisation of rice. If micropollutants (or the perception thereof) are a serious concern, struvite could possibly be used in agriculture outside food production, e.g. for the flower industry, animal feed production, plants used for starch production, etc. Although these might be small markets, it is not necessarily a disadvantage. The availability of nutrients from wastewater is also relatively small and between 10–30% of the total nutrient flow in many societies. Re-use of sewage sludge presents many problems, such as heavy metal content or organic pollutants. One solution is that of phosphorus recovery from sewage sludge, with supercritical water (Stendahl and Jäfverström, 2003). This is impressive technology, removing all organic matter from sludge and allowing phosphorus recovery. Although it is claimed that energy consumption is similar to sludge incineration, the technique would probably have limited application. Capital and operational costs are high and some technical problems could be expected, e.g. corrosion. Many regions will simply apply treated sludge directly to land.

Treatment of source separated urine could make it a good fertiliser. Udert *et al.* (2003) show that oxidation and partial nitrification of urine reduces the pH sufficiently to prevent ammonia evaporation. This is an interesting option, and hopefully more work will be done to compare the fertilising potential of treated urine with that of industrial fertiliser or struvite.

Conclusion

Dilution is never a solution for pollution. In fact, dilution destroys exergy and makes the treatment of wastewater costly. Waste streams must therefore be managed in ways that keep them as concentrated as possible. Concentrated streams also enable easier recovery of energy and minerals.

Another important aspect is the fact that sanitation is not intrinsically linked to sewer systems and end-of-pipe wastewater treatment. They proved to be an efficient and powerful solution, however they are also costly and contain many severe disadvantages.

Wastewater engineers are solving problems created elsewhere. All societies (wealthy and developing) should be made conscious of the fact that wealthy consumption patterns are not sustainable and that technology alone can not solve all technical problems. Highly expensive removal/recovery techniques might not be sustainable. Solutions have to be found with respect to the integrated system: E.g. sanitation ensures hygiene and comfort, while waste treatment protects water bodies and enables recycling of resources. The wastewater engineer of the future should rather be a resource engineer, concerned with both water management and minimisation of exergy losses.

The conference showed that many developments are taking place related to making waste and water management more sustainable. More innovative concepts are clearly needed. Concepts based on source separation and separate handling/treatment are expected to create the most important breakthroughs in water and waste treatment. Such concepts are now becoming feasible. It is also clear that existing wastewater treatment plants would benefit from partial source separation. This provides good opportunities for a gradual change in the systems.

References

Ayres, R.U. (1995). Life cycle analysis – A critique. *Resources conservation and recycling*, **14**(3–4), 199–223.

Boers, P.C.M. (1996). Nutrient emissions from agriculture in The Netherlands, causes and remedies. *Water Science and Technology*, **33**(4–5), 183–189.

Burns, R.T., Moody, L.B., Celen, I. and Buchanan, J.R. (2003). Optimization of phosphorus precipitation from swine manure slurries to enhance recovery. *Water Science and Technology*, **48**(1) 139–146 (this issue).

Driver, J., Lijmbach, D. and Steen, I. (1999). Why recover phosphorus for recycling and how? *Environmental Technology*, **20**(7), 651–662.

Gajurel, D.R., Li, Z. and Otterpohl, R. (2003). Investigation of the effectiveness of source control sanitation concepts including pre-treatment with Rottebehaelter. *Water Science and Technology*, **48**(1) 111–118 (this issue).

Hao, X.-D. and van Loosdrecht, M.C.M. (2003). A proposed sustainable BNR plant with the emphasis on recovery of COD and phosphate. *Water Science and Technology*, **48**(1) 77–85 (this issue).

Hellström, D. (2003). Exergy analysis of nutrient recovery processes. *Water Science and Technology*, **48**(1) 27–36 (this issue).

Kvarnström, E., Schönning, C., Carlsson-Reich, M., Gustafsonand Enocksson, E. (2003). Recycling of wastewater-derived phosphorus in Swedish agriculture – a proposal. *Water Science and Technology*, **48**(1) 19–25 (this issue).

Larsen, T.A. and Gujer, W. (2003). Separate management of anthropogenic nutrient solutions (human urine). *Water Science and Technology*, **34**(3–4), 87–94.

Larsen, T.A. and Gujer, W. (1997). 'The concept of sustainable urban water management', *Water Science and Technology*, **35**(9), 3–10.

Larsen, T.A. and Gujer, W. (2001). Waste design and source control lead to flexibility in wastewater management. *Water Science and Technology*, **43**(5), 309–318.

Lesjean, B., Gnirss, R., Adam, C., Kraume, M. and Luck, F. (2003). Enhanced biological phosphorus removal process implemented in membrane bioreactors to improve phosphorous recovery and recycling. *Water Science and Technology*, **48**(1) 87–94 (this issue).

Lienert, J., Haller, M., Berner, A., Stauffacher, M. and Larsen, T.A (2003). How farmers in Switzerland perceive fertilizers from recycled anthropogenic nutrients (urine) *Water Science and Technology*, **48**(1) 47–56 (this issue).

Maurer, M., Muncke, J. and Larsen, T. (2002). Technologies for nitrogen recovery and reuse. In: Lens, P., Hulshoff Pol, L., Wilderer, P. and Asano, T. (2002). *Water and Resource Recovery in Industry*, 491–510. IWA Publishing, ISBN-1843390051.

Maurer, M., Schwegler, P. and Larsen, T.A. (2003). Nutrients in urine: Energetical aspects of removal and recovery. *Water Science and Technology*, **48**(1) 37–46 (this issue).

Muraviev, D. (2003). Conversion of pollutants to fertilisers: Ion exchange synthesis of potassium sulphate from acidic mine waters. *Water Science and Technology*, **48**(1) 199–206 (this issue).

Pahl-Wostl, C., Schönborn, A., Willi, N., Muncke, J. and Larsen, T.A. (2003). Investigating consumer attitudes towards the new technology of urine separation. *Water Science and Technology*, **48**(1) 57–65 (this issue).

Rauch, W., Brockmann, D., Peters, I., Larsen, T.A. and Gujer, W. (2003). Combining urine separation with waste design: an analysis using a stochastic model for urine production. *Water Research*, **37**, 681–689.

Roeleveld, P.J., Klapwijk, A., Eggels, P.G., Rulkens, W.H. and van Starkenburg, W. (1997). Sustainability of municipal wastewater treatment. *Water Science and Technology*, **35**(10), 221–228.

Shimamura, K., Tanaka, T., Miura, Y. and Ishikawa, H. (2003). Development of a high-efficiency phosphorus recovery method using a fluidized-bed crystallized phosphorus removal system. *Water Science and Technology*, **48**(1) 163–170 (this issue).

Stendahl, K. and Jäfverström, S. (2003). Phosphate recovery from sewage sludge in combination with supercritical water oxidation. *Water Science and Technology*, **48**(1) 185–191 (this issue).

Szargut, J., Morris, D.R. and Steward, F.R. (1988). *Exergy Analysis of Thermal, Chemical, and Metallurgical Processes*. Hemisphere – New York, ISBN 0-89116-574-6.

Tawfik, A., Zeeman, G., Klapwijk, A., Sanders, W., El-Gohary, F. and Lettinga, G. (2003). Treatment of domestic sewage in a combined USAB/RBC system: Process optimization for irrigation purposes. *Water Science and Technology*, **48**(1) 131–138 (this issue).

Udert, K.M., Fux, C., Münster, M., Larsen, T.A., Siegrist, H. and Gujer, W. (2003). Nitrification and autotrophic denitrification of source-separated urine, *Water Science and Technology*, **48**(1) 119–130 (this issue).

Ujang, Z. and Buckley, C. (2002). Water and wastewate in developing countries: present reality and strategy for the future, *Water Science and Technology*, **46**(9), 1–9.

Van Dongen, U., Jetten, M.S.M. and van Loosdrecht, M.C.M. (2001). The SHARON-Anammox process for treatment of ammonium rich wastewater. *Water Science and Technology*, **44**(1), 153–160.

Wilsenach, J.A. and van Loosdrecht, M.C.M. (2003). Impact of separate urine collection on wastewater treatment systems. *Water Science and Technology*, **48**(1) 103–110 (this issue).

Yoshino, M., Yao, M., Tsuno, H. and Somiya, I. (2003). Removal and recovery of phosphate and ammonium as struvite from supernatant in anaerobic digestion. *Water Science and Technology*, **48**(1) 171–178 (this issue).

Potential societal and economic impacts of wastewater nutrient removal and recycling

C.W. Randall

Environmental Engineering Rm. 418 Durham Hall, Virginia Tech, Blacksburg, VA 24061-0246, USA

Abstract Because adequate nutrient controls were not established when there were past opportunities to do so, nutrient pollution of estuaries and coastal waters has resulted in the impairment of ecosystems and major reductions or collapse of fisheries at numerous sites around the world, resulting in major economical and societal impacts. The root of the problem is that the political policies and processes have permitted municipalities, developers, industries and farmers to expand and operate without paying the full cost of their activities, and this has been done at the expense of those who rely on the productivity and recreational value of our estuarine and coastal waters. Some governments have developed remedial nutrient control programs, but most of them have been under funded and inadequately enforced, resulting in small increments of progress that tend to be lost because of inadequate land use and immigration controls. It is believed that nutrient recovery and controlled reuse can provide a major tool for the control of nutrient pollution and should be widely implemented. Plans are currently being developed to promote widespread use of nutrient recovery and reuse in the Chesapeake Bay region of the USA. An example of phosphorus reuse is presented.

Keywords Economic impacts; ecosystem impacts; estuarine and coastal fisheries; nutrient pollution; nutrient recovery and reuse; societal impacts

Introduction

Shortly after passage of the Clean Water Act of 1972, the U.S. Environmental Protection Agency (EPA) developed a construction grants program that did not include nutrient removal as a wastewater treatment requirement. The political decision also was made that nutrient pollution generated by farmers could not be regulated, even though the use of manufactured fertilizers was rapidly increasing at that time. The long term results of political decisions made and policies established by EPA during the 1970s are that a large majority of the estuaries in the USA are excessively eutrophic and this has resulted in deterioration of water quality, including the creation of large "dead zones" of low dissolved oxygen (DO) during the growing season, and a decline in fisheries (National Ocean Service, 2000). Notable examples of highly eutrophic estuarine USA waters and impacted fisheries are the Chesapeake Bay, the Albemarle and Pamlico Sounds of North Carolina, New York Bight, Long Island Sound near New York City, Mobile Bay, Alabama, and Tampa Bay, Florida. Furthermore, there is a large hypoxic (< 2 mg/L DO) area on the Louisiana Shelf of the Gulf of Mexico that seasonally exceeds the combined areas of the states of Rhode Island and Connecticut (Figure 1). The nutrient loads carried into the Gulf by the Mississippi and Atchafalaya Rivers have caused this hypoxic area, and it has grown in size over the past decade, from 11,000 km^2 in 1992 to 17,500 km^2 in 1993 to more than 18,000 km^2 in 1995.

The low DO area of the Louisiana Shelf of the Northern Gulf of Mexico is not exceptional. The DO conditions of many major coastal ecosystems around the world have been adversely affected through the process of eutrophication (Diaz and Solow, 1999), with annual summertime hypoxia the most common form of low DO event (Diaz and Rosenberg, 1995). A few of these systems are listed in Table 1, with comparison of the hypoxia-related ecological and economic effects. It can be seen that the degree of obvious ecological and economic effects of the combined problems of eutrophication and hypoxia

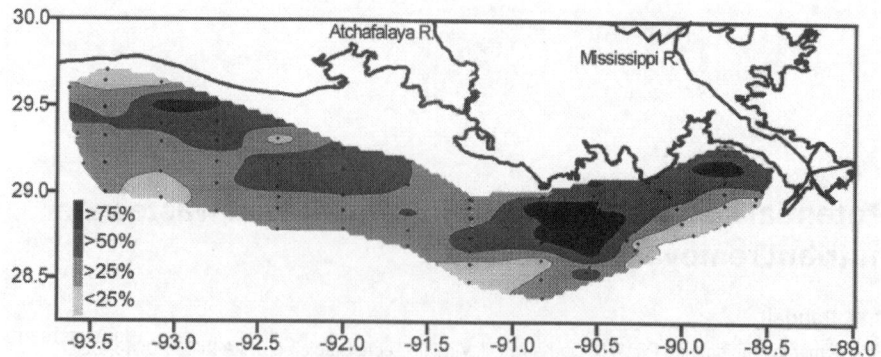

Figure 1 Distribution of frequency of occurrence of mid-summer hypoxia from 1985 to 1997 (data from Rabalais, Turner, and Wiseman hypoxia monitoring cruises). Rabalais, 2000

varies from system to system. The most serious effects are seen in the Black Sea and Baltic Sea where demersal trawl fisheries have been either eliminated or severely stressed (Mee, 1992; Elmgren, 1984). Hypoxia in the Kattegat, the sea between Denmark and Sweden, caused mass mortality of commercial and noncommercial species. Large-scale migrations and/or mortality among demersal fish and the Norway lobster (*Nephrops*) continue, resulting in a changed species composition and reduced growth and biomass. Hypoxia is believed to be partly responsible for the overall decline in the stock size, recruitment and landings of commercial fish over the last two decades. Two other stress factors are eutrophication (Caddy, 1993) and harmful algal blooms (Karup et al., 1993), both resulting from nutrient over-enrichment. In the Baltic Sea, declining DO levels were noted as early as the 1930s, and hypoxia was reported in the 1950s (Fonselius, 1969), resulting in loss of demersal fisheries, with hypoxia as a bottleneck for cod recruitment.

Economic and societal consequences

Hyper-eutrophication will inevitably lead to destruction of habitat and changes in the food web that will result in species changes, most of which will be detrimental to the established fisheries. As the fisheries collapse, there are major societal impacts such as the loss of employment and the migration of workers and their families. The relationship between fishery production and yield, and nutrient loading has been discussed by Diaz and Solow

Table 1 Comparison of ecological and economic effects of anthropogenic hypoxic zones from coastal seas around the globe similar to the northern Gulf of Mexico hypoxic zone (Diaz and Solow, 1999)

System	Area affected (km²)	Benthic response	Benthic recovery	Fisheries response
Louisiana Shelf	15,000	Mortality	Annual	Stressed, but still highly productive. No reports of mortality, except 'jubilees'.
Kattegat, Sweden-Denmark	2,000	Mass mortality	Slow	Collapse of Norway lobster, reduction of demersal fish. Hypoxia prevents recruitment of lobsters.
Black Sea North-west Shelf	20,000	Mass mortality	Annual	Loss of demersal fisheries; shift to planktonic species.
Baltic Sea	100,000	Eliminated	None	Loss of demersal fisheries, shift to planktonic species. Hypoxia is bottleneck for cod recruitment.

Note: Data from various sources cited in text

(1999). They state that there is an initial increase in production and yield as nutrient loading increases, followed by the final collapse of the ecosystem after nutrient loading exceeds the assimilative capacity of the system.

Chesapeake Bay ecosystem

The Chesapeake Bay is a prime example of a fragile estuarine ecosystem that has been brought to near collapse by nutrient pollution. It is the largest estuarine bay in the USA, and has a watershed that includes Washington, D.C. and parts of six states in the mid-Atlantic region, Delaware, Maryland, New York, Pennsylvania, Virginia and West Virginia. The population of the Bay region has increased substantially in recent years, resulting in large increases in nutrient pollution from all sources, i.e. wastewater treatment plant effluents, urban/suburban runoff pollution, automobile exhaust emissions, agricultural pollution and power plant stack emissions. In 1985 the total nutrient loads to the Bay were 163 million kilograms (358 million pounds) N, and 13 million kilograms (28.7 million pounds) P. Of that total, agricultural sources contributed 42 and 40% of the N and P loads, respectively, wastewater treatment plant effluents 24 and 32%, respectively, and direct water surface air pollution deposition plus forest and urban runoff most of the rest (Chesapeake Bay Program, 2000). It has been determined that, currently, 25% of the N load to the Bay comes from atmospheric sources, i.e. originates as air pollution, and that 62% of the airborne N originates from outside the watershed. About 66% of the airborne N consists of oxidized forms (Kerchner *et al.*, 2000). It is also estimated that 8% of the P load is directly deposited on the water surface from airborne sources (Chesapeake Bay Program, 2000).

The nutrient pollution has been especially detrimental to the Bay ecosystem because the increased algal blooms cause shading of the submerged aquatic vegetation (SAV) and also result in large hypoxic areas during the growing season. The SAV is essential habitat for the survival of newly hatched fish, crabs and oyster spat, but is very light sensitive. Algal shading has caused extensive reduction of SAV in the Bay and this has substantially reduced the fisheries productivity of the Bay. The potential SAV habitat in the Bay is 600,000 acres (242,820 hectares), but the area was only 40,000 acres (16,188 hectares) in 1984.

Chesapeake Bay fisheries

The hypoxia in the Gulf of Mexico has not yet resulted in fisheries declines sufficient to cause major societal changes in the region, but declines in fisheries sufficient to cause major societal changes have already taken place in the Chesapeake Bay region. Most notable are the declines in the American Shad and oyster fisheries. During the 19th century, the shad fishery was the largest in the Bay, with catches of 22,000 metric tons/yr. It declined for >80 years, and by 1992 had shrunk to only 700 metric tons. In the late 1800s, the oyster harvest from the Bay was more than a billion pounds (454 million kg) per year. It has averaged less than 2 million pounds (0.91 million kg) per year for the past decade. The tremendous decline in the oyster population has severely impacted the fishery and the cultural tradition of the watermen that ply the Bay bottom for this seafood product (National Marine Fisheries Service, 2002). During the 15 year period from 1970 to 1985, approximately 50% of all of the oyster processing facilities in Maryland ceased to function, putting thousands of employees out of work. Similar declines can be seen in other Bay fisheries. The catch of white perch has been declining since 1969. American eel landings have declined since 1981. Red drum has ceased to be an important commercial species although the catch was once as high as 180,000 pounds/year. The decline reaches into the coastal waters and bays beyond the Chesapeake Bay. Menhaden fishing is one of the most productive and important fisheries on the USA Atlantic Coast, but it, too, is in serious decline (NOAA CBO, 2002).

In 1955 there were 150 menhaden vessels in operation. This had shrunk to 31 in 1993. There were 23 shore-side reduction facilities in 1955 but only 5 in 1993.

Nutrient controls, recovery and recycling

The information given above makes it clear that excessive nutrient pollution of estuaries and coastal areas is now a worldwide crisis with major economic and social repercussions. It seems reasonable to say that this widespread crisis has occurred because our political and social structures have permitted population growth without adequate regulation of agricultural, municipal and industrial sources of nutrient pollution. The political policies and processes have permitted municipalities, developers, industries and farmers to expand and operate without paying the full cost of their activities, and this has been done at the expense of those who rely on the productivity and recreational value of our estuarine and coastal waters. In some cases, the fisheries have been permitted to fully collapse without any efforts at remedial actions, such as in the Baltic Sea. In other cases, the governments have developed somewhat belatedly, usually inadequate, and nearly always under funded remedial programs. For example, even when major remedial efforts such as the Chesapeake Bay Program are instituted and funded, the funding is inadequate for rapid reduction of nutrient inputs and no institutional efforts are made to establish land management and to control immigration into the region. The result is that established goals are not reached, and nutrient reductions achieved by the program are short-lived because, even with more effective point and non-point controls, the absence of land use and population controls results in increased pollution from population growth and increased pollution causing activities that overcome any reductions achieved through technology.

Given the political, social and demographic realities, what are some appropriate ways to establish and maintain effective nutrient controls? It is proposed that the most effective nutrient controls are those that follow "green engineering" principles. Green engineering can be defined as, "the utilization of technology that improves or is highly compatible with the environment, eliminates or minimizes secondary environmental impacts, and minimizes the costs of implementation". While all processes that remove nutrients from wastewaters and minimize the amounts that return to the aquatic environments can be characterized as green engineering, biological nutrient removal (BNR) more precisely fits the definition than chemical nutrient removal processes if the resulting sludges are simply sent to disposal. For example, both enhanced biological phosphorus removal (EBPR) and denitrification utilizing the wastewater as the organic carbon source reduce the amount of energy that must be consumed for the transfer of oxygen to complete COD removal. The reduction of energy requirements reduces the secondary environmental impacts of energy generation, i.e. fuel consumption, stack emissions, mining or extraction activities, etc. The reduction of chemical usage reduces the secondary impacts of chemical manufacturing and shipping. Additionally, denitrification and EBPR operation produce less waste sludge for the same operating conditions than aerobic COD removal and chemical addition processes, and this reduces the amount of secondary impacts related to waste sludge processing and disposal. Of course, if chemical processes can be used to reduce the lifetime costs of the treatment system when all costs including sludge dewatering and disposal are accounted for, then they can become the method of choice.

While biological processes are potentially more environmentally compatible than chemical processes, their advantages can be improved even more if they can be combined with recycle and reuse. The Hampton Roads Sanitation District (HRSD) of Virginia is an example of an organization that has implemented nutrient recycle through sludge processing, packaging and reuse. The District operates 11 WWTPs in the Tidewater Virginia region (Norfolk, Portsmouth, Newport News, Hampton, Virginia Beach, Williamsburg, etc.), two of which are three stage BNR plants. HRSD and the author worked together demonstrating and developing BNR wastewater treatment from 1985–1988, and phosphorus-rich sludge composting and recycle was implemented in 1986 in conjunction with the York River WWTP BNR demonstration project. The York River WWTP was designed as a 56.8 ML/d capacity conventional aerobic activated sludge facility. It was designed with aerated grit removal, a minimum of 30 minutes of pre-aeration, three primary clarifiers, six parallel aeration basins with coarse bubble aeration, three secondary clarifiers, sludge thickening by both gravity (primary) and dissolved air flotation (secondary), and two stage anaerobic digestion followed by belt filter press dewatering. The flow at the time of the BNR demonstrations was less than 26 ML/d, so only two of the parallel aeration basins were modified so that they could be operated for either two or three stage BNR operation, i.e. the other four were not used during BNR operation, which resulted in nominal hydraulic retention times (HRTs) of less than 6 hours in the activated sludge basins (Figure 2).

The percentage phosphorus in the York River activated sludge exceeded 15% during one phase of the two stage (EBPR) BNR experiments because of the low COD/TP ratio in the process influent, and averaged in excess of 10% throughout the first year once steady-state was achieved. Only a small fraction of the phosphorus in the waste activated sludge (WAS) was released during DAF thickening when it was properly operated. It was mixed with the thickened primary sludge immediately before entering the primary anaerobic digester, and the full stream entered the digester. Only one primary and one secondary digester were available at the WWTP, and because the flow was less than 50% of design flow, the HRT within the digesters was approximately 90 days. The digesters were always mixed and were never supernated to minimize the amount of phosphorus recycled to the BNR process.

The data in Table 2 compares inorganic chemical concentrations in the primary digester before and after EBPR was established and had come to steady state. The data show that the TP increased by 750 mg/L, but the soluble P increased by only 250 mg/L. This indicates that a large fraction of the phosphorus released in the digesters precipitated.

Potassium does not have an insoluble precipitate under the conditions in the anaerobic digesters, so the comparison of the increase in soluble K relative to the increase in the total K reveals the fraction of P release from the sludge because K is always taken up and

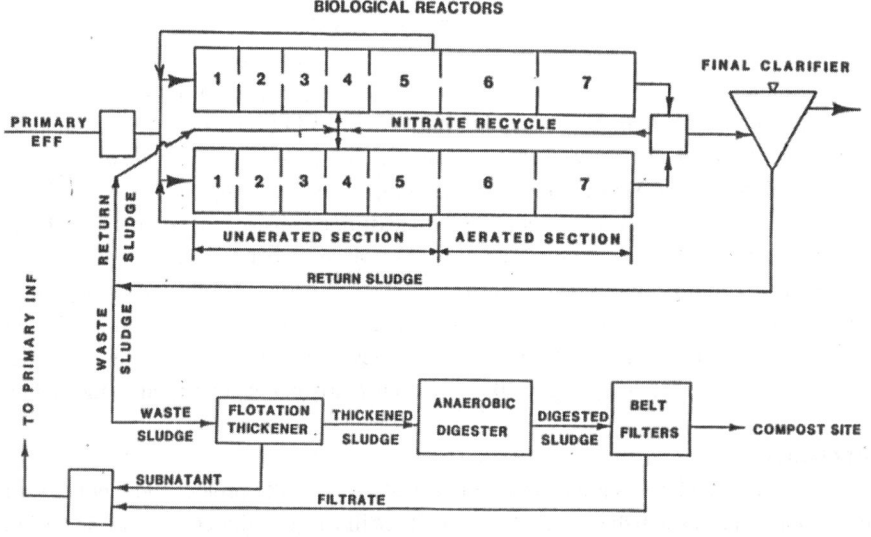

Figure 2 York River WWTP BNR modifications

Table 2 Inorganic chemical concentration changes in the York River digesters

Parameters	Chemical Concentrations, mg/L	
	Before EBPR	After 9 months of EBPR
Total phosphorus	350	1,100
Soluble phosphorus	50	300
Total potassium	60	300
Soluble potassium	55	230
Soluble magnesium	10	10
Soluble calcium	100	30
Ammonium	316	390

released with P during EBPR. Total K increased by 240 mg/L, whereas soluble K increased by 175 mg/L, indicated that 73% of the phosphorus in the cells was released during anaerobic digestion. Soluble magnesium did not increase at all indicating that all of the Mg released from the microbial cells precipitated upon release, and that Mg was the limiting chemical in the precipitation. Ammonium increased, but by only 74 mg/L, indicating that most of the released ammonium precipitated. The results indicate that both ammonium and Mg precipitated as struvite, i.e. $MgNH_4PO_4$, and that even more ammonium and phosphate would have precipitated if more Mg had been present. The formation of struvite was confirmed by the recovery of crystals from the dewatered digested sludge, and utilization of X-ray diffraction. In fact, the struvite crystals were so large and so numerous that the sludge sparkled when exposed to light.

The data show that the soluble calcium decreased by 70 mg/L after EBPR was instituted. The amount of calcium associated with P during EBPR reactions is typically small, i.e. less than 10% of the cations involved, and it may be near zero (Pattarkine, 1991). The reduction in soluble calcium clearly indicates that the high phosphate concentrations induced calcium phosphate precipitation within the anaerobic digesters. Mass balance calculations indicated that the reaction was non-stoichiometric. Regardless, most of the P released to solution during anaerobic digestion of the phosphate rich bacteria precipitated within the digesters and was retained in the sludge, not the supernatant. From mass balances it was determined that 30% of the P removed during EBPR was recycled back to the headworks in the belt filter press filtrate, and 70% of it was retained with the sludge cake. The retention of the struvite and calcium phosphate precipitates significantly increased the concentrations of P, ammonium, magnesium, and calcium in the sludge cake and increased its value as a fertilizer. Composting the waste sludge sterilized it and made it safer to handle.

The composted sludge is packaged in 40 lb (18.2 kg) bags and marketed locally as "NutraGreen". Each bag is currently sold for US$1.00. Additionally, it is sold in bulk for $14 per yd^3 (US$18.31 m^3). Currently, all of the sludge from the five HRSD operated treatment plants on the Virginia Peninsula is composted and completely recycled in this manner. The recycling effort has been very successful, and has been well received by the regional community. There are state regulations in place to mandate that the nutrient-rich composted material is appropriately utilized to minimize runoff pollution to the streams and estuaries in the region. There are plans underway to encourage similar nutrient recovery and recycle programs throughout the Chesapeake Bay region. This will begin with a workshop during the spring of 2003 that is currently being developed by the author.

Conclusions

Improved nutrient controls are needed in much of the world to protect and recover water quality and to preserve fisheries. Failure to do so will cause major economic and societal disruptions. The recovery and reuse of nutrients is recommended as a means of control.

References

Caddy, J. (1993). Toward a comparative evaluation of human impacts on fishery ecosystems of enclosed and semi-enclosed seas. *Review of Fishery Science*, **1**, 57–96.

Chesapeake Bay Program (2000). *State of the Bay Report*. USEPA Chesapeake Bay Office, 410 Severn Ave., Suite 109, Annapolis, MD 21403.

Diaz, R. and Rosenberg, R. (1995). Marine benthic hypoxia: A review of its ecological effects and the behavioural responses of benthic macrofauna. *Oceanography and Marine Biology: An Annual Review*, **33**, 245–303.

Diaz, R. and Solow, A. (1999). *Ecological and economic consequences of hypoxia: Topic 2 Report for the Integrated Assessment of Hypoxia in the Gulf of Mexico*. National Ocean Service, Coastal Ocean Program, NOAA National Centers for Coastal Ocean Science, 1315 East West Highway, Rm. 9700, Silver Spring, MD 20910, USA.

Elmgren, R. (1984). Trophic dynamics in the enclosed brackish Baltic Sea. *Rapports et Proces-Verbaux des Reunions*, **183**, 152–169.

Fonselius, S. (1969). *Hydrography of the Baltic deep basins III. Series Hydrography report no. 23*, 1–97, Stockholm: Fishery Board of Sweden.

Karup, H., Evans, S., Dahl, E., Klungsoeyer, J., Klungsoeyer, L., Reiersen, O., Gray, J., Iverson, P., Bokn, T. and Skjoldal, H. (1993). *Man's Impact on Ecosystems*. North Sea Subreg. 8 Assessment Report, 59–62.

Kerchner, M., Halloran, R., Thomas, J. and Hicks, B. (2000). The Significance of Ammonia to Coastal and Estuarine Areas. Airsheds and Watersheds III: A Shared Resources Workshop, November 15–16, 2000. NOAA National Center, Silver Spring, MD, USA.

Mee, L. (1992). The Black Sea in crisis: A need for concerted international action. *Ambio*, **21**, 278–86.

National Marine Fisheries Services (2002). *Chesapeake Bay Stock Assessment Committee Report, 2001*. NOAA National Centers for Coastal Ocean Science, Coastal Ocean Program, 1315 East West Highway, Rm. 9700 Silver Spring, MD, 20910, USA.

National Ocean and Atmospheric Administration, Chesapeake Bay Office (2002). *NCBO Fact Sheet*. 410 Severn Avenue, Annapolis, MD, 21403, USA.

National Ocean Service (2000). *Gulf of Mexico Hypoxia Assessment*. NOAA National Centers for Coastal Ocean Science, Coastal Ocean Program, 1315 East West Highway, Rm. 9700, Silver Spring, MD 20910, USA.

Pattarkine, V. (1991). The Role of Metals in Enhanced Biological Phosphorus Removal from Wastewater, Doctoral Dissertation, Virginia Polytechnic Institute and State University, Blacksburg, VA, USA.

Rabalais, N. (2000). *Characterization of hypoxia: Topic 1 Report for the Integrated Assessment of Hypoxia in the Gulf of Mexico*. National Ocean Service, Coastal Ocean Program, NOAA National Centers for Coastal Ocean Science, 1315 East West Highway, Rm. 9700, Silver Spring, MD 20910, USA.

Recycling of wastewater-derived phosphorus in Swedish agriculture – a proposal

E. Kvarnström*, C. Schönning**, M. Carlsson-Reich***, M. Gustafsson**** and E. Enocksson*****

* VERNA Ekologi AB, Malmgårdsvägen 14, SE-116 38 Stockholm, Sweden (E-mail: *elisabeth@verna.se*)
** Swedish Institute for Infectious Disease Control, SE-171 82 Stockholm, Sweden
(E-mail: *caroline.schonning@smi.ki.se*)
*** IVL Swedish Environmental Research Institute, Box 210 60, SE-100 31 Stockholm, Sweden
(E-mail: *marcus.carlsson@ivl.se*)
**** Skövde Municipality, SE-541 83 Skövde, Sweden (E-mail: *malin.gustafsson@skovde.se*)
***** Swedish Environmental Protection Agency, Blekholmsterrassen 36, SE-106 48 Stockholm, Sweden
(E-mail: *egon.enocksson@naturvardsverket.se*)

Abstract In 2001 the Swedish Government commissioned the Swedish Environmental Protection Agency (Swedish EPA) (i) to examine the need for stricter human health and environmental regulations governing the agricultural use of sewage sludge and (ii) to propose national targets for the agricultural recycling of phosphorus (P) originating from wastewater. The Swedish EPA may propose: (i) stricter regulations on sludge treatment to limit the risk of spreading pathogens in the environment and transmission of infectious diseases; (ii) recycling of 20–30% of wastewater-P to agriculture by 2015 and recycling of 35–50% of wastewater-P to agriculture by 2025.
Keywords Hygiene; nutrient recycling; phosphorus; regulation; sludge; wastewater

Introduction

Reducing the need for commercial phosphorus (P) fertilizers is one reason for recycling wastewater-P in agriculture. However, since wastewater is a heterogeneous and complex assembly of different elements, largely reflecting the society from which it is collected, recycling P from this kind of system poses difficulties. In fact, since only 10% of sludge produced in Sweden was used in agriculture in 2001, according to a Swedish Water and Wastewater Organization questionnaire, the P-cycle loop is broken in Sweden. One reason for the decline in sludge use in agriculture in Sweden is the variety of opinions held by stakeholders as to the risks involved and the sustainability of this use. To shed light on the sludge and P issue, the Swedish Government instructed the Swedish Environmental Protection Agency (Swedish EPA) in 2001 (i) to examine the need for stricter human health and environmental regulations governing the agricultural use of sewage sludge and (ii) to propose national targets for the agricultural recycling of phosphorus originating from wastewater. These targets are part of the national environmental quality objective "A good urban environment", one of 15 environmental quality objectives laid down in Swedish Government Bill 1997/98:145, enacted by Parliament in 2001. These objectives describe various environmental states necessary for ecologically sustainable development. The environmental quality objectives are to be achieved within a generation, i.e. by 2025.

The purpose of this paper is to present a selection of results from the study made by the Swedish EPA into the implications for human health when sludge is used in agriculture, as well as the national targets for recycling of phosphorus originating from wastewater. The study has resulted in a draft Action Plan that has been referred to stakeholders for their consideration. However, since the handling of the consultative input had not been concluded at the time of writing, the proposals presented in this paper may be subject to substantial change.

Methods
The Swedish EPA appointed a project committee to coordinate nine sub-projects;
1. Examination of the need to introduce new, stricter limit values for the **metal content** of sludge used in agriculture.
2. Examination of the need to introduce new, stricter limit values for the **organic pollutant** content of sludge used in agriculture.
3. Examination of the need to introduce new, stricter regulations for **hygiene** in relation to sludge use in agriculture.
4. Examination of whether there are legal obstacles to charging users for the increased costs associated to P recycling. Examination of whether municipalities should have greater powers to disconnect unwanted wastewater flows to wastewater treatment plants.
5. Description of P flows in Sweden and generally to discuss aspects of sustainability within the field of wastewater treatment systems.
6. Examination of the best available technology for phosphorus recycling/recovery in Sweden and abroad. A systems analysis was performed on six wastewater systems. The systems analysis included, among other things, energy consumption, nutrient recycling potential and financial aspects.
7. Transitional solutions for wastewater treatment.
8. Analysis of the stakeholders involved in the sludge debate in Sweden. The history of sludge use was also outlined. A welfare economic analysis was made of the systems defined under sub-project 6.
9. The production of an action plan including a synthesis of the sub-projects above and also national, time-defined targets for recycling/recovery of wastewater-P to agriculture. Appropriate economic and policy measures to achieve these targets were proposed and an environmental impact assessment for the proposed targets was included.

The project has been carried out in consultation with the Swedish Board of Agriculture, the National Food Administration, the National Chemicals Inspectorate, the Swedish Institute for Infectious Disease Control, the National Board of Health and Welfare, the National Veterinary Institute, and the National Board for Consumer Policies. The project has also involved other stakeholders, including organizations such as the Federation of Swedish Farmers, the Swedish Water and Wastewater Organization, the Swedish Food Federation, and various consumer organizations.

Each sub-project was carried out in more or less the same way. The bulk of the work was contracted either to consultants or researchers in Sweden. However, sub-project 2 was contracted to a Danish researcher. Sub-projects 1–3, 6, 8, and 9 have been monitored by various reference groups comprising the agencies involved, researchers and other stakeholders. Results from the various sub-projects were presented to the stakeholders continuously via a website conference and also at a meeting held in May, 2002.

Systems and welfare economic analyses (sub-projects 6 and 8)
Six wastewater systems suitable for urban settlements underwent a systems analysis as part of the work on the strategy for increased use of phosphorus and other wastewater nutrients in agriculture.

The six systems were:
1. Urine diversion

2. Collection of black water (urine, feces, and flush water)
3. Sludge use in agriculture
4. Recovery of P from wastewater
5. Recovery of P from sludge
6. Recovery of P from ashes after sludge incineration

The systems chosen had all been tested in full-scale implementation or near full-scale implementation. The systems analysis involved a reference system including conventional wastewater treatment using mechanical and biological treatment, chemical precipitation, nitrogen reduction and sludge incineration. It is worth noting that the sludge quality used in the reference system and for sludge use in agriculture was not the quality of average Swedish sludge, but a sludge quality one would expect from a wastewater system fed only by households and services such as offices, schools and shops.

Results and discussion
General targets and strategies for recycling wastewater nutrients to agriculture

As a result of the project, the Swedish EPA has formulated an overall objective for wastewater nutrients, viz., that wastewater nutrients should be returned to agriculture without risk to the environment or human health. This objective derives support both from the Swedish Environmental Code, which encourages reuse and recycling of resources and from the "good urban environment", which states that waste should be treated according to its properties and returned to the ecocycle in a balanced interplay between urban areas and their surroundings. That environmental quality objective underlines the importance of not exposing people to health and safety risks, whereas the "non-toxic environment" quality objective also states that levels of substances not occurring naturally must be close to zero and that naturally occurring substances must be close to background levels.

Four strategies have been defined to achieve the wastewater nutrient recycling target set by the Swedish EPA:
1. Increased use of phosphorus and other wastewater nutrients in agriculture by use of various wastewater fractions containing nutrients.
2. Reduced flow of pollutants to agricultural land
3. Reduced presence of pollutants in wastewater
4. Increased disease control

Proposed action for strategies 1 and 4 is presented below.

Increased use of phosphorus and other wastewater nutrients in agriculture by use of various wastewater fractions containing nutrients

A systems analysis was made of six wastewater systems to evaluate long-term recycling of phosphorus and other nutrients from wastewater to agriculture. Each system analyzed had its advantages and disadvantages, as may be seen in Table 1. The urine system and the black water system were by far the most expensive systems. The costs for the other P recovery systems were similar. This conclusion also held for the welfare economic assessment. The additional cost compared to the reference system per kg of recovered P, which can be seen as a form of objective fulfillment, still put the urine system and the black water system on a cost level far above the other systems. However, if the recovery of a broader perspective of nutrients was considered, kg of recovered NPKS, the additional costs for the urine system and the black water system were in the same order of magnitude as the other systems. However, the cost per kg nutrient was still far above the market value of approximately 10 SEK/kg. Additional costs for recovery of either P or NPKS were lowest for agricultural use of sludge. It is, however, of utmost importance to remember that the sludge quality used for the systems analysis is not the average Swedish sludge quality, but a sludge quality one

Table 1 A selection of results obtained in the systems analysis (Balmér et al., 2002), and for the welfare economic analysis (Carlsson-Reich, 2002)

	Urine	Black water	Sludge	P recovered from wastewater	P recovered from sludge	P recovered from ashes
Total recycled nutrient, %	P: 40 N: 65 K: 35 S: 55	P: 75 N: 90 K: 60 S: 70	P: 95 N: 20 K: 0 S: 35	P: 60 N: 0 K: 0 S: 0	P: 70 N: 0 K: 0 S: 0	P: 60 N: 0 K: 0 S: 0
Emission to water, basic and expanded system	Equal to reference	Lower than reference	Equal to reference	Equal to reference	Equal to reference	Equal to reference
Product quality, quantity of pollutants compared with current Swedish regulations on reuse of sludge in agriculture	Considerably lower	Considerably lower	Lower	Probably considerably lower	Probably considerably lower	Probably considerably lower
Energy consumption, expanded system	Lower than reference	Considerably higher than reference	Equal to reference	Slightly higher than reference	Higher than reference	Slightly higher than reference
Transport compared with reference system, tonne km/pe per year	+30	+130	−7			
Financial costs, SEK/pe,yr	1,045	2,145	409	464	488	459
Welfare economic assessment, SEK/pe,yr	1,108	1,789	660	705	736	703
Additional cost, compared to reference system, per kg P, SEK/pe,yr	2,098	3,112	−54	69	119	53
Additional cost, compared to reference system, per kg NPKS, SEK/pe,yr	151	275	−23	69	119	53
Disease control, including effluent	Slightly higher than reference	Slightly lower than reference	Equal to reference	Equal to reference	Equal to reference	Equal to reference
Ranking of systems according to the "good urban environment" quality objective	P: 3 NPKS: 1	P: 2 NPKS: 1	P: 1 NPKS: 2	P: 2 NPKS: 2	P: 2 NPKS: 2	P: 2 NPKS: 2
Ranking of systems according to the "non-toxic environment" quality objective	Metals: 1 Metals and organic pollutants: 2	Metals: 2 Metals and organic pollutants: 2	Metals: 3 Metals and organic pollutants: 3	Metals: 2 Metals and organic pollutants: 1	Metals: 2 Metals and organic pollutants: 1	Metals: 2 Metals and organic pollutants: 1

would expect from a wastewater system fed only by households and services such as offices, schools and shops. The costs for achieving this kind of sludge quality are not included into the analysis, and may prove to be very expensive.

However, of the systems studied, the urine and black water systems could recover the highest amount of the nutrients nitrogen, phosphorus, potassium, and sulfur, while simultaneously containing very small metal quantities. The urine and black water systems may therefore be said to be components of non-toxic ecocycles, which is one of three strategies proposed by the Swedish Government as guidance to achieve the 15 environmental quality objectives. Of the systems studied, these systems may therefore be said to be those best fulfilling the Swedish EPA's overall objective for nutrient recycling without risk to human health or the environment. However, the systems analysis indicated that these systems are in need of organizational improvement and technical development to reduce the financial costs. Energy-efficient hygienization of the black water system should also be developed. The systems analysis also showed that a narrow focus on recycling of phosphorus alone might be unfortunate from a sustainability viewpoint, with resulting sub-optimization of systems. Mean recycling of nitrogen, phosphorus, potassium and sulfur resources from a wastewater system might be a better way of analyzing the resource efficiency of different wastewater systems.

Sludge contained more nutrients than the fractions from the recovery techniques. Sludge use in agriculture may work in the long run when the "non-toxic environment" quality objective is achieved. In all probability it will take more than one generation to achieve this objective. Sludge use in agriculture may also be an alternative in the short term, provided that the sludge intended for agricultural use meets future quality requirements. Moreover, it is essential to gain the support of stakeholders in Sweden to overcome the diminishing acceptance of sludge use in agriculture. There are reasons to believe that the acceptance is higher for recovered P as compared with sludge, for example. However, the methods examined in the systems analysis recover only P and no other nutrients, which probably constitutes a less than optimal use of resources.

It may therefore be concluded that neither system can be said to be the obvious choice for full implementation in Sweden to achieve the Swedish EPA's overall objective for recycling nutrients from wastewater to agriculture.

Formulating the Action Plan involved setting time targets for recycling wastewater-derived phosphorus to agriculture. National P recycling targets will allow local and regional differences in choice of methods. Considering that a narrow focus on P recycling alone might be sub-optimal and that recycling of all nutrients should be encouraged, the Swedish EPA proposes that the P recycling targets for 2015 and 2025 be set so as not to hinder the recycling of other wastewater nutrients. The Swedish EPA proposes that 20–30% of wastewater-P be recycled to agriculture by 2015 and 35–50% by 2025. These recycling goals were calculated by using a variety of the systems studied, respecting a pace of refurbishment of wastewater systems of 2% per year (Hansson et al., 1993), and taking into consideration the results from the systems analysis.

Increased disease control
Present Swedish regulations (SNFS 1994:2), based on the EC directive 86/278/EEC, allow untreated sludge to be spread on agricultural land, as long as it is worked into the soil within 24 hours. It is now suggested that all sludge must be treated before it is used, irrespective of land use option chosen. The treatments suggested are divided into three categories (Table 2). In theory, these may reduce pathogens to the same extent, with the probable exception of parasite eggs in category C. The main difference between treatments is the safety level and possibility of controlling the process.

Table 2 Proposed treatment processes for sewage sludge, divided into categories A, B and C. All times are given as minimum values

Category	Treatment process	Parameters to be fulfilled
A	Thermal drying	Temperature: ≥80°C
		Exposure time: 10 minutes
		Moisture: <10%
A	Pasteurisation	Temperature: ≥70°C
		Exposure time: 60 minutes
A	Thermophilic anaerobic digestion	Temperature: ≥55°C
		Exposure time[a]: 6 hours
		Mean retention time[b]: 7 days at ≥55°C
A	Liquid composting	Temperature: ≥55°C
		Exposure time: 10 hours
		Mean retention time[b]: 7 days at ≥55°C
A	Thermophilic aerobic stabilisation (in-vessel composting)	Temperature – time requirements: 55°C – 2 weeks; 60°C – 6 days; 65°C – 3 days; or 70°C – 1 day
A	Conditioning with lime (quick lime)	pH: ≥12 *and*
		Temperature: ≥55°C
		Time: 4 hours
B	Thermophilic aerobic stabilisation in windrows or piles (open composting)	Temperature – time requirements[c]: 55°C – 2 weeks; 60°C – 6 days; 65°C – 3 days; or 70°C – 1 day
B	Conditioning with lime (slaked lime)	pH: ≥12
		Time: 3 months
C	Treatment and drying in reed beds[d]	1 year without addition of new sludge
C	Storage	1 year without addition of new sludge

[a] Exposure time is defined as the time when no sludge is added or withdrawn
[b] The mean retention time is calculated to include at least 95% of the material
[c] The temperature – time requirement should be met during three periods, the material being turned over between each period
[d] Methods involving dewatering and possibly biological degradation

The proposal also suggests additional requirements, e.g. that pasteurisation needs to be followed by anaerobic stabilisation and that the parameters must be fulfilled for the entire material.

Present regulations also include restrictions on the type of crops on which fertilizers may be used. Although it is proposed that treatment be compulsory, the restrictions remain with only small proposed amendments:

- Sewage sludge should not be used on pasture or on fields where forage crops, vegetables, berries, fruit, potatoes or root vegetables are grown, or will be grown during the same calendar year. An exception is fruit on trees, sugar beet and potatoes grown for the production of starch.
- If the sludge is treated according to category B or C, a two year period must elapse before the crops listed above can be harvested. These categories of sludge must also be worked into the soil the same day they are applied.
- In addition, sludge treated according to category C may not be used on green areas such as parks, sports fields and golf courses to which the public normally has access, or in gardens.

Alternative treatments may be allowed following approval by the competent authority. Restrictions have been proposed in addition to treatment mainly as a second barrier against exposure to pathogens.

Conclusions

The systems analysis of six various wastewater systems showed that a narrow focus on recycling of phosphorus alone might be unfortunate from a sustainability viewpoint, with resulting sub-optimization of systems. Mean recycling of nitrogen, phosphorus, potassium and sulfur resources from a wastewater system might be a better way of analyzing the resource efficiency of different wastewater systems. The Swedish EPA may propose: (i) stricter regulations on sludge treatment are proposed to manage the risk of pathogen spreading in the environment and transmission of infectious diseases and (ii) that 20 – 30% of wastewater-P be recycled to agriculture by 2015 and that 35–50% of wastewater-P be recycled to agriculture by 2025.

References

Balmér, P., Book, K., Hultman, B., Jönsson, H., Kärrman, E., Levlin, E., Palm, O., Schönning, C., Seger, A., Stark, K., Söderberg, H., Tideström, H. and Åberg, H. (2002). *System för återanvändning av fosfor ur avlopp ("Systems for reuse of phosphorus from wastewater")*. Swedish Environmental Protection Agency, Stockholm, Sweden. (In Swedish).

Carlsson-Reich, M. (2002). Swedish Environmental Protection Agency, Stockholm, Sweden. *Samhällsekonomisk analys av system för återanvändning av fosfor ur avlopp ("Welfare economics analysis of systems for reuse of phosphorus from wastewater")*. (In Swedish).

Hansson, R., Johansson, B.G. and Lindgren, S. (1993). *Avloppsstammars livslängd ("Life span of wastewater piping")*. Council for Building Research, Stockholm, Sweden. (In Swedish).

Exergy analysis of nutrient recovery processes

D. Hellström

Stockholm Water Co., SE-106 36 Stockholm, Sweden

Abstract In an exergy analysis, the actual consumption of resources in physical and chemical processes is calculated. Energy and chemical elements are not consumed in the processes – they are only transformed into other forms with lower quality. The principals of exergy analysis are illustrated by comparing different wastewater treatment systems for nutrient recovery. One system represents an end-of-pipe structure, whereas other systems include source separation of grey water, black water, and urine. The exergy flows analysed in this paper are those related to management and treatment of organic matter and nutrients. The study shows that the total exergy consumption is lowest for the system with source separation of urine and faeces and greatest for the conventional wastewater treatment system complemented by processes for nutrient recovery.

Keywords Black water; exergy analysis; nitrogen; phosphorus; urine separation

Introduction

The use of physical resources can be measured by using exergy analysis (Hellström, 1999). Exergy is energy of supreme quality, i.e. energy that is convertible into all other forms of energy (Wall, 1992). The quality of energy depends on how concentrated, ordered and structured the energy source is (Holmberg, 1995). Examples of quality factors for different energy sources are given in Table 1.

Energy and chemical elements are not consumed in the processes – they are only transformed into other forms with lower quality (order). An example of this is the use of gasoline for driving. The gasoline is consumed, i.e. the chemical structure is destroyed, but the chemical bound energy is transformed into mechanical work and heat and no energy is consumed. The chemical elements, e.g. carbon and hydrogen, are converted into other chemical compounds of much lower exergy content. Thus, low-entropy matter and energy has been converted into high-entropy matter and energy. Hence, exergy may be used to measure depletion and use of resources.

The difference between exergy and energy analysis can be illustrated by an example from Henriksdal's wastewater treatment plant (WWTP) in Stockholm (Figure 1). The treatment includes pre-treatment, chemically enhanced pre-sedimentation using Fe_2SO_4, activated sludge process with pre-denitrification, and sand filter (with some addition of precipitants). The treatment of sludge consists of anaerobic digestion and dewatering. The heat of wastewater is dominant as an energy input but not as an exergy input (Figure 1). However, organic matter is the dominant input if the quality of the different energy flows

Table 1 Example of quality factors for some energy sources. (Reference temperature 5°C)

Form of energy	Exergy/Energy (%)
Electrical energy	100
Hot steam (200°C)	~70
District heating	~30
Heat of wastewater (15°C)	< 5

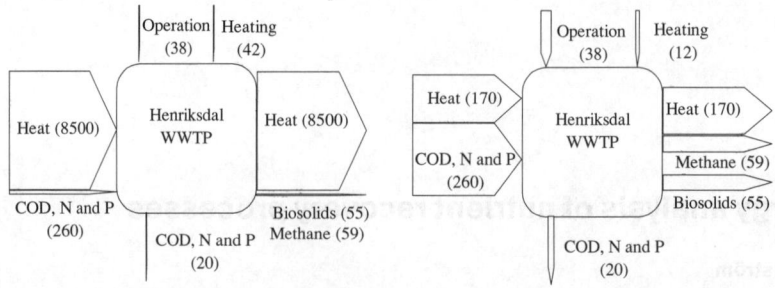

Figure 1 Major energy (left) and exergy (right) flows at Henriksdal WWTP, Stockholm (kWh/pe.yr). Data from 1999

are considered. It is noteworthy that only 25% of the exergy in organic matter is recovered as methane. Other significant exergy flows are due to electricity for operation. The method for analysis has been described by Hellström (1997).

Objectives and scope. The principals of exergy analysis are illustrated by comparing different wastewater treatment systems for nutrient recovery. One system represents an end-of-pipe structure, whereas the other systems include source separation of grey water, black water and urine. The comparison will focus on management and treatment systems for urban areas.

Description of studied systems

The studied systems have been chosen to illustrate different strategies for nutrient recovery. All systems are assumed to achieve a high degree of removal of organic matter, phosphorus (> 90%), nitrogen (> 90%), and nutrient recovery.

It is probably beneficial to treat organic household waste together with toilet waste in the blackwater systems described below. In the other systems, the organic waste is treated separately by anaerobic digestion. Hence, in all systems organic waste will generate methane gas and the residues can be used for nutrient recovery. However, the exergy consumption for management of organic waste is not included in this analysis.

A. Reference system – conventional treatment

The treatment of wastewater includes mechanical pre-treatment, chemical precipitation, activated sludge process (including nitrogen removal) and filtration. The treatment of sludge includes pasteurisation (70°C, 1 h), anaerobic digestion, and dewatering. Dewatered sludge will not be used for agricultural purposes, but can be used as landfill cover or for land reclamation projects.

B. Conventional treatment complemented by nutrient recovery processes

The treatment of wastewater includes mechanical pre-treatment, chemical precipitation, activated sludge process (excluding nitrogen removal), sand filtration, and reverse osmosis (RO) for nutrient recovery. The concentrate from the membrane is sent to an evaporator for further concentration. Sludge treatment involves dewatering, anaerobic digestion, sludge fractionation (SF) by hydrolysis and acidification for recovery of phosphorus.

C. Urine separation and sludge fractionation

The reference system is complemented with urine separation (US) and sludge fractionation. By implementation of urine separation, about 50 to 80% of nitrogen and 30 to 50% of phosphorus will be found in the collected urine. In this study, it is assumed that the degree of urine separation is 75% (Jönsson *et al.*, 1998). Hence, urine separation is complemented by conventional treatment to achieve 90% nitrogen and phosphorus removal. Sludge treatment involves dewatering, anaerobic digestion, sludge fractionation by hydrolysis and acidification for recovery of phosphorus.

D. Urine separation and blackwater system

Urine is separated by means of source-separation toilets and the degree of urine separation is assumed to be 75%. For further transport, urine is led by gravity to underground tanks near houses, then transported by truck to farms for use as a fertiliser. Brown water (toilet waste excluding urine) and organic household waste are collected by low-flush toilets and kitchen waste disposers, then transported to a treatment plant by a low-pressure system. Since most of the nitrogen is recycled as urine, the only nutrient recovery process in the treatment plant is chemical precipitation of phosphorus. The blackwater treatment plant consists of chemically enhanced sedimentation, UASB (Uplow Anaerobic Sludge Bed) or EGSB (Expanded Granular Sludge Bed) reactors and BNR (Biological Nitrogen Removal as described by Edström et al. (2001). Sludge treatment consists of anaerobic digestion, dewatering and hygienisation prior to transportation to farmland by trucks.

Greywater is treated in a conventional WWTP with mechanical, biological and chemical processes. Since the amount of nutrients in greywater is low, the sludge treatment includes anaerobic digestion, dewatering, and incineration.

E. Blackwater system

A system with separation of blackwater (toilet waste) and conventional treatment of greywater. Black water (toilet waste) and organic household waste are mixed and transported to an anaerobic digestion unit by a combination of vacuum and low-pressure system. Underwatered anaerobic residues are, after pasteurisation, transported to agriculture. Greywater treatment includes mechanical, biological, and chemical processes. The sludge treatment includes anaerobic digestion, dewatering, and incineration.

The potential for nutrient recovery for the systems is presented in Table 2. The potential has been calculated as the amount of nutrients that can be recovered for agricultural purposes compared with the inflow of nutrients to each system. Data have been collected from Hellström (1999), Stark (2002), Jönsson et al. (1998), Bark et al. (2001), and Vinnerås (2001).

Regarding the studied systems, conventional treatment in combination with RO and sludge fractionation has the highest nutrient recovery potential (Table 2). The amount of

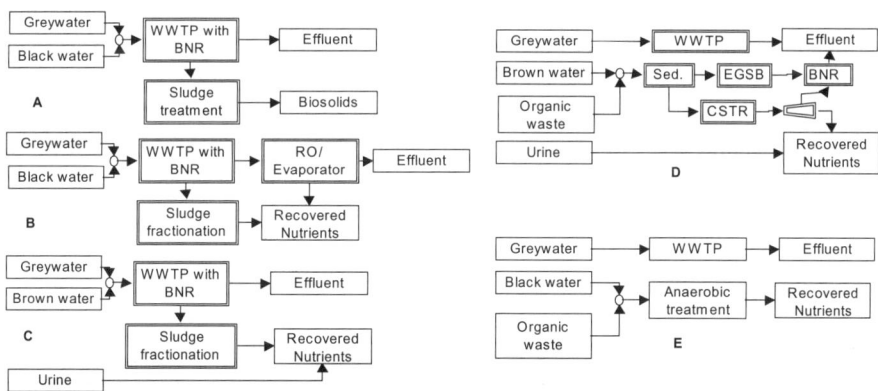

Figure 2 Schematic picture of: (A) reference system, (B) conventional system and nutrient recovery processes, (C) urine separation and sludge fractionation, (D) urine separation and blackwater system, and (E) blackwater system

Table 2 Potential nutrient recovery for different systems

System		% P	% N	% K
A.	Reference	0	0	0
B.	RO and sludge fractionation	> 80	> 80	> 90
C.	Urine Separation and sludge fractionation	> 80	> 50	> 30
D.	Urine Separation and blackwater	> 70[1]	> 60	> 30
E.	Blackwater system	> 70[1]	> 80	> 60

[1] > 80% if phosphorus containing detergents and washing powders are prohibited

recovered nutrients that can be used in agriculture without further treatment is also dependent on the quality of the different products. For example, concentrate from the RO may be contaminated by soluble toxic compounds from greywater, and blackwater may be contaminated by toxic compounds from cleaning (e.g. by emptying the scouring bucket into the toilet). Data for the quality of different fractions such as urine and blackwater is available (Table 3). To make a fair comparison it is necessary to study how these compounds will be affected and allocated by different nutrient recovery systems.

Methods used for exergy analysis

The system of interest is the sewerage system, including collection and transport of wastewater from households to treatment units, operation of treatment units, transport of recovered nutrients, and application of nutrients on farmland. The flows analysed are those related to management and treatment of organic matter and nutrients from domestic wastewater sources, i.e. wastewater from industrial sources and storm water are not included in the comparison of the different systems. Exergy flow caused by the heat of wastewater is not included in the calculations, but has been exemplified above and discussed elsewhere (Hellström, 1997; Hellström and Kärrman, 1997). The study considers only the operation of the sewerage system, and not the building and maintenance of the system.

An estimation of the *total* amount of exergy needed to operate a sewerage system should also include external costs such as the production of electricity and chemicals. However, the results from such an analysis will also include the "exergy efficiency" of the systems connected to the sewerage system. For example, production of electricity in (Swedish) nuclear power plants is very inefficient compared to production of electricity from water power (Wall, 1992). Thus, inefficiency in surrounding systems will influence the calculations, and the result will be less dependent on the efficiency of the sewerage system itself. Including the surrounding systems would probably give a more holistic picture, but the risk of arbitrariness would also increase. Thus, to avoid interference from inefficiencies in surrounding systems, the exergy consumed to produce the resources used in wastewater treatment is excluded from this analysis.

Water use and dilution by inflow of stormwater and drainage water are important parameters for the exergy consumption, especially for nutrient recovery by RO. The exergy consumption will also be affected by the water use for flushing of urine and blackwater. Thus, two different scenarios are considered in this study. The "best case" scenario is valid

Table 3 Quality indicators for different fractions (Vinnerås, 2001; Palmquist, 2001; Stockholm Water Co., 2002)

	Urine	Faeces	Blackwater	Biosolids
mg Ni/kg P	7	150	210	670
mg Pb/kg P	2	40	50	1100
mg Cd/kg P	1	20	7	34

for a rather new city-area where storm water and domestic wastewater are totally separated and where best available technology and behaviour are practised (e.g. low water consumption). The "worst case" scenario is valid for the existing sewer system and water use. Early experiences of sorting systems (Norin *et al.*, 2000; Malmén, 2001; Palmquist, 2001) also indicate that the use of water for flushing faeces and urine may be higher than expected. There is also a risk of dilution due to leakage (Hanæus *et al.*, 1997). The data used for different systems and scenarios are presented in Table 4.

Chemicals, organic matter and nutrients

Organic matter, nutrients as well as chemicals used in the treatment processes are regarded as different chemical compounds and the exergy for those species can be calculated by methods described by Szargut *et al.* (1988) and Hellström (1997). Standard chemical exergy of different compounds is presented in Table 5. Amount of organic matter and nutrients generated per person and day are presented in Table 6 with the assumed distribution between different fractions.

The main chemical for wastewater treatment is iron or aluminium salts for phosphorus precipitation. The average dosage of precipitants at the Swedish WWTP used was 1.5 mole Al or Fe/mole $P_{reduced}$ (Hellström, 1999) and the median exergy content in metallic salts and polymers was 2.4 kWh/(pe.yr). The installation of a source separation system will decrease the inflow of phosphorus to the WWTP and reduce the use of chemicals (Hellström, 1999).

An external carbon source is probably necessary to achieve 90% nitrogen removal if conventional nitrification/denitrification processes are used. For the reference system, it is assumed that 20% of all nitrogen is to be denitrified by an external carbon source and the exergy in the external carbon source for this alternative will be 15 kWh/(pe.yr) (Hellström, 1999).

The main chemical consumption for sludge fractionation is the use of acid or base to release phosphorus. If processes such as KREPRO or BioCon are applied to a treatment process using mainly chemical phosphorus removal (such as the Henriksdal process), about 800 kg acids and bases/ton TS are needed for phosphorus recovery, independent of the chosen technology (Stark, 2002). However, the necessary amount of acids and bases will be

Table 4 Data for water use, wastewater flow, and urine concentration for the studied systems based on data from VAV (1997), Jönsson *et al.* (1998), and Stockholm Water Co. (2002)

System	Water use (m³/pe,d)	Wastewater (m³/pe, d)	Toilet waste (lit/pe, d)	Urine concentration (g N/l)
A. Reference	0.16–0.24	0.18–0.36	–	–
B. RO, SF.	0.16–0.24	0.18–0.36	–	–
C. US, SF.	0.15–0.24	0.17–0.36	–	2.5–3.5
D. US, blackwater	0.15–0.24	0.13–0.32	30–40	2.5–3.5
E. Blackwater	0.13–0.23	0.13–0.32	12–25	–

Table 5 The standard chemical exergy of different compounds

Source	Exergy value	Source	Exergy value
Aluminium chloride	445 kJ/mole	Organic matter	13.6 kJ/g COD
Ferric chloride	230 kJ/mole	Polymer	15.2 kJ/g
Ferrous sulphate	173 kJ/mole	Phosphorus	134.1 kJ/mole HPO_4^{2-}
Magnesium oxide	67 kJ/mole	Sodium hydroxide	38 kJ/mole
Methane gas	832 kJ/mole	Sulphuric acid	108 kJ/mole
Nitrogen	322.1 kJ/mole NH_4-N		

reduced in proportion to the decrease of the amount of used precipitants and if only biological phosphorus removal is used, less than 300 kg/tonne TS are needed (Stark, 2002). In system B, with RO and sludge fractionation, it will be possible to recover phosphorus by both processes and the necessary amount of acids and bases could be reduced. For system C, the conditions for biological phosphorus removal are more favourable compared to the reference system because of a higher BOD/P-ratio. Thus, a different amount of chemicals is used in the scenarios "worst" and "best".

Chemicals for maintenance of the RO are necessary and an addition of acid may be necessary for adjustment of pH (Bergström *et al.*, 2002). Based on experiments from concentration of urine, it can be assumed that the use of acids is less than 0.5 mole H_2SO_4/mole N (Bergström *et al.*, 2002).

Electricity for pumping and treatment

Based on earlier studies it is assumed that 0.50 MJ/m^3 is needed for transport of wastewater and 2.0 MJ/m^3 are needed for the low-pressure system transporting blackwater (Hellstöm, 1999). Concerning vacuum systems, two levels of exergy demand are used, 30 kWh/(pe.yr) in the "best" scenario and 60 kWh/(pe.yr) in the "worst" scenario (Otterpohl, 1997; Balmér *et al.*, 2002).

Electricity consumption for treatment of wastewater and toilet waste is shown in Table 7. Data in Table 7 include the electricity needed for conventional sludge treatment, e.g. mixing of anaerobic digestion tank and centrifugation. For system B, the exergy consumption for RO and the evaporator is assumed to be 13 MJ/m^3 (Bark *et al.*, 2001; Bergström *et al.*, 2002).

Heating of sludge, sludge fractionation and incineration

In system A and E, pasteurisation (70°C, 1 h) is used to achieve pathogen reduction. In the other systems, pathogen elimination is achieved by the sludge fractionation process. However, heat is needed for the anaerobic digestion tanks. It is assumed that the energy demand could be reduced by 40% using heat exchangers and that the losses are 10–20%.

Data for sludge fractionation have been collected from the KREPRO process (Balmér *et al.*, 2002; Stark, 2002). The KREPRO process uses thermal hydrolysis and addition of acid to dissolve phosphorus and other compounds. Phosphorus is recovered as ferric phosphate. The process also generates a fibre fraction which in this study is incinerated in a heat power

Table 6 Assumed distribution of organic matter and nutrients in sewage and estimated amounts generated per capita and day based on data from Swedish EPA (1995a and 1995b)

Source	g/cap., day	Greywater, %	Urine, %	Feaces, %	Blackwater, %
COD	165	50			50
BOD	66.1	50			50
Tot-P	2.2	28	48	24	
Tot-N	13.3	7	82	11	

Table 7 Electricity for wastewater treatment (Hellström, 1999)

System	kWh/(pe.yr)	Comment
A. Reference	50	
B. RO, SF[1]	45	Biogical nitrogen removal not required. RO not included
C. US[2], SF[1]	47	Reduced nitrogen load, but nitrogen removal required
D. US[2], blackwater	45	Treatment of greywater and post-treatment of toilet waste
E. Blackwater	40	Including blackwater treatment

plant for generation of electricity and heat. Sludge generated from greywater treatment in system D and E will also be used as a fuel for production of heat and electricity.

Production of methane

Data from Swedish WWTP have been used to calculate the production of methane (Hellström, 1999). Data for WWTP with nitrogen removal have been used for the reference system, and data for WWTP without nitrogen removal have been used for conventional systems complemented with RO. For system D, blackwater system complemented with urine separation, it is assumed that about 50% of the collected organic matter is converted to methane gas. The corresponding value for system E is 70%.

Transport and spreading of recovered nutrients

The average distance for transport of recovered nutrients from WWTP to agriculture is 25 km in the "best" scenario and 50 km in the "worst" scenario, which probably will be valid for most of the Swedish WWTP (Hellström, 1999). Exergy consumption for transport will also be influenced by the concentration of nutrients. A source-separating system (C-E) will generate the largest volumes to be handled (Table 4). The amount of fuel needed, including empty return transport, is assumed to be 1.2 MJ/(ton, km) for both biosolids and urine (based on data from Malmqvist et al., 1995; Sonesson, 1996). Exergy necessary for spreading will depend on the technique and machinery used and also on factors such as soil type (Godwin et al., 1990). By using data applicable to Swedish conditions, it can be assumed that the fuel consumption for spreading of urine will vary between 5–15 MJ/m^3 urine (e.g. Dalemo, 1996; Nybrant et al., 1996). A value of 10 MJ/m^3 nutrient solution is used in this study.

Results and discussion

The results from exergy and energy calculations are shown in Figure 3 and Figure 4. The difference between exergy (or energy) needed for operation (including transport and spreading) and the exergy content in methane and recovered nutrients could be regarded as a net result. The "best" and "worst" scenarios are indicated as minimum and maximum values at each staple in Figure 3 and Figure 4.

System B, with RO and sludge fractionation (SF), has the greatest exergy demand. This is mainly due to the exergy consumption for RO and the evaporator (230–460 kWh/pe.yr).

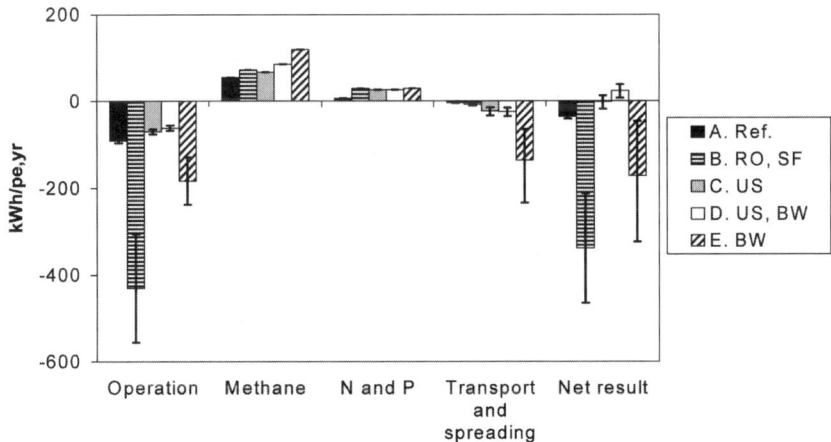

Figure 3 Exergy flows from operation of sewer system and treatment plants, transport and spreading of residuals, and exergy content in methane gas and recovered nutrients. Legend text refers to system descriptions

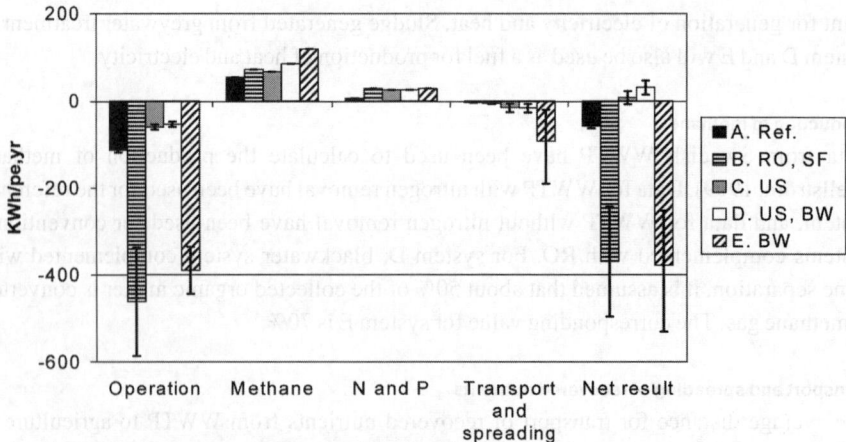

Figure 4 Energy flows from operation of sewer system and treatment plants, transport and spreading of residuals, and energy content in methane gas and recovered nutrients. Legend text refers to system descriptions

Concerning energy demand, there is a relatively small discrepancy between system B and E. The relatively large energy demand for the blackwater system (E) is due to heating of sludge (190–420 kWh/pe.yr) and transport (50–210 kWh/pe.yr). However, energy of lower quality can be used for the heating of sludge and the net exergy consumption for system E is about 170 kWh/(pe.yr) lower than the consumption for system B.

Both systems B and E are very sensitive to the amount of produced wastewater and blackwater. It is interesting to note that the exergy consumption for the blackwater system (E) is almost as low as for the reference system (A), if best available technology and practice is used. However, there is a substantial risk that technical failures and/or unmotivated users may cause a rather high exergy demand. The analysis also shows that it is important to separate storm water and drainage water from wastewater, if RO and the evaporator are used for nutrient recovery (system B). Another possibility is to use RO and the evaporator only for treated blackwater, that is, combining systems B and E.

System D, using urine separation and blackwater treatment, seems to be the most favourable alternative if only exergy and energy aspects are considered. The main reasons for this are relatively low exergy consumption for operation and relatively high production of methane. Compared to a conventional system, the need for aeration is reduced and a larger part of the organic matter is used for methane production. Compared to system E, the need for heating is substantially reduced by using sludge separation in combination with UASB or EGSB reactors for blackwater treatment.

System C, using urine separation, seems also to be an interesting alternative for nutrient recovery compared to systems B and E. However, the amount of recovered nutrients will probably be of the same magnitude as for system D, but lower than for system B and E (Table 2).

This study indicates that nutrient recovery to a certain extent could be achieved without using more exergy compared to conventional treatment systems. However, a high degree of recovery of nutrients will probably require a rather high exergy consumption. This result illustrates that nutrient recovery processes and strategies must be improved. It could also be discussed how much and which nutrients are important to recover (and for which purposes).

Conclusions

Source separation of blackwater and urine is a promising strategy for nutrient recovery if only consumption of natural resources in terms of exergy is considered. However, it is necessary to implement efficient treatment methods for blackwater.

Acknowledgements

This work is a part of the national Swedish research programme "Sustainable Urban Water Management" and the Swedish research programme "Systems for efficient management of resources in wastewater and organic household waste". The work has been financially supported by the Swedish Research Council for Environment, Agricultural Sciences and Spatial Planning (FORMAS), the Swedish Foundation for Strategic Environmental Research (MISTRA), and Stockholm Water Co. Finally, Lena Jonsson at Stockholm Water Co. is acknowledged for reading and checking the manuscript.

References

Balmér, P., Book, K., Hultman, B., Jönnson, H., Kärrman, E., Levlin, E., Palm, O., Schönning C., Seger, A., Stark, K., Söderberg, H., Tideström, H. and Åberg, H. (2002). *Wastewater systems for recovery of phosphorus*. Swedish EPA, (In Swedish).

Bark, U., Filipsson, S., Bjurhem, J.-E., Bergström, R., Isaksson, N., Nemeth, T. and Björlenius., B. (2001). Membrane techniques – a pre-study for local treatment plant in Hammarby Sjöstad (phase I), Stockholm Water Co. report R 31, Stockholm. (In Swedish).

Bergström, R., Bjurhem, J.-E., Ek, M., Björlenius, B., and Hellström, D. (2002). *Nutrient concentration processes for urine and reject water from anaerobic digestion*. Stockholm Water Co. (In Swedish).

Edström, E. Nordberg, Å., Olsson, L.-E., and Hellström, D. (2001). Anaerobic processes – a pre-study for local treatment plant in Hammarby Sjöstad (phase I), Stockholm Water Co. report R 30, Stockholm, (In Swedish).

Godwin, R.J., Warner, N.L. and Hann, M.J. (1990). Comparison of Umbilical Hose and Conventional Tanker-Mounted Systems. *Agricultural Engineer*, **summer**, 45.

Hanæus, J., Hellström, D. and Johansson, E. (1997). A Study of a Urine Separation System in an Ecological Village in Northern Sweden. *Wat. Sci. Tech.*, **35**(9), 153.

Hellström, D. (1997). An Exergy Analysis for a Wastewater Treatment Plant – an Estimation of the Consumption of Physical Resources. *Wat. Environ. Res.*, **69**, 44.

Hellström, D. (1998). Exergy Analysis: A Comparison of Various Treatment Alternatives for Nutrient Removal. In: *Chemical Water and Wastewater Treatment V*, H.H. Hahn *et al.* (Eds.). Springer, Berlin Heidelberg, 313–324.

Hellström, D. and Kärrman, E. (1997). Exergy Analysis and Nutrient Flows of Various Sewerage Systems. *Wat. Sci. Tech.*, **35**(9), 135.

Hellström, D. (1999). "Exergy Analysis: A Comparison of Urine Separation Systems and Conventional Treatment Systems". *Wat. Environ. Res.*, **71**, 1354–1363.

Holmberg, J. (1995). Socio-Ecological Principles and Indicators for Sustainability. Ph. D. Thesis, Institute of Physical Resource Theory, Chalmers University of Technology and Gothenburg University.

Jönsson, H., Burström, A., and Svensson, J. (1998). Measurements for Two Urine Separating Sewage Systems, Dep. Agricultural Engineering, Swedish University of Agricultural Sciences, Report 228. (In Swedish).

Malmén, L. (2001). Batchwise Aerobic Thermophilic Sludge Treatment in a Recycledbased System for Organic Waste and Waste water in Bomarsund. PROGRESS REPORT III. Project No. LIFE98 ENV/FIN/000575.

Malmqvist, P.-A., Björkman, H., Stenberg, M., Andersson, A.C., Tillman, A.-M. and Kärrman, E. (1995). *Alternatives to Conventional Wastewater Treatment Systems in Bergsjön and Hamburgsund*. VA-FORSK report 1995-03. (In Swedish).

Norin, E., Gruvberger C. and Nilsson, P.-O. (2000). *Handling of black water from the school of Tegelviken – recycling system with liquid composting*. VA-FORSK report 2000-03. (In Swedish).

Nybrant, T., Jönsson, H., Sonesson, U., Frostell, B., Mingarini, K., Thyselius, L., Dalemo, M. and Sundqvist, J.-O. (1996). *Systems Analysis of Organic Waste – The ORWARE Model, Case Study, Part One*. AFR-Report 109. Swedish Environmental Protection Agency. Stockholm, Sweden.

Otterpohl, R., Grottker, M. and Lange, J. (1997). Sustainable water and waste management in urban areas. *Wat. Sci. Tech.*, **35**(9), 121–133.

Palmquist, H. (2001). Hazardous substances in wastewater systems – a delicate issue for wastewater management. Licentiate thesis 2001:65, Div. Sanitary Engineering, Luelå University of Technology, Luleå, Sweden.

Sonesson, U. (1996). *Modelling of the Compost and Transport Processes in the ORWARE Simulation Model*. Dept. of Agricultural Engineering, Swedish University of Agricultural Sciences. Report no. 214. Uppsala.

Stark, K. (2002). Phosphorus Release from Sewage Sludge by Use of Acids and Bases, Licentiate Thesis TRITA-AMI LIC 2005, Dep. Land and Water Resources Engineering, KTH, Stockholm.

Stockholm Water Co. (2002). *Environmental report 2001*. Report 240–439. (In Swedish).

Stockholm Water Co. (2002). *Environmental accounts and annual report*.

Swedish EPA (1995a). *Large Wastewater Treatment Plants – Biosolids and Wastewater 1993*. Swedish Environmental Protection Agency, Report 4423. (In Swedish).

Swedish EPA (1995b). *What does Household Wastewater Contain?* Swedish Environmental Protection Agency, Report 4425. (In Swedish).

Szargut, J., Morris, D.R. and Steward, F.R. (1988). *Exergy Analysis of Thermal, Chemical and Metallurgical Processes*. Springer-Verlag, New York, N.Y.

VAV (1997). *Statistics for Water and Wastewater system*, Report S 95. (in Swedish).

Vinnerås, B. (2001). Faecal separation and urine diversion for nutrient management of household biodegradable waste and wastewater, Licentiate thesis, report 244, Dep. Agricultural Engineering, Swedish University of Agricultural Sciences, Uppsala.

Wall, G. (1992). *Exergy – an holistic approach applied in Gothenburg* (In Swedish). Report "Underlagsmaterial 6:92", Stadsbyggnadskontoret, Göteborg, Sweden. (In Swedish).

Nutrients in urine: energetic aspects of removal and recovery

M. Maurer*, P. Schwegler and T.A. Larsen

EAWAG, Environmental Engineering, Überlandstrasse 133, CH-8600 Dübendorf, Switzerland
* Author to whom all correspondence should be addressed

Abstract The analysis of different removal and recovery techniques for nutrients in urine shows that in many cases recovery is energetically more efficient than removal and new-production from natural resources. Considering only the running electricity and fossil energy requirements for the traditional way of wastewater treatment and fertiliser production, the following specific energy requirements can be calculated: 45 MJ $kg^{-1}{}_N$ for denitrification in a WWTP, 49 MJ $kg^{-1}{}_P$ for P-precipitation in a WWTP, 45 MJ $kg^{-1}{}_N$ for N-fertiliser and 29 MJ $kg^{-1}{}_P$ for P-fertiliser production. These numbers are higher than the values derived for thermal volume reduction of urine (35 MJ $kg^{-1}{}_N$ for eliminating 90% water) or production of struvite (102 MJ $kg^{-1}{}_N$, including 2.2 kg P). Considering only the electricity and fossil energy for the traditional way of wastewater treatment and fertiliser production, the energy value of 1 PE urine is 0.87 MJ $PE^{-1}d^{-1}$ (fertiliser value: 0.44, wastewater treatment: 0.43 MJ $PE^{-1}d^{-1}$).

A more detailed life cycle assessment (LCA) of the entire urine collection system, including the required materials and the environmental burden, support the energy analysis. The LCA compares conventional denitrification in a wastewater treatment plant with collecting urine in households, reducing the volume by evaporation and using it as a multi-nutrient fertiliser. The primary energy consumption for recovery and reuse of urine, including the nutrients N, P and K, is calculated with 65 MJ $kg^{-1}{}_N$, compared with 153 MJ $kg^{-1}{}_N$ derived for the conventional 'recycling over the atmosphere'.

Keywords Energy requirements; LCA; nitrogen; phosphorus; urine source separation

Introduction

In regions with sensitive surface waters, the costs for wastewater treatment are dominated by the conversion and elimination of nitrogen and phosphorus. The introduction of nitrification, a prerequisite for enhanced nitrogen elimination, into an activated sludge plant increases the reactor volume significantly and leads to higher energy consumption of approximately 60 to 80%. The elimination of phosphorus requires either the addition of chemicals and subsequent disposal of inorganic sludge or an increase of reactor volume for enhanced biological phosphorus removal (EBPR). At the same time, the elements N and P are essential for agriculture and have to be produced technically from natural resources. Phosphorus is gained from rock phosphates, which deplete in quantity and quality (USGS, 2002). Nitrogen fixation is determined by the cost and availability of fossil fuels. The idea is compelling that by recovering the nutrients from wastewater streams and reusing it for agricultural purposes, the costs of extensive wastewater treatment could be avoided and resources saved.

Several removal and recovery techniques have been investigated (Brett *et al.*, 1997; Maurer *et al.*, 2002). However, currently only very limited information is available to decide which route, removal or recovery, should be taken. This article gives an overview over the possible technologies for nutrient recovery from liquid wastes, compares them with the current practice, and discusses the feasibility of recovery. Energy data for the various technologies are compiled and give an important decision guideline for removal or recovery.

Urine

In municipal wastewater, urine is the dominating source for the major agricultural nutrients nitrogen, phosphorus and potassium, and holds 50% to 90% of these essential elements (Larsen and Gujer, 1996). Urine is therefore a prime target, for achieving a more sustainable handling of nutrients from urban wastewater. Of the fertiliser consumption in the EU, 12 % of N, 6% of P and 10% of K could be recovered from urine maximally (EFMA, 1999). These relatively small amounts indicate that currently there is a substantial excess of nutrient supply in agriculture.

Nutrient loads in wastewater and their source are subject to uncertainty. For nitrogen, a load of 11 $g_N PE^{-1} d^{-1}$ seems well established (ATV, 2000). For P and K, we assume a load of 1.1, and 2.7 $gPE^{-1} d^{-1}$ respectively. In Table 1, we present the basic data for the calculations performed in this paper.

Nitrogen

In principle there are two ways to prevent the eutrophication of aquatic systems by wastewater and to make its nitrogen content available for farming:
1. Conversion of the nitrogen compounds into elemental nitrogen gas and technical production of ammonia by the Haber-Bosch process ('recycling over the atmosphere').
2. Direct recovery from liquid waste and conversion into a reusable form.

In western Europe, northern America and Japan, the first pathway is common practice in wastewater treatment, where microbial denitrification is used to convert nitrate into elemental nitrogen gas. For pathway 2, large-scale applications are scarce and many approaches are still under development.

Nitrogen recovery techniques

Except aquacultures, all recovery techniques require relatively concentrated solutions. Some of them have already been tested and used on a full-scale base, such as struvite formation or ammonia stripping. Others, such as volume reduction by evaporation, partial freezing or reverse osmosis have only been tested in laboratories (Maurer *et al.*, 2002). In municipal wastewater handling, the only liquid waste streams with high concentrations of nitrogen are sludge dewatering liquid (digester supernatant), containing about 15–20 % of the total nitrogen load in wastewater, and human urine, containing about 80 % of the total nitrogen load (Table 1).

Besides the costs for installation and operation, the main question remains whether direct recovery and reuse technologies are more sustainable than denitrification combined with technical fertiliser production. One important indicator is the running energy requirement of a specific process. Energy considerations do not replace a careful evaluation of all relevant environmental factors, e.g. with life cycle assessments. Nevertheless, consumption of energy in the form of electricity, fuel and chemicals is a very important guideline to compare two processes.

Table 1 Typical values for the major nutrients in urine. Based on numbers in text and Ciba Geigy (1977) with characteristic concentrations in urine from adult

Element	In urine %	In urine $g_x PE^{-1} d^{-1}$	Typical conc. $g_x m^{-3}$	Typical conc. mmol l^{-1}
N	80	8.8	8180	584
P	50–80	0.7	670	21.6
K	80–90	2.3	2160	55.2

Energy

Tables 2 and 3 summarise the energy requirements for the various techniques to remove, produce and recover nitrogen. The calculations do not represent a complete life cycle assessment (LCA), but focus only on the running energy consumption (electricity, fuel and chemicals); so called 'grey energy' contained in hardware such as reactors and pumps are not included. The values are mostly derived from the literature, except the data for thermal volume reduction of urine, which is based on a detailed feasibility study. In the following the underlying calculations are explained in detail.

Denitrification: Technically there are several ways to achieve denitrification. To get an idea about the energy requirements for denitrification we distinguished three principal types: i) Heterotrophic denitrification without the use of an external carbon source. ii) Heterotrophic denitrification with addition of an external carbon source (e.g. methanol). iii) Autotrophic denitrification (Sharon / Anammox). The calculations only consider nitrogen that is converted into N_2-gas. The N incorporated into biomass is regarded as an intermediate that still needs further treatment. The oxygenation efficiency (energy efficiency of the aeration process), where needed, was assumed to be 1.7 $kg_{O_2} kWh^{-1}$ (= 2.2 MJ $kg^{-1}_{O_2}$).

i) A common way to eliminate nitrogen in a WWTP is pre-denitrification (Henze *et al.*, 2002). Typical for this process is that an excess of ammonium needs to be converted into nitrate (nitrified) to eliminate a specific amount of nitrogen. The calculations assume that from the nitrified ammonia 60% is finally converted to N_2 and the rest of the nitrate is 'lost' over the effluent. The electricity demand for the nitrification (aeration) is 17 MJ $kg^{-1}_{N,eliminated}$ and 3 MJ $kg^{-1}_{N,eliminated}$ is required for the internal recycling of water. However 6 MJ $kg^{-1}_{N,eliminated}$ is recovered in the form of oxidised COD in the denitrification process (Table 2).

ii) For denitrification with an external carbon source we assumed that typically methanol (3.4 $kg_{Methanol} kg^{-1}_{N,eliminated}$, Purtschert *et al.*, 1996) is used. The calculations consider the thermal energy content of methanol (77 MJ $kg^{-1}_{N,eliminated}$) and the electricity consumption for nitrification (10 MJ $kg^{-1}_{N,eliminated}$ for aeration; Table 2).

iii) Autotrophic denitrification (Van Dongen *et al.*, 2001; Udert *et al.*, 2003) is only reasonably applicable in concentrated solutions. The aeration in the Sharon process requires 4 MJ $kg^{-1}_{N,eliminated}$ electricity and approximately 1 MJ $kg^{-1}_{N,eliminated}$ for pumping, mixing, etc. The elimination of 1 $kg_{COD} kg^{-1}_{N}$ COD from urine requires additional 1 MJ kg^{-1}_{N} electricity. No heating for increased reaction temperature was considered (Table 2).

Table 2 Energy requirements for nitrification/denitrification and N-fixation with the Haber-Bosch process. Only running costs, e.g. for aeration, are considered (energy use for building materials and engines are neglected). The conversion of electricity into primary energy is based on an average European electricity mix (efficiency of 0.31) according to UCPTE (1994). See text for calculations and references

Process	Electricity MJ kg^{-1}_N	Other MJ kg^{-1}_N	Total UCPTE MJ kg^{-1}_N	Total UCPTE W PE^{-1}
Nitrification/pre-denitrification in WWTP	14	–	45	4.6
Nitrification/denitrification with methanol as substrate	10	77	109	11.1
Sharon/Anammox	5	–	16	1.6
Sharon/Anammox with urine (+COD removal)	6	–	19	1.9
Ammonia production (best available technology)	–	37	37	3.8
Average N-fertiliser production Europe	0.8	42	45	4.6
Average ammonia production Europe	–	43	43	4.4
Average urea production Europe	1	51	54	5.5

Table 3 Energy requirements for various urine treatment options to recover nitrogen. See also caption text in Table 2

Process	Electricity MJ kg$^{-1}_N$	Other MJ kg$^{-1}_N$	Total UCPTE MJ kg$^{-1}_N$	Total UCPTE W PE^{-1}
Thermal volume reduction of stabilised urine (10 fold concentration), one step distillation	–	389	389	39.6
Thermal volume reduction of stabilised urine (10 fold concentration) with vapour compression	7	11	34	3.5
Volume reduction of stabilised urine with reverse osmosis (5 fold concentration)	9	–	29	3.0
Struvite production (quantitative P-fixation)	6	6	25	2.5
Struvite production (quantitative N-fixation)	9	72	102	10.4
Stripping with air and (NH$_4$)$_2$SO$_4$ production	26	6	90	9.2

Fertiliser production: The best available technology for ammonia production requires 37 MJ kg$^{-1}_N$ with natural gas and 45 MJ kg$^{-1}_N$ with heavy oil (EFMA, 2000). Based on a careful investigation of the technologies actually used in Europe, Patyk and Reinhard (1997) reported 43 MJ kg$^{-1}_N$ for ammonium production and 54 MJ kg$^{-1}_N$ for urea production. In addition, they used fertiliser sales from the years 1993/94 to calculate the energy requirements of an 'average' fertiliser with 45 MJ kg$^{-1}_N$. Pimentel *et al.* (1990) used a value of 87.9 MJ kg$^{-1}_N$ for their estimation of energy use in agricultural production. However, it is not clear where they derived this value from (Table 2).

Volume reduction: Several techniques for volume reduction are described in the literature. For reverse osmosis of urine, Dalhammer (1997) reported an electricity demand of 5 to 10 kWh m^{-3} for a fivefold increase of concentration. At a nitrogen concentration of 3,100 g$_N$m^{-3} this gives a specific electricity consumption of 9 MJ kg$^{-1}_N$ (Table 3).

To estimate the cost and energy requirements for evaporation, a detailed feasibility study, based on experimental results, was performed (Mayer, 2002). Hydrolysis of the collected urine was inhibited by addition of sulphuric or acetic acid (Hellstrom *et al.*, 1999). Subsequently, urine was evaporated at 200 mbar and 78°C and neutralised. The nitrogen concentration increased from 9,606 to 96,522 g$_N$m^{-3} and the total solids from 43 to 445 kg$_{TS}$m^{-3}. The concentrated urine solution with a TS concentration of 573 kg$_{TS}$m^{-3} showed a viscosity of 6.42 mPa s (57 °C) and compared to pure water an increased boiling point of 10.1°C. Based on these data, a technical vapour condensation evaporation plant was laid out for a tenfold volume reduction of urine. The calculated energy requirements are: 7 MJ kg$^{-1}_N$ electricity and 11 MJ kg$^{-1}_N$ fuel for steam production (Table 3). This corresponds well with the values for seawater desalination given by Wood (1982), who reports 150–180 MJ m^{-3} distilled water (equivalent to 17–20 MJ kg$^{-1}_N$).

Another possibility is concentrating urine by partial freezing described by Lind *et al.* (2001), however no energy data are available for this technique.

Struvite production: In order to eliminate nitrogen quantitatively in the form of struvite (NH$_4$MgPO$_4$·6 H$_2$O), 2.1 kg of magnesium and 2.1 kg of phosphorus need to be added.
The energy requirements for the process itself are reported as 5.6 MJ kg$^{-1}_N$ (Siegrist, 1996). The following energy requirements were assumed (Patyk and Reinhard, 1997): magnesium oxide (MgO) 13.4 MJ kg$^{-1}_{Mg}$ and for phosphoric acid 21.1 MJ kg$^{-1}_P$ fossil fuel and 3.7 MJ kg$^{-1}_P$ electricity. No purification and granulising step was included in the calculation and it was assumed that no pH adjustments are needed (Table 3).

Stripping and production of ammonium sulfate: Air is used to strip out ammonia, which is thereafter absorbed in an acid solution. This process requires 2.4 kg kg$^{-1}_N$ calcium oxide for increasing the pH and 7.0 kg kg$^{-1}_N$ sulphuric acid for absorbing the gaseous ammonia (Siegrist, 1996). It produces a 40% ammonium sulfate solution. The electricity requirement for the aeration is an estimated 26.3 MJ kg$^{-1}_N$ and the following energy data are used for the calculations (Patyk and Reinhard, 1997): sulphuric acid (H_2SO_4) -1.1 MJ kg$^{-1}_{H2SO4}$ and calcium oxide (CaO) 5.8 MJ kg$^{-1}_{CaO}$ (Table 3).

Phosphorus

Recycling of phosphorus from conventional wastewater treatment plants is very limited. Although there are several ways to efficiently incorporate P into the excess sewage sludge, reuse of the latter is restricted. Reasons are the low plant availability of chemically precipitated phosphorus and the contamination of sludge with pollutants such as heavy metals and organic micropollutants (Renner, 2000). No energy data were found for the extraction of phosphorus from different sludge streams.

P-precipitation in WWTP: For chemical phosphorus elimination in WWTP we assumed that simultaneous precipitation with iron(II)-sulfate is performed. The stoichiometric ratio of iron to P-precipitated (β) is typically 1.8 and produces 6.8 kg$_{TS}$kg$^{-1}_P$ (Henze *et al.*, 2002). The energy requirements in total 49 MJ kg$^{-1}_P$ include the production of the precipitant (24 MJ kg$^{-1}_P$), the transport of the sludge to an incinerator (2 MJ kg$^{-1}_P$) and the energy loss for burning inorganic material in a modern incinerator with off-gas treatment (23 MJ kg$^{-1}_P$; Table 4).

Enhanced biological P elimination (EBPR): Under favourable conditions a WWTP can eliminate phosphorus biologically. It is assumed that the carbon substrate requirements can be covered by the wastewater and that the additional sludge production is 3.3 kg$_{TS}$kg$^{-1}_P$ (= 1 MJ kg$^{-1}_P$ for transport and 11 MJ kg$^{-1}_P$ for incineration and off-gas treatment). The electricity requirements for mixing and internal recycling is assumed to be 5 MJ kg$^{-1}_P$ (value depends strongly on the flow scheme; Table 4).

Struvite production. Focussing on P- instead of N-elimination, struvite production only requires the addition of magnesium, typically in the form of MgO. For details, see the same process described in the section 'Nitrogen' (Table 4).

Fertiliser production: For triple superphosphate production, based on a careful investigation of the technologies used in Europe, Patyk and Reinhard (1997) reported 3.6 MJ kg$^{-1}_P$

Table 4 Energy requirements for various urine treatment options to recover phosphorus. See also caption text in Table 2

Process	Electricity MJ kg$^{-1}_P$	Other MJ kg$^{-1}_P$	Total UCPTE MJ kg$^{-1}_P$	Total UCPTE W PE^{-1}
P-precipitation in WWTP (with FeSO$_4$), incl. energy for sludge transport and incineration	–	49	49	0.4
EBPR in WWTP	5	12	28	0.2
Thermal volume reduction of stabilised urine (10 fold concentrated) with vapour compression (corresponding values from)	85	134	408	3.3
Struvite production (quantitative P-fixation)	3	13	21	0.2
Triple superphosphate fertiliser production	4	23	36	0.3
Average P-fertiliser production Europe	4	16	29	0.2

electricity and 22.7 MJ kg^{-1}$_P$ fossil energy consumption. Taking the fertiliser sales 1993/94 into account, the production of an 'average' P-fertiliser consumed 3.8 MJ kg^{-1}$_P$ electricity and 16.1 MJ kg^{-1}$_P$ fossil energy. These numbers correspond well with the value of 26.4 MJ kg^{-1}$_P$ used by Pimentel *et al.* (1990) for their estimation of energy use in agricultural production (Table 4).

LCA of denitrification and urine source separation

A more detailed analysis of the entire urine collection system, including the required materials and the environmental burden, was performed to verify the conclusions drawn from the comparison of the running energy requirements alone (Schwegler, 2002). The life cycle assessment (LCA) compared two alternatives for Switzerland to keep away nitrogen (functional unit: mass of nitrogen eliminated) from the river Rhine:

1. DENI: upgrade of existing wastewater treatment plants (WWTP) for nitrification and denitrification and use of artificial fertiliser in agriculture. Base-line for the upgrade of the WWTP is their technical status in Switzerland in 1993, with a high number of high loaded plants and only partial nitrification. To achieve nitrification and denitrification for the entire year, additional reactor volume (0.071 m^{3}kg^{-1}$_N$yr^{-1}) and electricity (20.4 MJ kg^{-1}$_{N,eliminated}$) for the higher oxygen consumption (11.3 kg$_{O2}$kg^{-1}$_{N,eliminated}$) is required. In addition, chemical fertiliser production and application in Swiss agriculture is also included into this scenario.

2. USS (Urine Source Separation): decentralised collection of urine, centralised thermal volume reduction and use as fertiliser in agriculture (recovery and recycling). The scenario assumes that slightly diluted urine is collected locally in the households, collected by a truck twice a year and delivered to a centralised evaporation plant, where the urine is concentrated by a factor 10. The decomposition of urea in the storage tanks is prevented by adding sulphuric or nitric acid. The concentrated urine is applied to agriculture mainly as a urea-fertiliser. It is assumed that per PE a total tank volume of 0.356 m^3 (= 0.11 m^{3}kg^{-1}$_N$, incl. storage volume in work areas) and about 11.7 m tubing (= 3.7 m kg^{-1}$_N$) are required, having a life span of 30 years (material: polyethylene). The average transport distance is assumed as following: consumer → evaporation plant: 60 km and evaporation plant → farmers: 100 km.

The calculations were done with the software EMIS (version 3.5, 2001), based on data sets from various environmental inventories. The main results of the study are (Table 5):

- The primary energy consumption for recovery and reuse of urine, including the nutrients N, P and K, is 65 MJ kg^{-1}$_N$, compared with 153 MJ kg^{-1}$_N$ derived from the conventional 'recycling over the atmosphere'.
- Regarding energy requirements and greenhouse effect, USS performs markedly better than denitrification. Although thermal volume reduction and extensive transport is included, USS only yields 42% of the energy consumption and 52% of the greenhouse potential, compared with the conventional way of handling nitrogen from liquid waste.
- The results for the potentials to form environmentally relevant acids and atmospheric ozone is less clear. Here, the outcome depended very strongly on the chosen technology for applying the concentrated urine in agriculture and how experts assess the environmental consequences (scenario 1 and 2). The values given in Table 5 represent the worst cases, using field sprayers for the application of concentrated urine (scenario 1). Another scenario assumed that the farmers dilute urine and use trailing hoses (scenario 2). Using data from Johansson *et al.* (2000), the estimation for the potential acid formation drops from 112 for scenario 1 to 25 g$_{SOx-equivalent}$kg^{-1}$_N$ for scenario 2. These uncertainties indicate that the field application of urine, even if N is stabilised as urea, requires specific attention in order to quantify any ecological benefit or drawback.

- The results also show that the household collection of urine contributes significantly to the energy balance. The calculation is based on a relatively conservative estimation of tank volume and pipe length in households and workplaces, but also assumes that it will be possible to collect urine almost undiluted (5% tap water). Figure 1 shows that 22 MJ kg^{-1}_N in the form of plastic are required, which is a third of the net energy consumption of the whole process.

Discussion

At present, nitrification/denitrification and the subsequent recovery of nitrogen by the Haber-Bosch process is the prevalent way of dealing with nitrogen. These processes together have typical operating energy demand of totally 90 MJ kg_N^{-1} (45+45, Table 2). The advantage of the nitrification/denitrification technology is that it can be implemented relatively easily into the existing system. Energetically more favourable is the elimination of nitrogen in a Sharon-Anammox process, combined with fertiliser production (19 + 45 = 64 MJ kg_N^{-1}, Table 2). This process requires concentrated N-solutions like digester supernatant or urine, and enables the elimination of nitrogen in large amounts.

The high energy requirements for the N-fixation by the Haber-Bosch pathway makes many recovery and reuse techniques interesting. From an energetic point of view, the evaporation of the water content of urine seems to be competitive. Although farmers are used to handling liquid fertilisers with a high water content (liquid manure), a recent survey shows that a solid product would be much preferred and would be more readily accepted (Lienert et al., 2003). The elimination of water is one possibility to enable transportation and storage of nitrogen, hereby making it comparable with conventional N-fertiliser. Laboratory experiments with urine showed that a 90% elimination of water is relatively easy to achieve

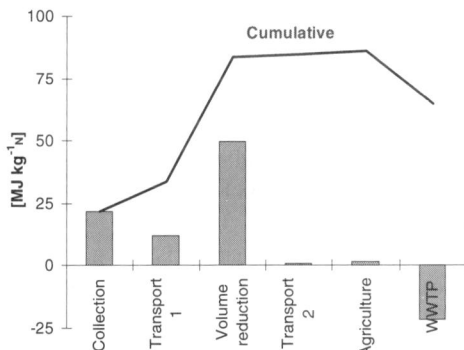

Figure 1 Energy consumption of the different units for recovery and reuse of urine in agriculture. 'Transport 1' is collecting and bringing the urine to the evaporation plant (60 km); 'Transport 2' is delivering the concentrated urine to the farmers (100 km). The energy savings in the WWTP includes the reduced oxygen consumption for nitrification and the decreased sludge

Table 5 Main results of the LCA to compare the two options to treat urine: USS – urine source separation, treatment and use as a fertiliser and DENI – denitrification in a WWTP and production of artificial fertiliser from elemental nitrogen (Schwegler, 2002)

Item	USS	DENI	Units
Energy consumption (primary resources)	65	153	MJ kg^{-1}_N
	7	16	W PE^{-1}
Ozone formation	15	12	$g_{ethen-equivalent} kg^{-1}_N$
greenhouse potential	4,093	7871	$g_{CO2-equivalent} kg^{-1}_N$
Formation of acids, scenario 1	112	39	$g_{SOx-equivalent} kg^{-1}_N$
Formation of acids, scenario 2	25	39	$g_{SOx-equivalent} kg^{-1}_N$

(Mayer, 2002). The last step towards a solid multi-nutrient fertiliser is technically possible, but seems to require specific know-how. Currently it is not clear whether this is feasible for small scale applications (less than 10^5 $kg_N d^{-1}$).

Another interesting product with similar properties as artificial fertiliser, is struvite recovered from urine (Lind et al., 2000). Compared with the 90 MJ kg_N^{-1} for the conventional 'recycling over the atmosphere', the energy demand of 102 MJ kg^{-1}_N for the production of struvite seems high (Table 3). However, in struvite every kilogram nitrogen is accompanied by 2.2 kg phosphorus. The production of equivalent amounts of artificial P-fertiliser requires 64 MJ (Table 4), which totals (90 + 64 =) 154 MJ kg_N^{-1} for the corresponding conventional treatment route. Table 3 shows that only the stripping of ammonia and the evaporation of water without energy recovery seems to be less favourable.

A recent investigation by Richard and Johnston (2001) indicate that recycled struvite has a potential to substitute for current fertiliser on the market. Together with the potential benefits of urine source separation for urban water management (Larsen and Gujer, 1996), this recovery route looks like a prime candidate for a sustainable improvement of the current wastewater handling.

The conclusion that from an energy point of view recovery and reuse of nutrients from urine seems to be more favourable than removal is also supported by a live cycle evaluation. In addition, the more detailed LCA revealed two critical points in the recovery and reuse scenario:
- The urine collecting system on a household level can consume large energy resources in the form of tanks and piping and therefore deserves further attention.
- The environmental consequences of applying urine in agriculture were ambiguous. The impacts depended strongly on the application technology and on differing opinions of experts. Manure and artificial fertiliser are known to cause negative impacts depending on their application. However, urine as a fertiliser is scientifically not very well investigated and more experience needs to be compiled.

The conclusions for phosphorus are very similar to nitrogen. The elimination of phosphate and the new fertiliser production requires between 57 MJ kg^{-1}_P (EBPR, Table 4) and 78 MJ kg^{-1}_P (P-precipitation). Struvite production only consumes 21 MJ kg^{-1}_P. Table 4 also indicates that the P-content of urine is too small to make evaporation worthwhile for the recovery of phosphorus only.

The only way to recover potassium or sulfur is using concentrated or diluted urine directly as a fertiliser. For K-fertiliser production Patyk and Reinhardt (1997) reported 0.5 MJ kg^{-1}_K electricity and 9.6 MJ kg^{-1}_K fossil energy consumption (a total of 11 MJ kg^{-1}_K primary energy according to UCPTE, 1994).

By combining the electricity and fossil energy consumption for the main elements nitrogen, phosphorus and potassium, we can calculate an 'energy value' for collected and untreated urine. Per PE we get an intrinsic fertiliser value of 0.44 MJ $PE^{-1}d^{-1}$ (5.1 W PE^{-1}) and a wastewater treatment energy bonus of 0.43 MJ $PE^{-1}d^{-1}$ (5.0 W PE^{-1}). This means that the total energy value (= intrinsic energy content) of 1 PE urine is: 0.87 MJ $PE^{-1}d^{-1}$ (10.1 W PE^{-1}). A majority (91%) of this energy content is due to nitrogen, only 6% to phosphorus and the rest to potassium.

Conclusions
- The detailed analysis of different removal and recovery techniques shows that in many cases recovery is energetically more efficient than removal and new-production from natural resources. This is especially valid for recovery and recycling of P and N by using urine directly as a fertiliser or by producing struvite as an end-product.
- Pathways working with concentrated solutions, such as urine, are in general

energetically more favourable than conventional processes in wastewater treatment plants.
- Considering only the electricity and fossil energy consumption for the traditional way of wastewater treatment and fertiliser production, the energy value of 1 PE urine is: 0.9 MJ $PE^{-1}d^{-1}$ (fertiliser value: 0.44, wastewater treatment: 0.43 MJ $PE^{-1}d^{-1}$). 91% of this energy content is due to nitrogen, 6% due to phosphorus and the rest is in potassium.
- Additional purification steps (e.g. for elimination of micropollutants) could be integrated into a recovery and recycling application and could energetically still be more favourable than the current traditional pathway for nutrient elimination in wastewater treatment.

References

ATV (2000). *Bemessung von einstufigen Belebungsanlagen.* Arbeitsblatt ATVL-DVWK-A131, GFA, Hennef, Germany. ISBN 3-933707-41-2.

Brett, S., Guy, J., Morse, G.K. and Lester, J.N. (1997). *Phosphorus Removal and Recovery Technologies.* Selper Publications, ISBN 0-948411-10-0.

BUWAL (1996). *Stickstoff-Frachten aus Abwasserreinigungsanlagen – Istzustand, Eliminationspotential und Kostenwirksamkeit von Massnahmen.* Schriftenreihe Umwelt, Nr. 276, Bern.

Ciba Geigy (1977). *Wissenschaftliche Tabellen Geigy, Teilband Körperflüssigkeiten,* 8. ed., Basel.

Dalhammar, G. (1997). Behandling och koncentrering av humanurin. Report, personal communication.

Dongen van, L.G.J.M., Jetten, M.S.M. and van Loosdrecht, M.C.M. (2001). *The Combined Sharon/Anammox Process.* IWA Publishing, London, ISBN 1-84339 000 0.

EFMA (1999). *Forecast of Food, Farming and Fertiliser use in the European Union – 2000 to 2010.* European Fertilizer Manufacturers' Association, B-1160 Brussels, Belgium [http://www.efma.org/publications/index.asp]. main@efma.be.

EFMA (2000). Best Available Techniques for Pollution Prevention and Control in the European Fertilizer Industry, European Fertilizer Manufacturers' Association, B-1160 Brussels, Belgium [http://www.efma.org/publications/index.asp]. main@efma.be.

EMIS (2001) LCA-Software by Carbotech AG, CH-4051 Basel [http://www.carbotech.ch/umweltmanagement/emis.htm]. f.dinkel@carbotech.ch.

Hellstrom, D., Johannson, E. and Grennberg, K. (1999). Storage of human urine: acidification as a method to inhibit decomposition of urea. *Ecological Engineering,* **12**, 253–269.

Henze, M., Harremoes, P., Jes la Cour, J. and Arvin, E. (2002). *Wastewater Treatment – Biological and Chemical Processes.* 3rd ed, Springer-Verlag Berlin, ISBN-3-540-42228-5.

Johansson, M., Jönsson, H., Höglund, C., Richert-Stintzing, A. and Rodhe, L. (2000). *Urine Separation – Closing the Nutrient Cycle.* Stockholm Vatten, Stockholmshem.

Larsen, T.A. and Gujer, W. (1996). Separate management of anthropogenic nutrient solutions (human urine). *Wat. Sci. Tech.,* **34**(3–4), 87-94.

Lienert, J., Haller, M., Berner, A., Stauffacher, M. and Larsen, T.A. (2003). How do farmers in Switzerland perceive fertilizers from recycled anthropogenic nutrients (urine)? *Wat. Sci. and Tech.,* **48**(1) 47–56 (this issue).

Lind, B.B., Ban, Z. and Byden, S. (2000). Nutrient recovery from human urine by struvite crystallization with ammonia adsorption on zeolite and wollastonite. *Bioresource Technology,* **73**(2), 169–174.

Lind, B.B., Ban, Z. and Byden, S. (2001). Volume reduction and concentration of nutrients in human urine. *Ecological Engineering,* **16**, 561–566.

Maurer, M., Muncke, J. and Larsen, T. (2002). Techniques for nitrogen recovery and reuse. In: Lens, P., *et al.* (2002) *Water and Resource Recovery in Industry.* IWA Publishing – in press.

Mayer, M. (2002). Thermische Hygienisierung und Eindampfung von Humanurin. Diplomarbeit des Institut für Umwelttechnik der Fachhochschule beider Basel, Muttenz, Schweiz (Thermal disinfection and evaporation of human urine. Diploma work of the Institute for Environmental Technology, Fachhochschule beider Basel, Muttenz, Switzerland).

Patyk, A. and Reinhardt, G.A. (1997). *Düngemittel – Energie- und Stoffstrombilanzen.* Vieweg, Braunschweig, ISBN-3-528-06885-X.

Pimentel, D., Dazhong, W. and Giampietro, M. (1990). Technological changes in U.S. agricultural energy use. In: Gliessmann, S.R. (ed.): *Agroecology: Researching the Ecological Basis for Sustainable Agriculture*, Springer-Verlag, New York, 305–321.

Purtschert, I., Siegrist, H. and Gujer, W. (1996). Enhanced denitrification with methanol at WWTP Zurich-Werdholzli. *Wat. Sci. Tech.*, **33**(12), 117–126.

Renner, R. (2000). Sewage sludge: pros and cons. *Env. Sci. Tech.*, **34**(19), 430A-435A.

Richard, I.R. and Johnston, A.E. (2001). Effectiveness of different precipitated phosphates as phosphorus sources for plants. Scientific research report. CEEP BIT, CEFIC Sector Group, B-1160 Brussel, Belgium.

Schwegler, P. (2002). Ökobilanzierung zur Human-Urineindampfung und Verwertung in der Landwirtschaft. Diplomarbeit des Institut für Umwelttechnik der Fachhochschule beider Basel, Muttenz, Schweiz (LCA of urine evaporation and use in agriculture. Diploma work of the Institute for Environmental Technology, Fachhochschule beider Basel, Muttenz, Switzerland).

Siegrist, H. (1996). Nitrogen Removal from Digester Supernatant – Comparison of Chemical and Biological Methods. *Wat. Sci. Tech.*, **34**(1–2), 399-406.

UCPTE (1994). *Yearly Report 1993*, UCPTE, Wien 1994.

Udert, K.M., Fux, C., Münster, M., Larsen, T.A., Siegrist, H. and Gujer, W. (2003). Biological nitrogen treatment of source-separated urine. *Wat. Sci. and Tech.*, **48**(1) 119–130 (this issue).

USGS, U.S. Geological Survey (2002) Phosphate Rock – Statistics and Information. [*http://minerals.usgs.gov/minerals/pubs/commodity/phosphate_rock/*]. Last update 05/13/02. *cdoud@usgs.gov*.

Wood, F.C. (1982). The Changing Face of Desalination – A Consulting Engineer's Viewpoint. *Desalination*, **42**, 17-25.

How farmers in Switzerland perceive fertilizers from recycled anthropogenic nutrients (urine)

J. Lienert*,****, M. Haller*, A. Berner**, M. Stauffacher*** and T.A. Larsen*

* Swiss Federal Inst. for Environmental Science and Technology (EAWAG), Überlandstrasse 133, CH-8600 Dübendorf, Switzerland (E-mail: *judit.lienert@eawag.ch*; *michelhaller@gmx.ch*; *larsen@eawag.ch*)
** Research Inst. of Organic Agriculture (FiBL), Ackerstrasse, CH-5070 Frick, Switzerland
(E-mail: *alfred.berner@fibl.ch*)
*** Swiss Federal Inst. of Technology (ETH), Dep. of Environmental Sciences, Natural and Social Science Interface, Haldenbachstrasse 44, ETH-Zentrum HAD, CH-8092 Zürich, Switzerland
(E-mail: *stauffacher@uns.umnw.ethz.ch*)
**** Correspondence address: Judit Lienert, EAWAG, Switzerland
(E-mail: *judit.lienert@eawag.ch/www.novaquatis.eawag.ch*)

Abstract We studied acceptance of a urine-based fertilizer product using a mail survey of 467 Swiss farmers. We distinguished among four production types: organic or IP farming, and with or without vegetable production. Considering that the idea of urine-based fertilizers is new, acceptance among the answering farmers was surprisingly high, with 57% explicitly stating that they thought it was a good or very good idea, and 42% willing to purchase such a product. The farmers of different production types did not differ strongly in their attitude towards urine-based fertilizers. Especially IP and vegetable farmers, who purchased additional fertilizers anyway, seem willing to accept urine-based fertilizers, hereby preferring a grainy, odorless ammonium nitrate fertilizer. Absolutely essential is a hazard-free product: 30% of all farmers had concerns regarding micropollutants. Based on fertilizer data, we demonstrate an existing demand for the nutrients N, P, and K in Switzerland, which could be partially substituted by a recycled urine product. Finally, we discuss methodological requirements of social science surveys. To obtain representative data on an entire population in a mail survey, multiple contacts with respondents are necessary. We argue that information and participation of stakeholders at an early stage is essential for successful technology transfer.
Keywords Farmers; mail survey; participation; sustainable nutrient recycling; technology transfer; urine separation

Introduction

Phosphorus is a limited resource, and the known worldwide phosphate rock reserves will be exploited in ca. 300 years, given the present phosphorus exploitation rate (calculations based on Jasinski, 2000/2002). Therefore, sooner or later some sort of recycling from urban areas must take place. Known forms of recycling are sewage sludge (the only remaining form in northern Europe), treated wastewater, or direct application of urine and feces to agricultural land. At least in Switzerland, the reuse of sewage sludge will be prohibited in the near future (Chassot and Mühlethaler, 2001).

For many different reasons, including the possibility of nutrient recycling, urine separation is propagated as an improvement of wastewater management (e.g. Larsen *et al.*, 2001). Consumer acceptance, information and participation of the population are essential elements of such a technological innovation. These non-technical aspects, however, are often neglected both by science and by developers of new technologies (Chopyak and Levesque, 2002). An important aspect of NOVAQUATIS, a larger research project on urine separation (www.novaquatis.eawag.ch) is to involve stakeholders in an early phase of technology development. Accordingly, we defined a number of sub-projects with this aim. These projects include focus groups on consumer acceptance (Pahl-Wostl *et al.*, 2003), studies on farmer acceptance, and ongoing sociological investigations in pilot projects, which generate more long-term data both in private and institutional settings with NoMix toilets.

The study presented here is an explorative rather than confirmatory study to identify trends concerning the acceptance of a urine-based fertilizer by Swiss German farmers. A main goal was to develop methods for a larger, quantitative investigation, and to determine whether research in this direction would be feasible and effective. In Switzerland, agriculture is heavily subsidized, federal requirements strongly regulating the production type of agriculture receiving subsidies. Today, two production types are eligible: organic agriculture and the so-called 'integrated production', IP. The latter may be described as a 'best management practice' and has less stringent regulations than organic farming. Important features of IP farming are for instance balanced nutrient budgets. Farmers increasingly convert to one of these standards, and less than 40% practiced conventional farming in the year 2000. Different management practices have different nutrient requirements. Although the use of synthetic mineral fertilizer is prohibited in organic farming, especially vegetable farmers are dependent on additional nutrients. IP farmers are allowed to use synthetic mineral fertilizer, provided that the nutrient balance on the farm is in equilibrium. Livestock farmers often face a nutrient surplus from farm manure and may be less interested in any new fertilizer product.

Provided acceptance by farmers, a market for urine-based fertilizer exists undoubtedly. In Table 1, we compare data from different regions on the use of artificial fertilizer with data on the maximum possible production of nutrients from human urine. In Switzerland, the trend towards IP and organic farming, and the current Swiss efforts to lower nutrient losses from agriculture (Spiess, 1999) will presumably reduce the gap between nutrient demand from agriculture and the possible supply from separate urine handling.

Previous studies have shown that a farmer, as any other human being, is open to new ideas if there are no essential drawbacks to be expected. Farmers have often been criticized of being responsible for environmental problems, especially in rural areas. At least in Switzerland, this led to skepticism against or even rejection of environmental research by many farmers. Therefore, intensive and continuous communication between scientists and farmers is essential (Pongratz, 1992). However, sociocultural aspects of agriculture have only rarely been investigated (Lobao and Meyer, 2001). Therefore, an additional objective of our investigation is to gain experience in conducting social research studies with this population group.

Finally, we explore methodological requirements of such sociological investigations, thereby contributing to a knowledge transfer among engineers and scientists on sociocultural aspects of technological innovations. We hope to pave the way for more extensive

Table 1 Yearly per capita consumption of artificial fertilizer and potential production from urine

	Consumption of artificial fertilizer		
	N	P	K
	$kg_N/p/y$	$kg_P/p/y$	$kg_K/p/y$
Switzerland[1]	8.7	1.5	5.4
Western Europe[2]	26.3	4.0	8.4
World[2]	14.0	2.4	2.9
Max. potential production of fertilizer from urine[3]	3.2	0.3	0.8

[1] Year 1995 for 6.7 Mio. Swiss inhabitants. Fertilizer data are taken from Spiess (1999)
[2] Year 2000. Fertilizer data are taken from ifa (2001). Data on populations are taken from United Nations Population Division (2000)
[3] European data; based on Maurer and Larsen (2003)

studies on stakeholder involvement in technology development, which among others could increase the acceptance of waste-based fertilizers in European agriculture.

Materials and methods

Data collection

At the time of our study, 39,270 farmers in the German speaking part of Switzerland – which is the largest Swiss region – were registered as organic (10.2%) or IP (89.8%) farmers (Bundesamt für Statistik, 2000). We also distinguished among farmers with and without vegetable production. To achieve sufficient group size to allow comparisons, we used a stratified random sampling procedure: from the national farmer register, we randomly sampled 155 organic farmers with vegetable production (29% of all organic farmers with vegetables), 154 IP farmers with vegetables (6%), 79 organic farmers without vegetables (2%), and 79 IP farmers without vegetables (0.2%). We did not sample conventional farmers, as they are all believed to convert to IP or organic farming in the near future.

In January 2000, the study farmers received information on the urine separation project NOVAQUATIS together with the questionnaire. In February 2000 a reminder was sent again to the full sample, since the fully anonymous procedure we used did not allow tracking those who had not returned their answer. We asked for four groups of data: (1) personal, (2) farm details, (3) data on the acceptance of urine-based fertilizers, and (4) nutrient demands and type of fertilizer product needed (Table 2). Most questions were closed questions in a multiple-choice format, so that respondents had only to tick the appropriate answer.

Data analysis

First, we summarized responses over all farmer groups. Then we compared organic and IP farmers; farmers with and farmers without vegetable production; vegetable production (with/without) within the group of only organic and only IP farmers; and younger farmers (<45 yrs) with older ones (>45 yrs; hypothesizing that younger farmers are more progressive than older ones). We used χ^2-tests to determine differences among groups. Samples being small and most variables of ordinal scale only, we analyzed group differences in continuous variables (area of farm and vegetable fields/livestock units) with Mann-Whitney U tests. To assess correlations between different variables, we used Spearman's rho. We used the statistical software SPSS (release 10.1; SPSS Inc, Chicago, Illinois, USA).

Table 2 Overview of questions from questionnaire concerning urine-based fertilizer that was sent to 467 farmers in the German part of Switzerland

Data group	Description
Personal data	Sex, age, childhood spent on farm?
Farm details	Swiss canton, is agriculture main income?, production type (organic/IP)
	Reason for choice of production type (financial, practical, ethic)
	Date of change from conventional to current production type
	Size of farm (ha: crops, pastures, vegetable, others), livestock (cattle, pigs)
Acceptance of urine-based fertilizer	How do you like the idea of re-using urine as fertilizer?
	Should regulations change to allow urine-based fertilizer in organic farming?
	Would you use urine-based fertilizer?
	Additional comments and concerns
Nutrient demands/ product needed	Use of fertilizer (manure, artificial, or other additionally purchased)
	How large is nutrient demand (N, P, K, B, others)
	Preferred form of nitrogen fertilizer (urea, NH_4^+, NO_3^-, organically bound)
	How much would you pay for a urine-based fertilizer product?
	Would you use fertilizer with urine odor (or cattle/hog manure)?
	Preferred type of urine-based fertilizer (liquid or grainy)

Response

Of the 467 questionnaires, 127 (27%) were returned. Three returned questionnaires were empty and contained a remark that the idea of urine separation was completely ridiculous. Two farmers wrote that they did not wish to participate in the study. Given the short time available to carry out the survey (1 month), the response rate was acceptable and comparable to similar surveys using mail questionnaires (see below). The percentage of returned questionnaires differed highly significantly among the farmer groups, being highest among organic farmers (all: N = 467; df = 1; P<0.001; Table 3). This means, generalizations to the overall population of Swiss farmers should be avoided. However, comparisons between different groups are well possible, but also require some caution, since the returned sample could be biased (see below). The group of farmers that returned the questionnaire after receiving the reminder differed only in personal data (growing up on farm, main income from farm), but not in any of the questions concerning fertilizer acceptance.

Sample description

Most heads of farms were male (95%), 35–55 years old (67%), grew up on a farm (87%), and gained their main income with farming (90%). The size of farms ranged from 0.3 to 70 ha (mean 18.6; median 16.8 ha), and the area of vegetable land from 0.02 to 50 ha (mean 3; median 1 ha). The 102 farms with livestock (85%) had 1 to 60 animal units (mean 22; median 20).

The time of and reasons for converting to the current production type differed among organic and IP farmers: 39% of organic, but only 19% of IP farmers had converted to this production type before 1990 (N = 119; df = 2; P<0.1). Likewise, 51% of farmers with vegetables had converted before 1990 to their current production type (organic or IP), but only 15% of those without vegetables (N = 199; df = 2; P<0.001). A majority of organic farmers expressed ethical reasons for conversion (84%, and 92% of organic farmers with vegetable production), but only 44% of IP farmers (N = 109; df = 1; P<0.001). Financial motives for conversion were mentioned by 55% of IP, but only by 26% of organic farmers (N = 111; df = 1; P<0.01).

Results

Acceptance of a urine-based fertilizer

Over all groups, 57% of farmers had an explicitly positive attitude towards the idea of a urine-based fertilizer, and 33% thought it was a bad idea (Table 4A). Changing the regulations to allow the use of urine-based fertilizer in organic farming was welcomed by 43%, while 38% were against it (Table 4B). Of all farmers, 42% would purchase a urine-based fertilizer (Table 4C). We tried to track down reasons for not wanting to buy this fertilizer: of those that would not buy the fertilizer, 76% indicated that they had no need for it. Fifty-four % had indicated in question A that they regarded re-using urine in agriculture as a bad idea. Concerns about the quality of the fertilizer were mentioned in additional remarks by 46% of those that would not buy it (mostly fear of micropollutants, rarely hygienic aspects).

Table 3 Number and percentage (in parentheses) of returned questionnaires of the different farmer groups

	Organic farmers	IP farmers	Total with/without vegetables
With vegetable production	37 (24%)	18 (12%)	55 (18%)
Without vegetable production	41 (52%)	24 (30%)	65 (41%)
Total of organic/IP	78 (33%)	44[1] (19%)	

[1] Two questionnaires did not contain information on vegetable production

Table 4 Acceptance of a urine-based fertilizer over all farmers. We asked whether farmers regarded urine recycling as a good idea (A), whether regulations should change to allow the fertilizer in organic farming (B), and whether they would buy it (C). Of those that would not buy it, we show how many indicated no need for it, regarded it as a bad idea, or indicated concerns of some sort. We also present how many farmers wrote additional remarks and of those, how many had concerns regarding urine-based fertilizer (D)

		Percent	N[a]
A	**Idea**		
	Bad idea	33%	
	No opinion	10%	
	Good idea	46%	
	Very good idea	11%	125
B	**Change of regulations**		
	No	38%	
	No opinion	19%	
	Yes	43%	122
C	**Market chances**		
	Would buy fertilizer	42%	123
	No: no need for it	76%	67
	No: bad idea	54%	67
	No: quality concerns	46%	67
D	**Concerns**		
	Wrote additional remarks	59%	125
	Indicated concerns	82%	74

[a] Different sample sizes (N), because not all farmers answered all questions. Here, we did not include the missing answers

Of all 125 farmers, 59% wrote additional remarks, and the majority of these were concerns (Table 4D). Micropollutants such as hormones and other impurities caused by far the biggest worry, even though we had mentioned that the urine-based fertilizer would be treated to such a degree that it should not pose any toxicological or hygienic risks: doubts concerning micropollutants were mentioned in 51% of 74 remarks, which equals 30% of all 125 farmers. Other frequent concerns were feasibility (9% of 74), hygiene (7%), high costs (5%), missing acceptance by consumers of agricultural products (5%), and a negative attitude towards ideas that come 'from the city' and are burdened on farmers (9%). Of those with remarks, 9% explicitly wrote that they thought it was a ridiculous idea altogether (6% of all 125).

Opinions rarely differed among groups; here we present the few significant results only. The farmers older than 45 years were less indifferent towards the idea of urine-based fertilizer and the change of regulations than the younger ones: more older farmers had a distinctly negative or positive opinion (Figures 1A, B). More farmers without vegetables would change regulations to allow urine-based fertilizer in organic farming (Figure 1C). More organic than IP farmers indicated that they would not buy a urine-based fertilizer because they had no need for it (Figure 1D). Finally, more farmers that would not buy a urine-based fertilizer wrote additional remarks (Figure 1E). Concerns were mentioned more often in the remarks, when the farmers also indicated that they would not buy a urine-based fertilizer – in contrast, the farmers that would buy the fertilizer indicated fewer concerns (Figure 1F).

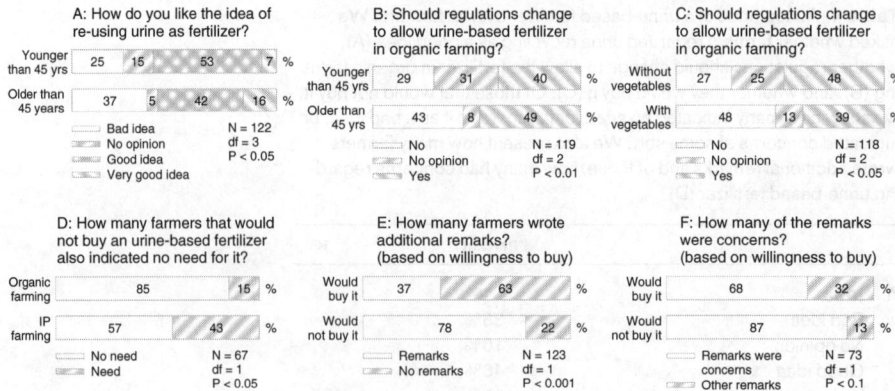

Figure 1 Significant differences among farmer groups concerning the acceptance of a urine-based fertilizer. We present the questions, the groups tested against each other (i.e. younger or older than 45 yrs; Figure 1A), responses to the possible answers (%), sample sizes (N), degrees of freedom for the χ^2 tests (df), and significance values (P)

Current use of fertilizer, nutrient demands and type of product needed

Over all groups, 38% of farmers bought additional fertilizers (Table 5A), with large significant differences among groups: more IP than organic farmers, more farmers with than without vegetables, and more vegetable farmers within only organic or only IP farmers purchased additional fertilizers (Figure 2A). Forty-six % of all farmers indicated a medium to large demand for N (Table 5B), the IP farmers indicating significantly more demand for N

Table 5 Use of additional fertilizers (A), nutrient demands (B), preferred form of nitrogen (C), acceptance of fertilizer with urine odor (D), and type of fertilizer product needed (E; percentage over all farmers)

		Percent
A	**Additional fertilizer used?**	
	Uses additional fertilizers	[a]49%
	Buys additional fertilizer	38%
B	**Nutrient demand**	
	Medium to large need for N	46%
	Medium to large need for P	28%
	Medium to large need for K	22%
	Medium to large need for B	18%
	Medium to large need for others	14%
C	**Preferred form of nitrogen**	
	Urea	14%
	NH_4^+	18%
	NO_3^-	18%
	Organically bound	25%
D	**Fertilizer with urine odor**	
	Would use it near house	12%
	Would use it on open field	31%
E	**Preferred type of fertilizer**	
	Has need for liquid fertilizer	11%
	Has need for grainy fertilizer	38%

[a] Sample size for all (N = 125); percentages inclusive missings; e.g., (A) 49% of all farmers indicated that they used additional fertilizer, and 51% that they did not or they gave no answer

than the organic farmers (Figure 2B). More IP farmers with vegetables than IP farmers without vegetables indicated a medium to large demand for K and B (Figure 2B).

The preferred form of nitrogen differed among some farmer groups: IP compared with organic farmers preferred NH_4^+ (Figure 2C). NO_3^- was preferred by more IP than organic farmers, by more farmers with than without vegetables, and also by many more IP farmers with vegetables than IP farmers without vegetables (Figure 2C). Organic farmers, and farmers without vegetables seemed to slightly prefer an organically bound slow-release fertilizer, but these preferences were not significant.

Of all farmers, only 35% would use a fertilizer with urine odor (near houses, in fields, or both), IP farmers being more willing to use it (especially in fields) than organic farmers (Figure 2D). Acceptance of cattle or hog manure was much higher: up to 100% of vegetable or IP farmers would use hog manure (with rather unpleasant odor) on fields. Farmers clearly preferred a grainy to a liquid fertilizer product (Table 5E; Figure 2D).

However, the price of a urine-based fertilizer would have to be moderate: only 4% of the farmers would pay more than for the fertilizer they are currently using, 34% would pay the same, 37% would pay 20% less, and 25% would pay 50% less. Regarding price, there were no differences among groups. Moreover, the older farmers did not differ from the younger ones in any of these variables.

Discussion

Attitude of farmers towards a urine-based fertilizer product

Considering that the idea of a urine-based fertilizer is new and possibly startling, acceptance among the answering farmers was surprisingly high, with 57% explicitly stating that

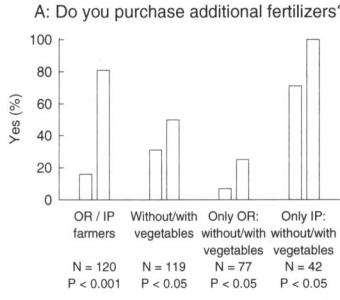
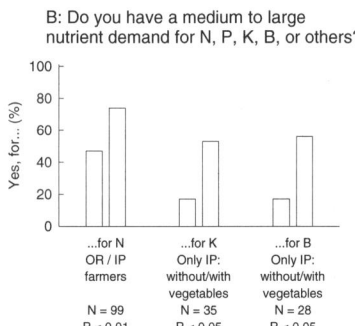
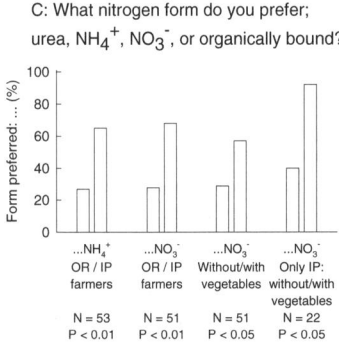
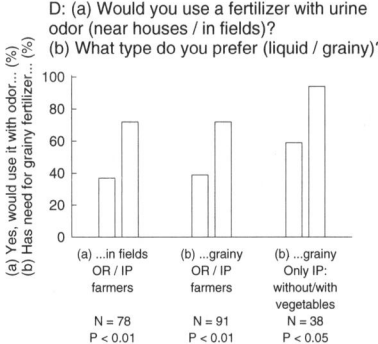

Figure 2 Significant differences among farmer groups concerning nutrient and fertilizer demands. We present the questions, the answers of the groups tested against each other (i.e., 16% of OR [organic], and 81% IP farmers purchased additional fertilizers; Figure 2A, 1st group), sample sizes (N), and significance values (P). The degree of freedom for the χ^2 test was always 1. Sample sizes differ, because not all farmers answered all questions

they thought it was a good or very good idea, and 42% willing to purchase such a product (Table 4). The farmer groups with different production types did not differ significantly in their attitude towards urine-based fertilizers. However, this might be an artifact of the biased sample, since probably primarily interested farmers answered (see below). Given certain quality criteria, farmers with a real nutrient demand (i.e. that purchased additional fertilizers anyway) would also buy a urine-based fertilizer product. Since a large majority of IP farmers – especially with vegetable production – purchases additional fertilizers, market chances would presumably be largest among this group. IP farmers were also less reluctant to apply fertilizer with urine odor compared with organic farmers. They strongly preferred a grainy nitrogen fertilizer in form of NH_4^+ or NO_3^-. Farmers were not willing to pay more than they paid for their current fertilizer product. Asking for the price can give an indirect estimation of attractiveness. Since only 34% of the farmers would even pay the same they currently paid, a urine-based fertilizer does not seem to be a highly desirable product at the moment. However, if one succeeds in producing a good quality, hazard free fertilizer, market chances seem to be good.

Concerns regarding micropollutants in urine
Concerns regarding the safety of a urine-based fertilizer were prominent, with 30% of all farmers remarking that they had doubts regarding residues such as hormones and pharmaceuticals. Such concerns were also mentioned by ca. 50% of farmers of any group that was not willing to purchase a urine-based fertilizer and seemed to be an important reason for rejecting the idea. Only 4% of all farmers mentioned hygiene as a problem. Apparently, farmers did not question the possibility of hygienizing a urine-based fertilizer, but did not believe in the technical feasibility of removing micropollutants. We hypothesize that this is due to a lack of positive experiences. Farmers in Switzerland – and Europe – were confronted with a paradigm change, especially regarding sewage sludge. Early recommendations strongly promoted the use of sewage sludge as fertilizer (Candinas, 1989). However, increasing awareness of environmental problems is now resulting in the total ban of re-use of this waste in Switzerland. Farmers are sensitized, and have been made responsible for various environmental problems, ranging from eutrophication to landscape destruction (Pongratz, 1992). This may result in total rejection of new ideas. Indeed, 6% of all farmers mentioned strong resentments against being burdened with problems 'from the cities'. Therefore, positive communication is essential. Moreover, the concerns regarding micropollutants are understandable. Without process engineering experience, it is not obvious why it should be possible to remove micropollutants from urine, but not from sewage sludge. Therefore, future communication with farmers must also emphasize technical knowledge transfer.

Nutrient demand in Swiss agriculture
Today, urine separation in Switzerland could substitute around 37% of N, 20% of P, and 15% of K from artificial fertilizers (Table 1). With increasing conversion from conventional to organic or IP farming, nutrient demands in agriculture will diminish, because organic agriculture applies by far less additional fertilizers. Therefore, a urine-based fertilizer could account for an even larger percentage of substitution. A popular argument against re-using nutrients in agriculture is the idea that agriculture generally faces a nutrient surplus. This applies to certain countries (e.g. The Netherlands; van Bruchem et al., 1999), and to certain production types (livestock farming; Hall, 1999). Our survey clearly indicates that there *is* a nutrient demand for N (46% of all farmers), P (28%), and K (22%; Table 5), and that fertilizers *are* imported into Switzerland (Table 1). Surplus of nutrients is a regional problem due to transportation limitations. With liquid urine, we would face similar prob-

lems. In contrast, purified and solidified urine could easily be transported. We therefore believe that the existing nutrient demand could be partially supplied by a urine-based fertilizer product, which is additionally a sustainable alternative to artificial fertilizers.

Methodological considerations

A drawback of our study is that the project NOVAQUATIS is still in a very early phase. It is possible that some farmers did not take our idea of urine recycling seriously, which could be responsible, in part, for the relatively poor response rate. With ongoing implementation of larger pilot projects and increasing publicity, this should change. However, possibly farmers with currently indifferent opinions (because they did not take the idea seriously), might then also adopt a negative attitude.

Given that the response rate was rather low (27%) and differed significantly among farmer groups (Table 3), our study is not representative for Swiss farmers in general. Possibly, only farmers with increased interest in the very particular topic answered our questions (Dillman, 1991). In a study on farm work satisfaction, response rates also differed among conventional (60% returned) and organic farmers (80%), which was attributed to the relative salience of the study to the two groups (Rickson et al., 1999). Work satisfaction is certainly much more relevant to all farmers than our topic, and is known to achieve better response (Heberlein and Baumgartner, 1978). Nevertheless, a better response would have been possible. In social research methodology low response to mail surveys is discussed extensively (Babbie, 2001; Schutt, 2001). Data collection with questionnaires can be implemented with three modes: (1) face-to-face interviews, (2) via telephone, or (3) with mail surveys. All three have their specific strengths and optimal application framework. In our study, due to restricted resources, neither telephone nor face-to-face interviews were possible, since the researcher needs even more time for each respondent than the respondent needs to answer the questions, especially in geographically dispersed areas (travel time). Hence, here mail survey is still the most promising method. Dillman (2000) postulates five important elements to achieve high response in mail surveys: (1) a respondent-friendly questionnaire, (2) up to five contacts with the recipient, (3) inclusion of stamped return envelopes, (4) personalized correspondence, and (5) a token financial incentive sent with the survey request. We followed (1), (3) and (4); (5) was not possible. Most importantly, increased response is possible with multiple contacts (Dillman, 1991). These need to be distinctive to attract different respondents at different times. Therefore, Dillman (2000) proposes: (1) a brief prenotice letter on the study, (2) a questionnaire mailing with detailed cover letter, (3) a postcard a few days later to thank for responding and reminding others to do so, (4) a second questionnaire to all non-respondents 2–4 weeks later, and (5) a phone reminder to the rest of the non-respondents. Evidently, steps (4) and (5) require an identification of each questionnaire, something that we did not want, because it could cause distrust among farmers. In short, increasing response in a mail survey is feasible, but requires a full set of measures and sufficient resources.

Conclusions

We conclude that there *is* a demand for the nutrients N, P, and K in Switzerland, which could be substituted, to a certain degree, by a recycled urine product. Those Swiss farmers that have a real need for additional fertilizers will most likely accept a urine-based fertilizer. Therefore, market chances would be especially high among IP and vegetable farmers. The fertilizer would have to be relatively cheap, odorless, of desired type (i.e. grainy ammonium nitrate), and – most importantly – free of micropollutants, since concerns regarding micropollutants seemed to be a prominent motive for rejection. Before actually introducing such an innovative product, stakeholder demands need to be assessed in further detail. Our study further shows that additional measures are necessary to obtain representative data for an entire population (all Swiss German farmers in our case). Therefore, we propose that future surveys either use more time-consuming telephone interviews, or increase response rates of a mail survey by establishing multiple and varying contacts with the respondents. Finally, we believe that acceptance of a urine-based fertilizer will only be successful with ongoing information and participation of farmers and other stakeholders from the start.

References

Babbie, E. (2001). *The Practice of Social Research*, 9th edn, Wadsworth, Belmont.
Bundesamt für Statistik (2000). Stichprobe aus *Betriebs und Unternehmensregister des 1. Wirtschaftssektors (BUR)*. Bundesamt für Statistik, Neuchâtel (in German).
Candinas, T. (1989). Anforderungen und Kriterien für die landwirtschaftliche Verwertung von Klärschlamm. Umwelt-Information VGL. **2**/89, 25–29 (in German).
Chassot, M. and Mühlethaler, B. (2001). Klärschlamm hat bald ausgedüngt. Magazin Umwelt des BUWAL (Bundesamt für Umwelt, Wald und Landschaft, Schweiz) **3** (Wasser Spezial), Chapter 14 (in German).
Chopyak, J. and Levesque, P. (2002). Public participation in science and technology decision making: trends for the future. *Technol. Soc.*, **24**, 155–166.
Dillman, D.A. (1991). The design and administration of mail surveys. *Annu Rev. Sociol.*, **17**, 225–249.
Dillman, D.A. (2000). *Mail and Internet Surveys: the Tailored Design Method*, 2nd edn, Wiley, New York.
Hall, J.E. (1999). Nutrient recycling: The European experience – review. *Asian-Australasian J. of Animal Sciences*, **12**(4), 667–674.
Heberlein, T.A. and Baumgartner, R. (1978). Factors affecting response rates to mailed questionnaires: a quantitative analysis of the published literature. *Am. Sociol. Rev.*, **43**, 447–462.
ifa, International Fertilizer Industry Association (2001). Total fertilizer consumption statistics by region from 1970/71 to 1999/2000 [*http://www.fertilizer.org/ifa/statistics.asp*]. Last updated 6th November 2001. *ifa@fertilizer.org*.
Jasinski S.M. (2000/2002). Phosphate rock. U.S. Geological Survey, Minerals Yearbook – 2000/Mineral Commodity Summaries – Jan. 2002 [*http://minerals.usgs.gov/minerals/pubs/commodity*]. Last updated 6th June 2002. *sjasinski@usgs.gov*.
Larsen, T.A., Peters, I., Alder, A., Eggen, R., Maurer, M. and Muncke, J. (2001). The toilet for sustainable wastewater management. *Env. Sci. Tech.*, **35**(9), 193A–197A.
Lobao, L. and Meyer, K. (2001). The great agricultural transition: crisis, change, and social consequences of twentieth century US farming. *Annu. Rev. Sociol.*, **27**, 103–124.
Maurer, M. and Larsen, T.A. (2003). Nutrients in urine: energetical aspects of removal and recovery. *Wat. Sci. Tech.*, **48**(1) 37–46 (this issue).
Pahl-Wostl, C., Schönborn, A., Willi, N., Muncke, J. and Larsen, T.A. (2003). Investigating consumer attitudes towards the new technology of urine separation. *Wat. Sci. Tech.*, **48**(1) 57–65 (this issue).
Pongratz, H. (1992). *Die Bauern und der ökologische Diskurs – Befunde und Thesen zum Umweltbewusstsein in der bundesdeutschen Landwirtschaft*. Profil Verlag, München (in German).
Rickson, R.E., Saffigna, P. and Sanders, R. (1999). Farm work satisfaction and acceptance of sustainability goals by Australian organic and conventional farmers. *Rural Sociology*, **64**(2), 266–283.
Schutt, R.K. (2001). *Investigating the Social World. The Process and Practice of Research*, 3rd edn, Pine Forge Press, Thousand Oaks.
Spiess, E. (1999). Nährstoffbilanzen der schweizerischen Landwirtschaft für die Jahre 1975 bis 1995. Schriftenreihe der FAL **28**. Eidg. Forschungsanstalt für Agrarökologie und Landbau (FAL), 8046 Zürich-Reckenholz (in German).
United Nations Population Division (2000). United Nations Population Information Network, Table 1 [*http://www.un.org/esa/population/publications/wpp2000/wpp2000at.pdf*]. Last updated 28th September 2001. 2 United Nations Plaza, Rm. DC2-1950, New York 10017.
van Bruchem, J., Schiere, H. and van Keulen, H. (1999). Dairy farming in the Netherlands in transition towards more efficient nutrient use. *Livestock Production Science*, **61**, 145–153.

Investigating consumer attitudes towards the new technology of urine separation

C. Pahl-Wostl*, A. Schönborn***, N. Willi**, J. Muncke** and T.A. Larsen**

* Institute of Environmental Systems Research, University of Osnabrück, Albrechtstrasse 28, D-49069 Osnabrück
** Swiss Federal Institute of Environmental Science and Technology (EAWAG), Überlandstrasse 133, CH-8600 Dübendorf
*** Armadillo Webworks, Lucerne, Switzerland
Corresponding author (E-mail: pahl@usf.uni-osnabrueck.de)

Abstract The technology of urine separation and the recycling of anthropogenic nutrients as fertilizer in agriculture are considered as major innovations to improve the sustainability of today's urban wastewater management. The acceptance of consumers will be key for the introduction of the new technology. Citizens will have to make important decisions in their role as tenants and owners of houses and as consumers buying products fertilized with urine. Consumer attitudes towards the new technology were explored in a number of citizen focus groups in Switzerland. Focus groups are deliberate, moderated group discussions with informed citizens on a certain topic. The information was provided by a computer based information system specifically designed for this purpose. The acceptance of individual citizens for the new technology proved to be quite high. The majority of the citizens expressed their willingness to move into an apartment with NoMix toilets and to buy food fertilized with urine. However, they were not willing to accept additional financial costs or efforts. Arguments related to long-term sustainability (closing nutrient cycles) were of less importance than arguments that relate directly to the effects of micropollutants on human and ecosystem health. For the introduction of the new technology on a wide scale it will thus be crucial to explore the fate and effects of micropollutants.
Keywords Citizen focus groups; consumer attitudes; micropollutants; sustainable nutrient recycling; urine separation technology

Introduction

Historically, wastewater treatment technology has been developed exclusively by engineers. The public was not involved in the whole process of planning and implementation. Today's situation is characterized by a quite complex centrally controlled system about which most people have only little knowledge. However, with the introduction of decentralized technologies at the household scale (see e.g. Larsen and Gujer, 2001) citizens are increasingly taking an important role. Citizens will have to make decisions as voters, as house-owners, as tenants, and as consumers. It is therefore essential to invoke citizens at an early stage in the process of technology development and application, when there is still ample room for taking into account their wishes and concerns. This corresponds to an overall increasing awareness of the importance of public participation in environmental decision making and the assessment of environmental risks (Renn *et al.*, 1995; Connor, 1999; Glicken, 2000).

An example of a technological innovation at the household scale is given by urine separation and the recycling of anthropogenic nutrients (nitrogen, phosphorus, and potassium) as fertilizer in agriculture (Larsen *et al.*, 2001). The two main arguments for introducing urine separation are the improvement *and* at the same time simplification of water pollution control (because most nutrients and an important fraction of the micropollutants are contained in urine) and the possibility of closing the nutrient cycle – especially for phosphorus, a limited, essential resource. In the context of the interdisciplinary project NOVAQUATIS (at EAWAG, Switzerland, www.novaquatis.eawag.ch), consumer attitudes towards NoMix (urine separating) toilets and the idea of anthropogenic nutrient recycling were investigated using the IA-focus group methodology. This paper introduces the methodology, which is a novelty for technology assessment, especially for urban water management, and reports the results.

Materials and methods

The method employed was citizen focus groups. Focus groups are deliberate, moderated group discussions with informed citizens on a certain topic. The focus group method for integrated assessment (IA-focus groups) is a participatory method that draws on elements of both public opinion research and marketing studies (Dürrnberger *et al.*, 1999; Jaeger *et al.*, 1999; Schlumpf *et al.*, 2001). They differ from both fields in their explicit goal of providing ordinary citizens as well as various other stakeholders with an opportunity to articulate their voice in a debate. Citizens are provided with information that draws on the state-of-the-art in scientific research and the assessments aim at receiving from citizens information that is suitable for shaping actual decision making.

With IA-focus groups, the following questions may be clarified (Schlumpf *et al.*, 2001):
- How do citizens perceive and translate scientific input into shaping their opinion and assessments?
- How do they cope with uncertainties?
- How do informed citizens judge the risk of an environmental issue?
- What policy options do they prefer?
- What are their policy recommendations?

Typically, IA focus groups are used for a comprehensive assessment of an environmental problem such as climate change or sustainable development of a region. This implies giving recommendations for a range of possible options. This study, however, focused on a technology assessment. The intention was to involve relevant stakeholders in an early phase into a research process, where they could still have some influence on the other, more technical-scientific research projects of NOVAQUATIS (see also Lienert *et al.*, 2002 on a survey amongst Swiss farmers).

The main issues that were addressed in the study are:
- Assessment of citizens' perception of nutrient recycling, production of urine-based fertilizer, and its use in organic farming as well as its contribution to sustainable development.
- Assessment of citizens' perception of risks arising from pharmaceuticals in urine for the environment today. Furthermore, we also assessed perception of potential risks in the new system, for instance of micropollutants contained in urine-based fertilizer products.
- Elicitation of consumer preferences regarding product attributes such as use, maintenance or design of NoMix toilets in comparison to current technology.

In general IA focus groups consist of eight to ten participants. The input is provided by specifically designed computer tools, fact sheets that summarize in one or two pages the most important facts on one issue, or expert statements.

Meetings with ten gender specific groups of citizens with eight to ten participants were conducted during 2000/2001. The focus group participants were recruited both by phone and by a newspaper advertisement in the local press. They were chosen to be a representative sample of the population regarding age, profession, political attitude, and environmental awareness. The groups met twice for two hours in two successive weeks. Information about the NoMix technology was provided by a specifically designed interactive computer-based information system, the NoMix Tool (*www.novaquatis.eawag.ch*; see below). The focus group participants had the opportunity to visit a NoMix toilet (Dubbletten, from BB Innovation, Sweden). The state of knowledge and citizen attitudes was assessed using questionnaires. The state of knowledge was assessed at the beginning of the meetings, and attitudes at the end. During each meeting the most important statements were recorded on a protocol. In addition, each session was recorded on a tape recorder and partly transcribed to check individual statements.

The focus group methodology allows exploration of the range of arguments and perceptions that could arise about an environmental issue in a well-defined setting. Ideally, it consists of a representative sample of the informed public. However, since the number of people that can be addressed in such detailed studies is limited, sample-groups are often not representative for the entire population, and the data generated are qualitative, rather than quantitative.

An interactive computer-based citizen information tool (ICIT) was developed for the focus group discussions. The purpose of ICITs is to make decision oriented expert knowledge on complex problems accessible and utilizable for citizens (Schlumpf *et al.*, 2001). To fulfill their task of informing citizens about scientific results and making the process of research visible, ICITs need to match the following criteria:

- scientific results must be described comprehensibly and clearly
- the presentation ought to be trustworthy and balanced
- the information needs to be up-to-date
- ICITs have to enable and support independent study
- and they should be fun to work with.

The scientific information in the NoMix Tool is based on literature surveys and on subjective expert judgements, which were gained by an iterative exchange with scientists. This scientific information represents the state of the art of knowledge in the field of urine separation, and is relevant for a lay public. Figure 1 shows the interface of the NoMix Tool, where the main idea of closing nutrient cycles and of preventing remnants from hormones and drugs to enter the aquatic environment is visualized. The users can choose any one of the different issues simply by scrolling over it with the computer mouse to receive more in

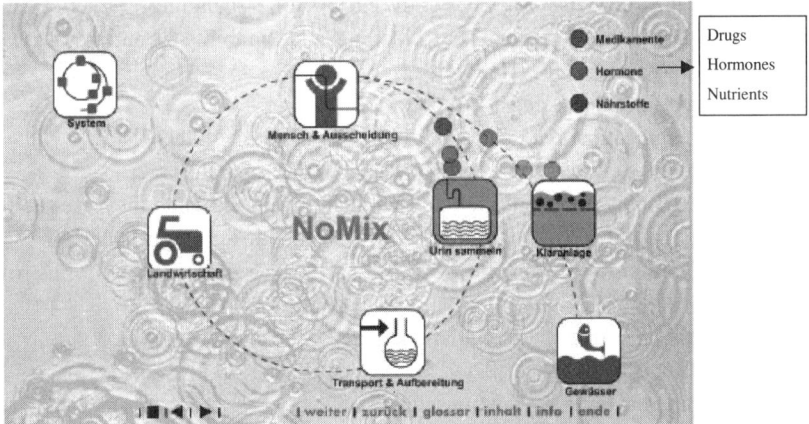

Figure 1 Graphical user interface of the interactive citizen information tool (ICIT). The flows of drugs (red) hormones (blue), and nutrients (green) are symbolized by animation. The user can get access to additional information from this information tool (*www.novaquatis.eawag.ch/NoMix Tool*). The symbols within the circle denoting the new system refer to urine collection (Urin sammeln), transport and processing (Transport und Aufbereitung), agriculture (Landwirtschaft), human beings and excretions (Mensch und Ausscheidung). The symbols in the path leaving the cycle refer to waste water treatment plant (Kläranlage) and aquatic environment (Gewässer). Additionally, higher-level information is given under the symbol system (System)

depth information. In the tool, information on the following topics is given: the content of nutrients and micropollutants (pharmaceuticals and hormones) in urine, the problems of collecting and storing urine in households, the possible technologies for transport and treatment of urine, and the concerns of agriculture. Moreover, the important role of phosphorus as a limiting and essential nutrient was discussed as well as the advantages for treatment plants and receiving waters which can be expected from the introduction of NoMix technology.

Tables 1 and 2 summarize the questions of the questionnaires. The results are given in the next section.

Table 1 Questionnaire 1 – general information (prior to group discussions)
1. How do you judge the quality of the drinking water in Switzerland today?
2. Has the quality of drinking water increased or decreased over the past years?
3. Which type of substances are removed in wastewater treatment? Please indicate for each substance in the list to which extent they are removed (a list of substances followed).
4. Please indicate if the following statements are appropriate (a list with statements about the fate and effects of hormones and drugs, their removal in waste water treatment).
5. A number of questions to judge environmentally friendly behavior.
6. To which extent do you consider yourself to be environmentally friendly?
7. How often do you buy food produced in organic farming?
8. Gender, Age, Education
9. Size of the household and number of children.
10. Are you tenant or owner?

Table 2 Questionnaire 2 – about the NoMix toilet (after group discussions)
1. What do you think in general about the NoMix toilet?
2. What do you find particularly good and what do you find particularly bad regarding the NoMix toilets?
3. How would you explain your neighbor the idea of the NoMix toilet?
4. Please judge the following properties of the NoMix toilet (handling, aesthetics, ecological aspects, smell, waste management, cleaning, and suitability for children).
5. Do you have suggestions for improvement?
6. Could you imagine purchasing a NoMix toilet?
7. Could you imagine moving into an apartment with a NoMix toilet?
8. Could you imagine collecting the used toilet paper in a separate bag?
9. NoMix toilets imply that both men and women always sit when they use the toilet. Could you imagine to always sitting down when using the toilet to urinate?
10. What do you prefer – vegetable fertilized with urine or vegetable grown with artificial fertilizer?
11. Could you imagine to purchase regularly vegetable fertilized with urine?
12. Should all farmers use urine fertilizer?
13. Who is cleaning the toilet in your household?
14. Could you imagine that all of Switzerland would have NoMix toilets?
15. What could be the societal benefits of a Switzerland with NoMix toilets only?

Results

State of knowledge and representativity of the groups

When asked to classify their own environmental awareness on a scale between 1 (no environmental awareness) and 10 (very high environmental awareness) the focus group participants classified themselves on average at 7.45. The answers to the more specific questions about their behavior in daily life showed as well that the participants had a high environmental awareness and were prepared to translate this into environmentally friendly behaviour.

Most citizens are not aware of the complexity of the urban water system, but in general the confidence in the technical system is high. To assess the overall knowledge about the current system a number of questions were asked before the focus groups started. Most participants were convinced that the quality of the drinking water in Switzerland today was good and had improved over the past years. If more specific questions were asked – e.g. regarding the efficiency of wastewater treatment plants – it was evident that citizens were quite uncertain. We conclude that there is a high public confidence in the quality of drinking water supply and waste water treatment, which is however based on little detailed knowledge. The average citizen is not interested in a technology that is invisible and outside of the realm of decisions made in daily life. The confidence in the reliability of the technical system is high given the fact that failures are rarely realized by the public.

General evaluation of acceptance

The overall acceptance for the new technology was high. Eighty per cent of the focus group participants made a positive judgement of the idea of the NoMix toilet (Figure 2), and more than 60% expressed a willingness to introduce the new technology in their households and to purchase "urine-fertilized" products from organic farming.

The focus group participants were also asked what they regarded as particularly positive or negative about the NoMix technology. For women, the main arguments in favor of the new technology were environmental friendliness and the idea of water saving. For men, the main arguments in favor of the new technology were the removal of hormones and drugs. The main arguments against the new technology for women were the considerable additional effort required for maintenance, urine collection and transport, and for men the unsolved technical problems.

Participants expressed a high willingness to decide in favor of the new technology as far as their own decisions were concerned (Figures 3A, B). The results showed a clear gender difference. Whereas 47% of the women were clearly willing to purchase a NoMix toilet, only 17% of the male participants were ready to do so. The willingness to move into an apartment with a NoMix toilet was higher for both men and women. 71% of the male participants expressed their clear willingness to move into an apartment with a NoMix toilet, whereas 58% of the women were clearly willing to do so. The higher overall acceptance for question 3B can partly be explained by the fact that most people in Switzerland are tenants

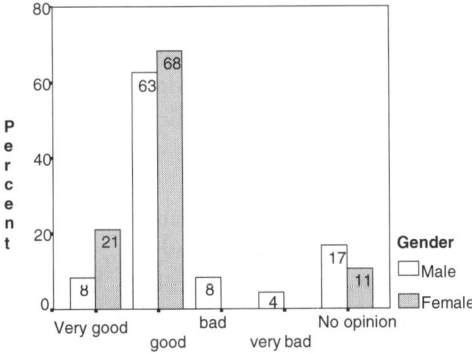

Figure 2 Distribution of answers to the question "What do you think in general about the NoMix toilet?" The percentages are given on the top of each bar

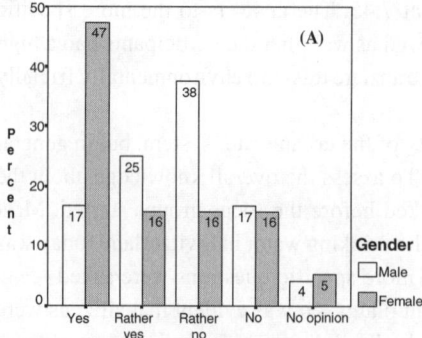
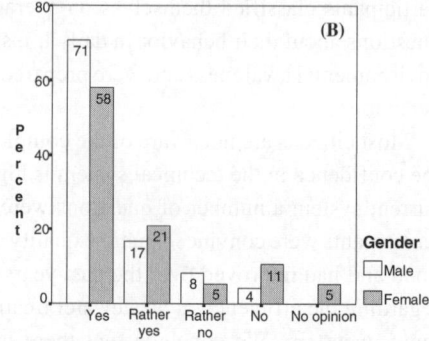

Figure 3 Answers to the question (A) Could you imagine purchasing a NoMix toilet? (B) Could you imagine moving into an apartment with a NoMix toilet? The percentages are given on the top of each bar

and not owners of their house or apartment. Hence, they are not able to make the decision to purchase a NoMix toilet. Further, citizens hold the subjective perception that the purchase and installation of the NoMix technology is associated with considerable additional efforts. In particular the male participants justified their negative attitude to buying a NoMix toilet with the premature state of the whole technology. If additional costs and efforts are carried by the landlord the willingness to move into an apartment with NoMix toilet is high.

Focus group participants were less optimistic regarding the large-scale introduction of the NoMix technology in a country such as Switzerland. More than 50% of the participants could not imagine Switzerland having only NoMix toilets. This may be attributed to the fact that the whole idea is still quite remote for most people. A typical statement made during discussions in the plenary supports this:

"Overall I think that the separation technology is a good thing. But I have difficulties to imagine how to put the new philosophy into practice. I am interested to find out how to communicate the idea and how to translate it into action. How many generations will be required – the whole thing is very remote. To be honest, I do not give the whole new philosophy a real chance. Human beings are comfortable and complacent. Often a message is not communicated in the right fashion and to the right audience at the right moment (designer, age 56)."

Ideas for improvement of the NoMix technology and general recommendations

Participants made quite a few practical recommendations for improvements. Most suggestions from women were related to an improved and more esthetic design. The suggestions from men were mainly related to technical improvements. Overall, the suggestion was repeatedly made to introduce the NoMix technology into public buildings, hospitals and home for the retired where the input of drugs via urine is particularly high. This shows the importance attributed to the removal of pharmaceuticals.

Urine as fertilizer in agriculture

Participants were asked if they could imagine eating vegetables fertilized with urine. Figure 4 shows that a majority (72%) answered positively. Also 80% stated that they would prefer vegetable fertilized with urine to artificial fertilizer. Arguments in favor were related to the fact that urine fertilizer is more natural. Schmidtbauer (1996) reported similar results from a survey in a Swedish municipality.

People pointed out the need to find an attractive name and develop good marketing to overcome the possible bias towards urine as a basis for fertilizer production. Despite

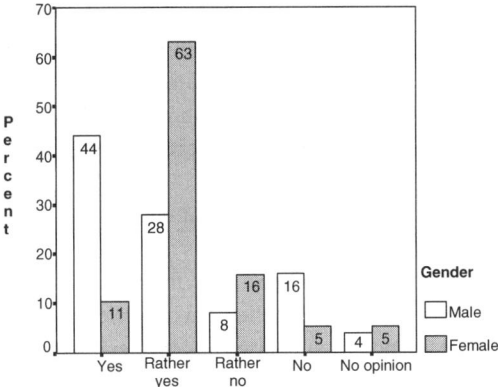

Figure 4 Answers to the question "Could you imagine buying regularly urine-fertilized vegetables". The percentages are given on top of each bar

this overall positive attitude participants emphasized that any health risks should be excluded:

> "As long as potential sources for disease and threats to human health cannot be excluded, I would not welcome the application of human urine even when artificial fertilizer is not a very attractive alternative (Environmental consultant, age 35)."

Discussion

Finding focus group participants proved to be rather difficult. This resulted in a biased composition of the focus groups, with a large level of environmentally friendly attitudes. A larger number of citizens could be addressed with a survey. This would allow putting the quantitative results on a stronger statistical basis. However, it can be expected that the bias towards environmentally friendly citizens would be at least as high or even higher unless the focus of the questionnaire were shifted towards human health. The issue of urban water management and wastewater treatment does not rank high on the agenda of the average Swiss citizen. The system seems to function perfectly and there is not much awareness for a need for change. This was also the message that was conveyed to the public for years. The complexity of the overall system is not anticipated by the broad public, who has only a little responsibility for the well-functioning of the system.

It proved to be quite difficult to convey the complex idea of nutrient recycling despite the efforts devoted to producing a comprehensive and easy to understand citizen information tool. Citizens were mainly interested in the practical aspects of household/bathroom technology and in the whole issue of drug removal. This supports the experience with citizen focus groups on climate change (Schlumpf et al., 2001; Pahl-Wostl et al., 2000; Kasemir et al., 2000). In particular the whole aspect of uncertainties is difficult to convey to a lay public. This was especially prevalent for climate change, but became also evident in the discussion of the focus groups on NoMix technology. Citizens need further a clear link to topics of relevance in their daily lives (see e.g. Schlumpf et al., 1999). In particular direct threats to environmental and human health proved to be the main focus of interest. Here citizens requested absolute certainty that potential threats to human health could be excluded for using urine-based fertilizer in organic farming. Considerable efforts have thus to be devoted to research into developing the technological basis of removing drugs and antibiotics in the process of fertilizer production and to explore methodologies to assess the ecotoxicological effects of drugs. These are important topics of the NOVAQUATIS project. Given the fact that absolute certainty cannot be obtained, much emphasis will have to be devoted to the communication of risks and possible uncertainties. One has further to keep in mind that the focus groups were run just after the BSE crisis, which was particularly strong in Switzerland. People were quite sensitive to potential health threats. The whole idea of closing cycles in the human chain was partly linked to negative associations given the experience with BSE. Given this fact the high general positive attitude is very encouraging.

However, one has also to take into account that citizens are hardly prepared to carry additional costs and efforts. If this attitude prevails already in groups where the environmental awareness is high, one has to expect that the average citizen will be even less prepared to invest time and money. This implies that in particular in the initial state the new technology will have to be financially supported and that technological development has to focus as well on comfort and the ease of handling.

The high individual willingness of citizens to accept a NoMix toilet in their own households and to buy urine-fertilized vegetable corresponds to the results from a survey amongst Swiss farmers presented by Lienert *et al.* (2002). 57% of the farmers thought the idea of using urine-based fertilizer was good or very good, and 42% were willing to purchase such a product. Farmers, however, raised similar concerns regarding micropollutants as found in this study: 30% indicated concerns regarding micropollutants, without being specifically asked to answer any question on this subject.

Despite the individual willingness to accept the NoMix technology, there was a general skepticism regarding the willingness of the majority in society or another stakeholder group to do the same. In informal discussions with different stakeholders, we often encounter the same phenomenon. Many people find the idea of urine separation convincing, but think that farmers will not accept or apply the fertilizer. Experts on sanitary devices accept readily that NoMix technology would be beneficial for the environment, but they are equally convinced that people will not want to buy them. It would be an interesting issue to explore further the reasons for the discrepancy between individual willingness and the skepticism regarding the willingness for the majority in society or another stakeholder group to accept the new technology. It would make also much sense to foster an exchange among the different stakeholder groups. A future project could involve an actors' platform with representatives from the major stakeholder groups (e.g. farmer association, consumer association, manufacturers of sanitary technology, etc.).

The study reported here is focused on a technology assessment. It provides a good base for a more comprehensive investigation of new scenarios for urban water management and socio-technical transformations, of a new role for citizens and households in a system with changed decision-making structures and institutional settings.

Conclusions

In summary, the technology or urine separation is well accepted as long as it offers the same level of comfort and is not more expensive for individuals than conventional technologies. For the participants in the focus groups, the topic of micropollutants proved to be the most important argument for introducing NoMix technology.

There was a high acceptance of urine as fertilizer, under the condition that risks connected to hygiene and micropollutants could be excluded. However, this seemed to be based on more intuitive arguments such as urine being more natural than artificial fertilizers. Arguments related to long-term sustainability (closing nutrient cycles) are still of less importance than arguments that relate directly to the effects of micropollutants on human and ecosystem health.

For a successful introduction of NoMix technology in Switzerland, removal of micropollutants and a high hygienic standard are essential elements. However, development will also have to focus on esthetic aspects and practical everyday handling as well as providing sound technical solutions for storage, transport, and process engineering.

The attitude towards the new technology is in general positive. However, more efforts need to be devoted to inform the public and to get the different stakeholder groups involved in a more in depth discussion.

References

Connor, D.M. (1999). *Constructive Citizen Participation; A Resource Book.* 7th edition.

Dürrenberger, G., Kastenholz, H. and Behringer, J. (1999). Integrated Assessment Focus Groups: Bridging the gap between science and policy? *Science and Public Policy*, **5**, 341–349.

Glicken, J. (2000). Getting stakeholder participation "right": a discusion of participatory processes and possible pitfalls. *Env. Sci. and Policy*, **3**, 305–310.

Jaeger, C., Schüle, R. and Kasemir, B. (1999). Focus Groups in Integrated Assessment: A Micro-Cosmos for Reflexive Modernization. *Europ. J. Social Sci.*, **3**, 195–219.

Kasemir, B., Schibli, D., Stoll, S. and Jaeger, C.C. (2000). Involving the Public in Climate and Energy Decisions'. *Environment*, **42**, 32–42.

Larsen, T.A., Peters, I., Alder, A., Eggen, R.I., Maurer, M. and Muncke, J. (2001). Reengineering the toilet for sustainable wastewater management. *Environ. Sci. & Technol.*, May 1, 192A–197A.

Lienert, J., Haller, M., Berner, F., Stauffacher, M. and Larsen, T.A. (2002). How farmers in Switzerland perceive fertilizers from recycled anthropogenic nutrients (urine). *Water Science and Technology*, **48**(1) 47–56 (this issue).

Pahl-Wostl, C., Schlumpf, C., Schönborn, A., Büssenschütt, M. and Burse, J. (2000). In Integrated Assessment. Models at the interface between science and society: *Impacts and Options. Integrated Assessment*, **1**, 267–280.

Renn, O., Webler, T. and Wiedmann, P. (Eds). (1995). *Fairness and Competence in Citizen Participation: Evaluating Models for Environmental Discourse.* Dordrecht, Kluwer.

Schlumpf, C., Behringer, J., Dürrenberger, G. and Pahl-Wostl, C. (1999). The personal CO_2 calculator: A modeling tool for Participatory Integrated Assessment methods. *Environmental Modeling and Assessment*, **4**, 1–12.

Schlumpf, C., Pahl-Wostl, C., Schönborn, A., Jaeger, C.J. and Imboden, D. (2001). IMPACTS – An Information Tool for Citizens to Assess Impacts of Climate Change from a Regional Perspective. *Climate Change*, **51**, 199–241.

Schmidtbauer, P. (1996). *Hinder och möjligheter för källsortering av humanurin – intervjuuntersökning bland lantbrukare, fastighetsförvaltare och boende i Ale kommun. Technical Report no. 5*, Swedish University of Agricultural Sciences, Division of Soil Management. ISSN 1400–7207 (In Swedish).

The quest for sustainable nitrogen removal technologies

A. Mulder

Amecon Environmental Consultancy, P.O. Box 606, 2600 AP Delft, The Netherlands
(E-mail: *arnoldmulder@amecon.nl*)

Abstract In this paper the sustainability of current available and future nitrogen removal systems has been investigated. For the assessment of the sustainability six indicators were used; sludge production; energy consumption, resource recovery; area requirement and N_2O-emission. For the evaluation of the position of the individual nitrogen removal systems in the anthropogenic nitrogen cycle a broad outline for a life-cycle analysis has been presented.
Keywords Algal ponds; Anammox; constructed wetlands; denitrification; duckweed ponds; life-cycle analysis (LCA); nitrification; nitrogen cycle; resource recovery; sustainable processes; wastewater treatment

Introduction

For human life and health a diet with sufficient protein is essential. This results in a central position of man in the anthropogenic nitrogen cycle (Figure 1). The supply of protein food for the global population by agriculture is nowadays largely dependent on the use of synthetic nitrogen fertiliser produced from atmospheric N_2 by the Haber-Bosch process. In the last century the world's annual industrial output of nitrogenous fertiliser increased from 10 Mt N in 1960 to about 90 Mt N in 1998 (Figure 2). The global estimate for biological nitrogen fixation is in the range of 200–240 Mt N which shows that the anthropogenic mass flows for nitrogen have a major impact on the global nitrogen cycle (Gijzen and Mulder, 2001). However, the consumption of proteins will ultimately result in the discharge of the protein nitrogen in wastewater (Figure 1).

In view of the substantial contribution of the inorganic nitrogen in wastewater in the global nitrogen balance it is relevant to consider the sustainability of the current available nitrogen removal and cycling processes. The processes and technologies that will be considered in this paper are physical–chemical processes, conventional and advanced nitrification and denitrification in activated sludge systems, algal ponds, duckweed ponds and constructed wetlands. This paper presents further a framework to assess the sustainability of the individual processes and technologies. The process properties and criteria which have been evaluated are energy consumption in relation to the total nitrogen cycle, area requirement, process stability, sludge production, global or regional applicability, N_2O emission and resource recovery. Further, the aspect of dilution of sewage in the widely applied flush toilet will be discussed.

Indicators for sustainable nitrogen removal

For the assessment of the sustainability of wastewater treatment systems for N-removal in this study the following indicators have been used: sludge production; energy consumption; resource recovery; area requirement and emission of nitrous oxide (N_2O). In the literature there is general agreement on the relevance of minimising sludge production and maximising resource recovery for sustainability of treatment systems (Henze, 1997; Roeleveld *et al.*, 1997; Van der Graaf *et al.*, 1997). However with respect to energy consumption some authors conclude that the contribution of wastewater treatment in the total energy consumption is relatively low and therefore minimisation of the energy demand was considered to be less relevant (Roeleveld *et al.*, 1997). However in this study the energy consumption for nitrogen removal is compared with the energy input in the total nitrogen cycle. For the assessment of the environmental impact of the alternative nitrogen removal processes a life-cycle analysis (LCA) has been made for the anthropogenic N-cycle. When the sustainability of total nitrogen cycle is discussed the aspect of dilution will be included.

Anthropogenic nitrogen balance

For a concise life-cycle analysis of the anthropogenic use of reactive nitrogen, the input and ultimate human output of nitrogen have been compared. At the end of the twentieth century the global annual input of nitrogen fertiliser in the human food chain was about 15 kg N per capita (Figure 3). A fraction of this nitrogen will be incorporated in vegetable or animal protein. The amount of protein in the human diet varies world-wide from 20 to more then 100 g protein per head daily. After human consumption ultimately a large fraction of the protein nitrogen will be excreted. Based on a nitrogen content in protein of 6.25 the daily human discharge is estimated at 5 to 16 g nitrogen per head per day. This is in agreement with reported typical values for human nitrogen excretion (Table 1).

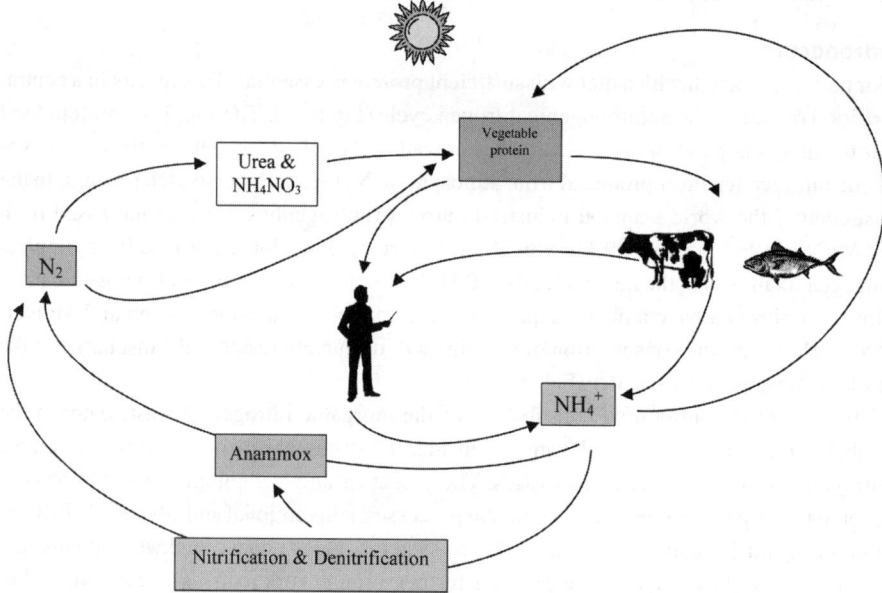

Figure 1 The anthropogenic nitrogen cycle

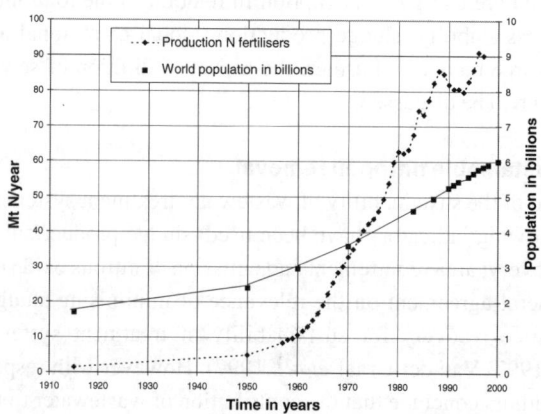

Figure 2 Global rise in population and production of nitrogen fertiliser (data FAO, 1966–1998)

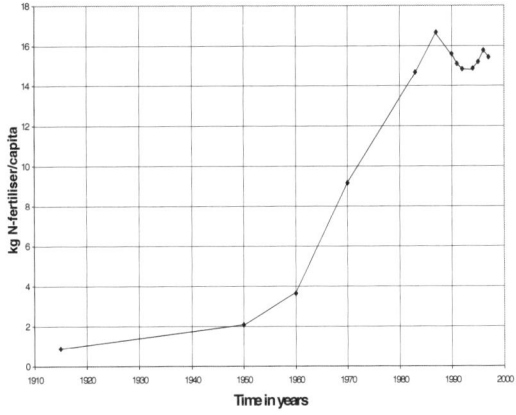

Figure 3 Development in the global annual production of nitrogen fertiliser per capita

Based on an average excretion of 13 g N per capita per day the annual excretion is 4.75 kg N per capita. Comparison of the global averages for input and output of N in the human food chain show that about 30% fertiliser nitrogen ends up in wastewater (Figure 4). However for Norway and The Netherlands the balance of fertiliser nitrogen input and human output through wastewater differ considerably from the global average (Figure 4). This confirms the presence of regional imbalances (Laegreid *et al.*, 1999).

In countries with intensive agriculture and factory farming (Norway and The Netherlands) only about 18% of the fertiliser nitrogen will end up in the wastewater. This means that in these European countries in theory with full nitrogen recovery from wastewater the agricultural demand for N-fertiliser can be met for only 18%. Due to the input of other sources such as imported cattle feed, wet and dry deposition and biological

Table 1 Typical values human nitrogen excretion

Faeces (g N/Capita.d)	Urine (g N/Capita.d)	Total (g N/Capita.d)	Reference
1.8–4.9	7.5–13.3	9.3–18.2	Gootaas, 1956
1.5	12.2	13.7	Pöpel, 1993
–	9.2–13.8	–	Larsen and Guyer, 1996

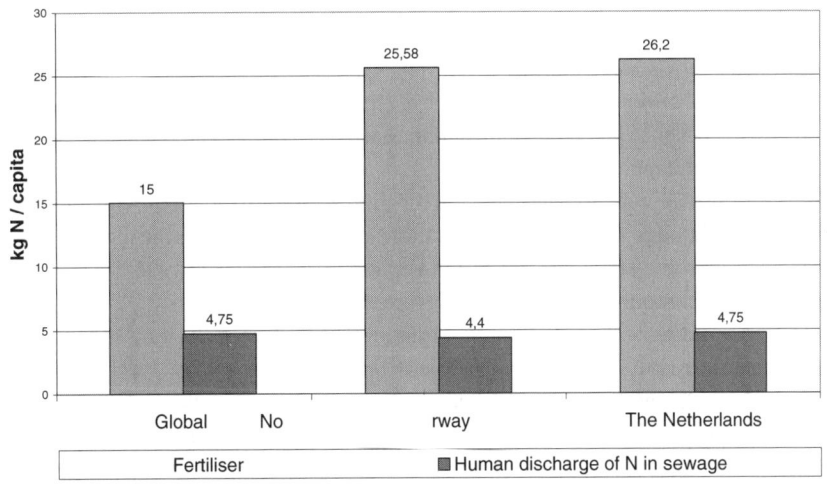

Figure 4 Comparison of the annual balance of N-fertiliser use and human N-discharge on global scale and on national scale in Norway (Bleken and Bakken, 1997) and The Netherlands (RIVM, 1991)

fixation, the contribution of anthropogenic nitrogen output via sewage in the total N-input is actually only about 10% (Bleken and Bakken, 1997). Measures aimed at an increase of the efficiency in the nitrogen cycle, such as the development of slow release fertilisers and improving the conversion efficiency in cattle feed are therefore considered to be more effective compared with resource recovery from sewage (Bleken and Bakken, 1997). On the other hand there will be regions with insufficient N-input where the contribution of N-recovery from sewage can be a relevant factor. This is in agreement with the suggestion of Larsen and Guyer (1996) to apply separate management of anthropogenic nutrient solutions in nutrient demanding nations.

Survey of nitrogen removal systems for wastewater

Wastewater concentration as factor in the selection of an optimal N-removal system

For the removal of ammonium from wastewater a wide variety of biological and physico-chemical removal systems are available (Henze et al., 1996; Priestley et al., 1995). For a specific application the available alternatives will be evaluated on cost aspects, chemical and energy requirements, operational experience and process reliability. The selection of the most optimal alternative is generally based on cost-effectiveness. However in practice the selection of either a biological or a physicochemical method is determined by the nitrogen concentration of the wastewater. Three concentration ranges can be distinguished:

1. Diluted wastewater with an ammonium concentration < 0.1 kg N/m^3. In this range biological N-removal is the preferred process based on cost-effectiveness. Domestic wastewater is within this range.
2. Concentrated wastewater with ammonium concentrations in the range 0.1–5 kg N/m^3. A typical example is sludge liquor for which after extensive investigations biological treatment was preferred (Janus et al., 1997). Although ammonia stripping and producing $MgNH_4PO_4$ were identified as interesting alternatives for resource recovery these options were not cost-effective (Priestley et al., 1995; Janus et al., 1997).
3. Concentrated wastewater with ammonium concentrations > 5 kg N/m^3. In this range physicochemical methods are technically and economically feasible. A successful example is the steam stripping of a wastewater with an ammonium concentration of 1.5% followed by ammonia recovery which has been in operation on industrial scale since 1985 (Harmsen et al., 1986).

The concentration of wastewater is generally < 2 kg N/m^3 thus biological treatment systems are preferred.

Survey of biological N-removal systems

For the biological N-removal systems which are investigated in this study the typical values of specific operational parameters are summarized in Table 2. The following N-removal systems are considered:
- Activated sludge with conventional nitrification and denitrification. Around the world the activated sludge system with nitrification and denitrification is the most widely used system for N-removal with many design variations (Henze et al., 1996). In the nitrification process ammonium is via nitrite converted into nitrate: $NH_4^+ + 2O_2 \rightarrow NO_3^- + 2H^+ + H_2O$. According to this reaction the oxidation of 1 kg N requires 4.57 kg O_2. For denitrification the required COD/N ratio varies from 3–6.
- Activated sludge with nitrification denitrification via nitrite. In this process ammonium is oxidised into nitrite: $2NH_4^+ + 3O_2 \rightarrow 2NO_2^- + 4H^+ + 2H_2O$. The oxidation of 1 kg N requires 3.42 kg O_2. Under anaerobic conditions nitrite is reduced into nitrogen gas and the required COD/N ratio is 2–4.
- Activated sludge with autotrophic N-removal. Recently the Anammox process in which

ammonium is oxidised under anaerobic conditions was discovered by serendipity (Mulder et al., 1995). This process can be considered as an advanced biological nitrogen removal process in which nitrification and denitrification are integrated. Initially it was unknown whether nitrite or nitrate was used as electron donor however later it was found that in the Anammox process ammonium and nitrite are reacting and release nitrogen gas into the N-cycle (Strous, 2000). First ammonium must be oxidised to nitrite and then with Anammox the overall reaction of the autotrophic nitrogen removal process is: $4NH_4^+ + 3O_2 \rightarrow 2N_2$. According to this reaction the removal of 1 kg N requires 1.71 kg O_2.

- Algal ponds. In algal ponds ammonia is assimilated into algal biomass according to the equation: $NH_3 + 5CO_2 + 2H_2O \rightarrow C_5H_7O_2N + 5O_2$. The energy use in the algal pond is required for mixing and pumping. The highest value of the N-load is corresponding with the lowest efficiency.
- Duckweed ponds. From the applied N-load 41–68% is assimilated in duckweed (Alaerts et al., 1996).
- Constructed wetlands. In constructed wetlands biomass is not recovered. The highest value of the N-load corresponds with the lowest removal efficiency.

Table 2 Typical values of specific operational parameters of biological N-removal systems

N-removal system	N-load (kg N/ha.d)	Energy consumption (kWh/kg N)	COD/ N-ratio	Sludge/biomass production (kg d.w./kg N)[1]	N_{total} removal efficiency (%)	Comments and references
Activated sludge conventional nitrification-denitrification	200–700[2]	2.3[3]	3–6	1–1.2	>75	Based on plants with typical load of 0,05–0,1 kg BOD/ kg d.w.d. Henze et al., 1996
Activated sludge nitrification-denitrification via nitrite	200–700[2]	1.7[3]	2–4	0.8–0.9	>75	Abeling and Seyfried, 1992; Balmelle et al., 1992
Activated sludge autotrophic N-removal	>200–700[2]	0.9	0	<0.1	>75	Feasibility demonstrated for concentrated wastewater. Mulder, 1995; Seyfried et al., 2002
Algal pond	15–30	0.1–1	6–7	10–15	23–78	El Hamouri et al., 1995; Oswald, 1995; Polprasert, 1996
Duckweed pond	3–4	<0.1	28	20–26	74–77	System operated at 17–33°C. Alaerts et al., 1996
Constructed wetland	3–26[4]	<0.1	2–7	–	30–70	Hammer and Knight 1994; Haberl, 1995; Wittgren and Tobiason, 1995; Meuleman, 1999

1. d. w. = dry weight
2. Calculated based on a conservative value of the aeration tank depth of 2 m and surface area aeration tank is 25% of total plant area
3. The given values are energy required for aeration and are exclusive energy in COD used for denitrification. Aeration efficiency is 2 kg O_2/kWh
4. An exceptional high load of 112 kg N/ha.d was reported by Van Oostrom (1995)

Assessment of the sustainability of N-removal systems
Energy use for N-fixation and N-removal

The energy use of the alternative N-removal processes has been compared with of the energy use for N-fertiliser production (Figure 5). In the Haber-Bosch process nitrogen fixation with natural gas proceeds according to the equation: $3CH_4 + 4N_2 + 6H_2O \rightarrow 3CO_2 + 8 NH_3$. For the currently applied N-fertiliser production processes the energy use is 44.5 MJ/kg N which can be reduced to 34.5 MJ/kg N in production plants where best available technology is applied (Laegreid et al., 1999). The energy required for nitrification was calculated from the data given in Table 2. The energy used for denitrification was calculated from the required COD. When resource recovery is applied the production of N-fertiliser will require an equivalent reduced amount of energy. The obtained results show that for conventional nitrification-denitrification the energy consumption for N-removal is 42.2 MJ/kg N. This is nearly similar with the energy use for N-fertiliser production. The energy required in the autotrophic N-removal process is 3.1 MJ/kg N which is only 7% of the energy required in the conventional process. By application of resource recovery the situation becomes more favorable, however it must considered that the recovery and processing of the nitrogen from urine will require energy.

Recovery of reactive nitrogen versus denitrification

From the biological N-removal systems only the algal and duckweed pond system apply resource recovery, by using the produced algae and duckweed respectively. In the activated sludge systems the ammonium nitrogen is recycled as nitrogen gas. Direct recovery of ammonium from human wastewater is only feasible when the flush toilet and the collection and transport system for urine are modified fundamentally (Stoner, 1977; Larsen and Guyer, 1996). For the assessment of the environmental advantage of the resource recovery for nitrogen the energy use for N-fertiliser production has been compared with the energy use of the alternative N-removal processes (Figure 5). The energy use of the autotrophic N-removal process and the 20% resource recovery will come on a similar level after addition of the energy used for recovery.

Sludge production

The autotrophic N-removal process has lowest sludge production (Table 2).

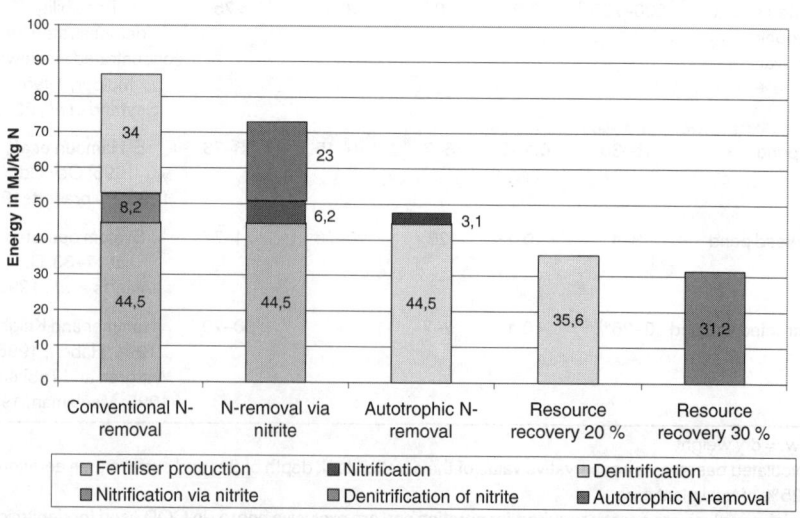

Figure 5 Comparison of the energy use for N-fertiliser production with the energy use of the alternative N-removal processes

Area requirement

For the individual N-removal systems the N-load, as measured for the area requirement, differs considerably (Table 2 and Figure 6). The wide range for the activated sludge systems is caused by differences in plant size. The average value of 450 kg N/ha.d is a factor 30–150 higher then for respectively the algal and duckweed pond systems. The higher loading rates are the result of faster biological processes which are possible due to the aeration. However because of this the energy use for the activated sludge systems is higher (Figure 6). In systems with the autotrophic N-removal process the energy use is reduced at similar N-load (Figure 6).

Emission of N$_2$O in the anthropogenic nitrogen cycle

At different stages of the anthropogenic nitrogen cycle emission of N$_2$O occurs. The global anthropogenic emission of N$_2$O is estimated at about 3 M ton N$_2$O-N/year (exclusive the contribution from combustion of fossil fuel, calculated from data in Pérez-Ramírez, 2002). Based on this figure it can be calculated that from the global annual production of nitrogen fertiliser of 90 Mton N about 0.3% will dissipate as N$_2$O. This value is lower then the emission of 0.5–2% given by Smil (2001). It is reported (Pérez-Ramírez, 2002) that this emission largely occurs in agriculture (50%), fertiliser production (6%) and wastewater treatment (21%). The emission of N$_2$O in wastewater treatment is largely determined by operational conditions and under unfavourable conditions like high load and low COD/N

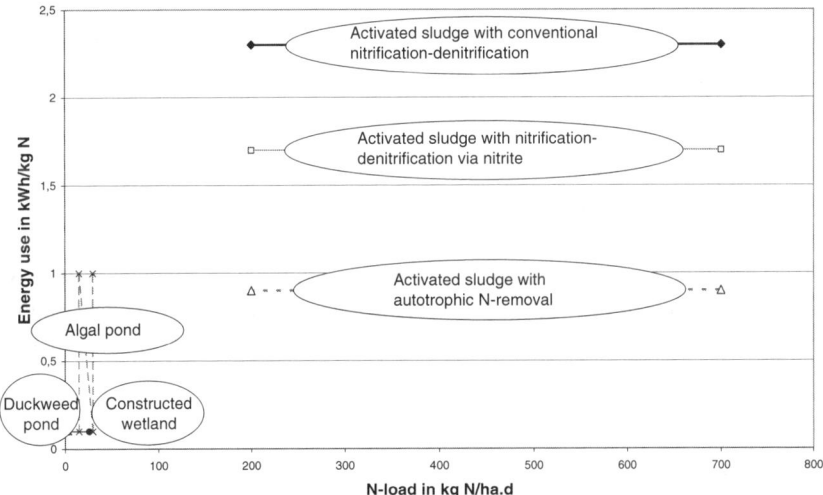

Figure 6 Comparison of energy consumption and area requirement of biological N-removal systems

Table 3 Matrix for the assessment of the sustainability of biological N-removal systems

N-removal system	Sustainability indicator				
	Sludge production	Energy consumption	Resource recovery	Area requirement	N$_2$O-emission
Activated sludge conventional nitrification-denitrification	– –	– – –	–	+++	–
Activated sludge nitrification-denitrification via nitrite	–	– –	–	+++	–
Activated sludge autotrophic N-removal	+	–	–	+++	–
Algal pond	+	–	+	– –	– –[1]
Duckweed pond	+	+	+	– –	– –[1]
Constructed wetland	+	+	–	– –	– –[1]

1. Based on prognosis because reliable data of N$_2$O emissions of these systems are not available

ratio's the emission of N_2 can contribute 10% in the total N-balance (Hanaki et al., 1992). In high loaded constructed wetlands the N-load is frequently much higher then the nitrogen input in intensive agriculture which is around 100 kg N/ha.year. Therefore it is likely that a considerable part of the N-load is not taken up in plant production systems and more information is required with respect to the potential emission of nitrous oxide (Bouwman, 1998). Also the N-balances of algal and duckweed ponds must be checked for the emission of nitrous oxide.

Final evaluation

The sustainability matrix of the investigated N-removal systems show that the activated sludge system with autotrophic N-removal is the most sustainable process (Table 3). The next best is the duckweed pond system. The executed analysis of the nitrogen balances disclosed that on global scale there are regional imbalances with respect to N-input and output. In regions with serious shortages in N-input resource recovery must developed and stimulated. On the other hand it is relevant to investigate expected future developments and to anticipate expected trends. With respect to the re-use of algae and duckweed for feed production the potential risk of accumulation of metals and organic micro-pollutants must be investigated.

Conclusions

- On a global scale regional conditions with respect to N-balance, climate, development differ considerably. As a result of this there is not one universal or paramount treatment system for N-removal from sewage. Regional conditions will determine which system is optimal under given conditions of climate, agricultural, environmental and economic development. This means that in a changing world the quest for sustainable N-removal techniques will continue.
- The process of autotrophic N-removal, which combines nitrification via nitrite and anaerobic ammonium oxidation is a very sustainable process because of: the minimal energy consumption, the absence of need for organic matter for denitrification, low sludge production and high N-load. However the autotrophic N-removal systems which are currently under development are applicable for ammonium concentrations of 100–500 mg N/l. For sustainable development it is relevant to investigate the feasibility of the autotrophic N-removal process for diluted wastewater.
- For high loaded constructed wetland systems, algal and duckweed ponds reliable data of the N-balance and the contribution of N_2O-emissions are not available. This information is required for a precise assessment of the sustainability of these systems.

References

Abeling, U. and Seyfried, C.F. (1992). Anaerobic-aerobic treatment of high-strength ammonium wastewater-nitrogen removal via nitrite. *Wat. Sci. Tech.*, **26**(5–6), 1007–1015.

Alaerts, G.J., Rahman Mahbubar, M.D. and Kelderman, P. (1996). Performance analysis of a full-scale duckweed-covered sewage lagoon. *Wat. Res.*, **30**(4), 843–852.

Balmelle, B., Nguyen, K.M., Capdeville, B., Cornier, J.C. and Deguin, A. (1992). Study of factors controlling nitrite build-up in biological processes for water nitrification. *Wat. Sci. Tech.*, **26**(5–6), 1017–1025.

Bleken, M.A. and Bakken, L.R. (1997). The nitrogen cost of food production: Norwegian society. *Ambio*, **26**(3), 134–142.

Bouwman, A.F. (1998). Nitrogen oxides and tropical agriculture. *Nature*, **392**, 866–867.

El Hamouri, B., Jellal, J., Outabiht, H., Nebri, B., Khallayoune, K., Benkerroum, A., Hajli, A. and Firadi, R. (1995). The performance of a high-rate algal pond in the Moroccan climate. *Wat. Sci. Tech.*, **31**(12), 67–74.

FAO (1966–1998). *Yearbook Fertiliser*. Vol. 16–48, FAO, Rome.

Gijzen, H.J. and Mulder, A. (2001). The nitrogen cycle out of balance. *Water21*, **8**, 38–40.

Gotaas, H.B. (1956). Composting: Sanitary disposal and reclamation of organic wastes. W.H.O., Geneva.

Haberl, R., Perfler, R. and Mayer, H. (1995). Constructed wetlands in Europe. *Wat. Sci. Tech.*, **32**(3), 305–315.

Hammer, D.A. and Knight, R.L. (1994). Designing constructed wetlands for nitrogen removal. *Wat. Sci. Tech.*, **29**(4), 15–27.

Hanaki, K., Hong, Z. and Matsuo, T. (1992). Production of nitrous oxide gas during denitrification of wastewater. *Wat. Sci. Tech.*, **26**(5–6), 1027–1036.

Harmsen, L.W.F., Lourens, P.A. and Van Leeuwen, H.J.M.L. (1986). Stikstofverbindingen verwijderen uit afvalwater. P.T./ Procestechniek 41, 27–29. (in Dutch).

Henze, M. (1997). Trends in advanced wastewater treatment. *Wat. Sci. Tech.*, **35**(10), 1–4.

Henze, M., Harremoës, P., La Cour Jansen, J. and Arvin, E. (1996). *Wastewater Treatment: Biological and Chemical Processes*. 2nd edn, Springer, Heidelberg.

Janus, H.M. and Van der Roest, H.F. (1997). Don't reject the idea of treating reject water. *Wat. Sci. Tech.*, **35**(10), 27–34.

Laegreid, M., Bockman, O.C. and Kaarstad, O. (1999). *Agriculture, Fertilisers and the Environment*. CABI Publishing, Wallingford, UK.

Larsen, T.A. and Guyer, W. (1996). Separate management of anthropogenic nutrient solutions (human urine). *Wat. Sci. Tech.*, **34**(3–4), 87–94.

Meuleman, A.F.M. (1999). Performance of treatment wetlands. PhD Thesis, Utrecht University.

Mulder, A., Van der Graaf, A.A., Robertson, L.A. and Kuenen, J.G. (1995). Anaerobic ammonium oxidation discovered in a denitrifying fluidized bed reactor. *FEMS Microbiol. Ecology*, **16**, 177–184.

Oswald, W.J. (1995). Ponds in the twenty-first century. *Wat. Sci. Tech.*, **31**(12), 1–8.

Pérez-Ramírez, J. (2002). Catalyzed N_2O activation: Promising (new) catalysts for abatement and utilization. PhD Thesis, Faculty of Applied Sciences, Technical University of Delft.

Polprasert, C. (1996). *Organic Waste Recycling*. John Wiley & Sons Ltd., Chichester, England.

Pöpel, F. (1993). Lehrbuch für Abwassertechnik und Gewässerschutz. Deutscher Fachschriften-Verlag, Wiesbaden.

Priestley, A.J., Cooney, E., Booker, N.A. and Fraser, I. (1995). Nutrients in wastewater's – ecological problem or commercial opportunity? *AWWA 17th Federal Convention*, 340–346.

RIVM (1991). Nationale Milieuverkenning 1990–2010. Samson, H.D. Tjeenk Willink bv, Alphen aan den Rijn. (in Dutch).

Roeleveld, P.J., Klapwijk, A., Eggels, P.G., Rulkens, W.H. and Van Starkenburg, W. (1997). Sustainability of municipal wastewater treatment. *Wat. Sci. Tech.*, **35**(10), 221–228.

Seyfried, C.F., Rosenwinkel, K.H. and Hippen, A. (2002). Deammonification: a cost-effective treatment process for nitrogen-rich wastewater's. *Proceedings WEFTECH*, Chicago (in press).

Smil, V. (2001). Nitrogen and food. *Papers of the Second International Nitrogen Conference*, 14–18 October 2001, Washington, D.C.

Stoner, C.H. (1977). *Goodbye to the flush toilet: Water-saving alternatives to cesspools, septic tanks and sewers*. Rodale Press, Emmaus, PA.

Strous, M. (2000). Microbiology of anaerobic ammonium oxidation. PhD Thesis, Technical University of Delft.

Van der Graaf, J.H.J.M., Meester-Broertjes, H.A., Bruggeman, W.A. and Vles, E.J. (1997). Sustainable technological development for urban-water cycles. *Wat. Sci. Tech.*, **35**(10), 213–220.

Van Oostrom, A.J. (1995). Nitrogen removal in constructed wetlands treating nitrified meat processing effluent. *Wat. Sci. Tech.*, **32**(3), 137–147.

Wittgren, H.B. and Tobiason, S. (1995). Nitrogen removal from pre-treated wastewater in surface flow wetlands. *Wat. Sci. Tech.*, **32**(3), 69–78.

A proposed sustainable BNR plant with the emphasis on recovery of COD and phosphate

X.-D. Hao*,** and M.C.M. van Loosdrecht**

* The R & D Centre for Sustainable Environmental Biotechnology, Beijing University of Civil Engineering and Architecture, 1 Zhanlanguan Rd., Beijing 100044, P. R. China (E-mail: *xdhao@hotmail.com*)
** Kluyver Laboratory for Biotechnology, Delft University of Technology, Julianalaan 67, 2628 BC Delft, The Netherlands

Abstract Water problems have to be solved in an integrated way, and sustainability has become a major issue. For this reason, developing more sustainable wastewater treatment processes is needed. New discoveries and good understanding on microbial conversions of nitrogen and phosphorus make more sustainable processes possible. New options for decentralized sustainable sanitation are generally compared to conventional sewage systems, we think that for a proper comparison also innovative centralized treatment schemes should be evaluated. In this article, a more sustainable WWTP is proposed for municipal wastewater treatment, mainly based on the principles of denitrifying dephosphatation and anaerobic ammonium oxidation (ANAMMOX). The proposed system consists of a first stage of the A/B process in which maximal sludge production is achieved. In this way, COD is regained as sludge for methanation. The following BCFS® and CANON processes can remove N and P with minimal or no COD need. As a potential fertiliser, struvite can easily be removed from the sludge water by adding magnesium compounds. A case study is done on the basis of the mass balance over the proposed plant. The effluent from the system has a good quality to be recycled. This could also make a contribution to meeting the world's water needs and lessening the impact on the world's water environment. Since all the separate units are already applied or tested on pilot-scale, no problems for technical implementation are foreseen.
Keywords ANAMMOX; BCFS®; CANON; denitrifying dephosphatation; methanation; struvite

Introduction

Nowadays, wastewater treatment deals with the main polluting components such as COD, ammonium and phosphate. Traditionally, COD and ammonium are eliminated by biological oxidation and nitrification coupled with denitrification. Phosphate is removed either by biological accumulation in the sludge or by chemical precipitation. The main problems of conventional processes include:
- energy consumption due to aeration for COD oxidation and nitrification, which actually results in considerable loss of chemical energy from COD oxidation (about 14 MJ/kg COD);
- COD requirement for denitrification and biological phosphate removal;
- sludge production;
- no recovery of nutrients;
- emissions of CO_2 into the atmosphere.

Wastewater treatment plants (WWTPs) have to cope with increasingly stricter effluent standards. The general EC effluent standards for nutrient discharge have been tightened; the total effluent nitrogen and phosphorus have to be below 10 g N/m^3 and 1 g P/m^3, respectively. In other continents of the globe, eutrophication is also a major problem in the water environment. For example, China has tightened its effluent standards for nutrient discharge ($N_{tot} \leq 15$ g N/m^3 and $P_{tot} \leq 0.5$ g P/m^3) in the last decade, even though the development of its WWTPs is still in the early stages. Complying with these stricter effluent standards will cause even more energy consumption (thus higher CO_2 emissions), more organic carbon consumption (even including additional COD for denitrification and/or dephosphatation) and larger sludge production if conventional processes have to be used. Therefore, improved effluent quality by conventional processes may easily lead to overall adverse effects on the environment (van Loosdrecht et al., 1997). From the point of view of an integrated approach to environmental aspects, conventional processes somehow seem to have an unsustainable feature in fighting eutrophication.

It is really needed to develop sustainable wastewater treatment processes. Sustainable processes can be characterised by a minimum of COD oxidation, a maximum of methanation (by COD conversion), a minimum of energy consumption, a minimum of CO_2 emissions, a minimum of sludge production, and recovery of phosphate. On the other hand, increasing attention has to be paid to recycling of purified effluent, especially in regions with water shortage, such as in large parts of China. Purified effluent is a valuable resource. Instead of being thrown away, this water after appropriate treatment can be reused to reduce the demand on the fresh water sources. Therefore, water recycling can make a sustainable contribution to meeting the world's water needs and lessening the impact on the world's water environment. A move, from the old "use once and throw away" approach to a new "conserve, use wisely and recycle" water economy, will benefit the whole world.

Effluents resulting from full biological treatment and nutrient removal (tertiary treatment) have a good quality to be recycled. By applying advanced treatment processes such as flocculating filtration, membrane filtration, UV and activated carbon to remove the remaining inert COD and bacteria, effluents from tertiary treatment can be further improved and reused for various purposes such as process water, household water, urban water, agricultural irrigation, groundwater supplement and so on (van der Graaf, 2001). Therefore, constructing sustainable WWTPs can simultaneously contribute to two key water problems: i) deterioration in quality, and ii) shortage in quantity. This is like killing two birds with one stone.

Developing novel biological processes for wastewater treatment has to depend on new discoveries and a good knowledge of microbiology and biochemistry. In the last decade, the discoveries of denitrifying dephosphatation (Vlekke et al., 1988; van Loosdrecht et al., 1992; Kerrn-Jespersen et al., 1994; Kuba et al., 1996) and anaerobic ammonium oxidation – ANAMMOX (Mulder, 1992) have made sustainable WWTPs possible. As a practical technique for denitrifying dephosphatation, the BCFS® process has been successfully applied in several Dutch WWTPs (Brandse and van Loosdrecht, 2002). Although still under research, ANAMMOX coupled with partial nitrification has been tested in experiments (Strous et al., 1997; Dijkman and Strous, 1999) and observed in several on-site applications (Hippen et al., 1997; Siegrist et al., 1998; Helmer and Kunst, 1998; Helmer et al., 1999).

It has been quantitatively evaluated that minimising use of COD in WWTPs can significantly contribute to the total environmental quality (Hao et al., 2001a). Moreover, recovery of phosphate will make WWTPs more sustainable. In consideration of these key aspects for sustainability, a sustainable WWTP is proposed for biological nutrient removal (BNR), based on the BCFS® process for denitrifying dephosphatation evaluated by Hao et al. (2001b), the CANON process for completely autotrophic ammonium removal in biofilm evaluated by Hao et al. (2001c and 2001d), the A/B process for COD recovery developed by Boehnke (1978), and potential struvite formation for recovery of phosphate tested by Battistoni et al. (1998), Li et al. (1999), Jaffer et al. (2001) and Battistoni et al. (2001).

Principles of the BCFS® and CANON processes
The BCFS® processes
The BCFS® (biological and chemical phosphorus and nitrogen removal) process is based on the modified UCT process, with the emphasis on developing denitrifying phosphorus-

removing bacteria (DPB). The BCFS® process was developed and has been applied in The Netherlands (van Loosdrecht et al., 1998; Brandse and van Loosdrecht, 2002). Unlike the standard UCT process, the BCFS® process is extended to five compartments, three internal recirculation flows and an integrated P-stripper in the first compartment (Figure 1).

A contact tank (DO = 0) is added as a second selector, to remove soluble hydrolyses products coming from the anaerobic tank with nitrate from return sludge. In this way, growth of filaments is strongly repressed. A mixed tank (DO = 0.5 mg/l) is specially designed for simultaneous nitrification and denitrification and to ensure a low nitrogen concentration in the effluent. This added tank is only aerated when necessary. When influent COD is not enough for bio-P removal or sludge age is too long to biologically accumulate phosphate, released phosphate can be easily extracted from the anaerobic stage. This option can also be used in order to recover phosphate.

Denitrifying phosphorus-removing bacteria (DPB) use nitrate as electron acceptor instead of oxygen. DPB can contribute towards a significant reduction in the need of COD for integrated nitrogen and phosphorus removal. Up to 50% less COD is required, compared to conventional aerobic P-uptake processes (Kuba et al., 1996; Hao et al., 2001a). Moreover, oxygen requirement and sludge production can be decreased up to about 30% and 50%, respectively (Kuba et al., 1996). Because of the reduced COD need, the saved COD can be used for methanation (energy production) in a digestion reactor. Both energy savings from aeration and energy production from methanation can finally contribute to a significant reduction of CO_2 emissions into the atmosphere.

The CANON process

A one-stage ammonium removal process is possible to be achieved in a biofilm reactor with simultaneous nitrification and ANAMMOX (Figure 2), which was tested and observed in experiments (Strous et al., 1997; Dijkman and Strous, 1999) and in practical applications (Hippen et al., 1997; Siegrist et al., 1998; Helmer and Kunst, 1998; Helmer et al., 1999). This process has been entitled "CANON" (completely autotrophic N-removal over nitrite, Strous, 2000).

In a biofilm supporting nitrification and ANAMMOX, there can generally be three autotrophic organisms: ammonium oxidisers, nitrite oxidisers and ANAMMOX organisms. They compete with one another for oxygen, ammonium and nitrite. Due to different affinity constants for oxygen between ammonium and nitrite oxidisers as well as the mass transfer limitation, accumulation of nitrite is completely possible. Under this condition, ANAMMOX could occur with ammonium and accumulated nitrite. The overall stoichiometry of the CANON process can be represented by Eq. (1).

$$NH_4^+ + \frac{3}{4}O_2 \rightarrow \frac{1}{2}N_2 + \frac{3}{2}H_2O + H^+ \qquad (1)$$

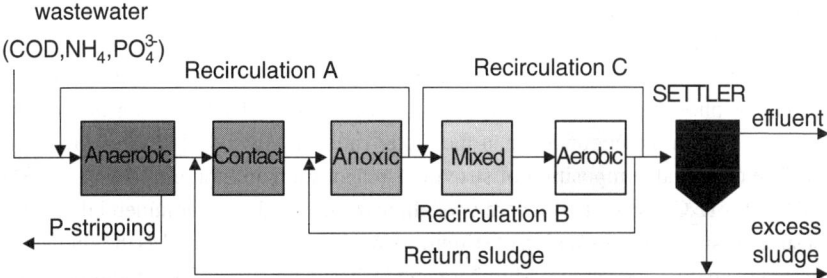

Figure 1 Schematic representation of the BCFS® process

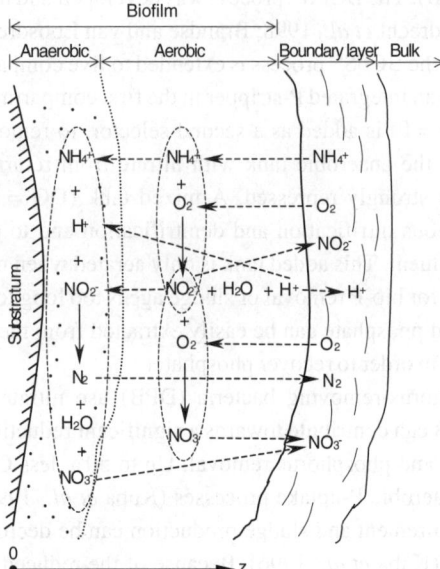

Figure 2 Reaction model of the CANON process

Proposed concept for a sustainable BNR plant

The regularly applied A/B process (Boehnke, 1978) was specially designed for recovery of COD from wastewater. The A-stage of the A/B process has a short solids retention time (SRT = 8–25 h), and a large part (up to 70–80%) of both suspended and soluble COD can be converted into biomass sludge. The sludge produced in this stage can be concentrated and used for methanation. This also maximises the assimilation of nitrogen and phosphorus in the sludge thereby concentrating them for potential recovery. The sludge produced in the A-stage has better digestion characteristics than normal secondary sludge, which results in a lower overall sludge production (Boehnke, 1978). The B-stage of the A/B process was designed for oxidation of the remaining ammonium.

Instead of the B-stage, the BCFS® process is proposed here to remove the remaining N and P with the remaining COD in the main stream (a large diluted cold stream). The CANON process is proposed to treat the sludge liquor (reject-water from sludge digestion, a small warm concentrated side stream). The relatively constant flow-rate with a constant warm temperature will be favourable for the CANON process.

Figure 3 depicts the proposed flow scheme of a sustainable BNR plant. In the proposed BNR plant, recovery of COD and phosphate is emphasised. The A-stage of the A/B process is only used to concentrate both suspended and soluble COD and no chemical P-removal is involved. After the separation and digestion of the sludge from the A-stage, two streams remain: (i) a main stream: the wastewater itself and (ii) a side stream: the sludge liquor. Parts of COD, N and P remain in the main stream, and the BCFS® process is designated to treat this stream.

The main components of the side stream are enriched ammonium and phosphate. Recovery of phosphate from this stream could be achieved by formation of struvite by adding magnesium compounds when the molar ratio of $Mg^{+2}:NH_4^+:PO_4^{3-}$ is greater than 1:1:1. The chemical composition of struvite is magnesium ammonium phosphate (MAP: $MgNH_4PO_4·6H_2O$). Struvite is a hard crystalline deposit and has a commercial value. The most suitable stream for formation of struvite in WWTPs was identified as the sludge liquor separated from digested sludge (Momberg and Oellermann, 1992; Battistonni et al., 1998, 2001; Jaffer et al., 2001). Experiments indicated that the solubility of struvite decreased with increasing pH and reached its minimum at pH = 8.5–9.0 (Li et al., 1999; Jaffer et al., 2001).

Following formation of struvite, the CANON process would be suited to remove the ammonium remaining in the side stream.

Case study of the proposed BNR plant

To demonstrate the proposed BNR plant, an assumed inflow (Q = 8,500 m^3/d, COD = 625 g/m^3, TKN = 60 g N/m^3 and P = 9.5 g P/m^3) is used here as a case study. According to the proposed flow scheme (Figure 3) and some experimental data, it can be assumed that 40% of N-load in the influent can be bound into biomass sludge in the A-stage. It is assumed that sludge thickening is performed by centrifugation. Therefore, 1,200 g $NH_4^+ - N/m^3$ can be expected in the side stream (Mulder et al., 2001). Before making a mass balance, it is further assumed that ordinary biomass sludge contains 8% N and 2% P (maximal 7–8% for P-bacteria sludge) and that about 50% of the sludge-COD can be converted into methane (CH_4). This means that maximum 300 g P/m^3 could exist in the side stream, leading to a necessity and possibility of phosphate recovery by formation of struvite.

Based on the above assumptions and a temperature of 15°C, the mass balance of COD, N and P in the proposed BNR plant can be calculated, which is shown in Figure 4. Practical removal efficiencies in every treatment unit have been considered in calculating the mass balance. If 74% of the influent COD-load is removed in the A-stage, 40% of the influent N-load and 63% of the influent P-load will be assimilated, and 26% of the influent COD-load (based on a sludge yield of 0.65 g biomass-COD/g substrate COD) will be oxidised. This actually results in a total COD removal efficiency of 74% in the A-stage.

The remaining COD (26%) is introduced in the BCFS® process to remove the remaining nitrogen (60%) down to 8 g N/m^3 in the effluent and the remaining phosphorus (37%) down to 0.5 g P/m^3 in the effluent. The conversions are based on a full dynamic simulation model (Hao et al., 2001b). A SRT of 20 days is set for the BCFS® process, leading to a small part of excess secondary sludge (6.6% of the influent COD load). This sludge will be combined with the sludge from the A-stage for digestion.

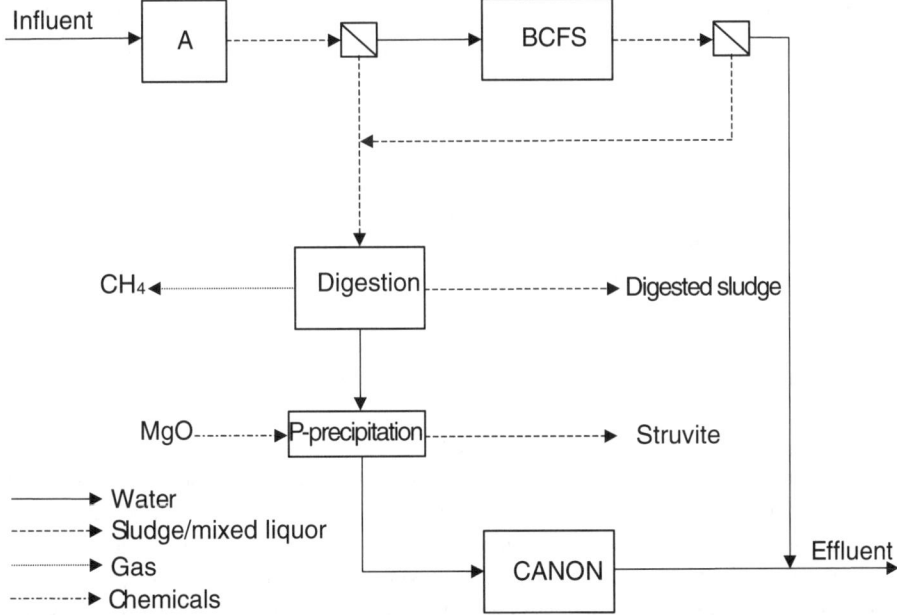

Figure 3 Flow scheme of a proposed sustainable BNR plant

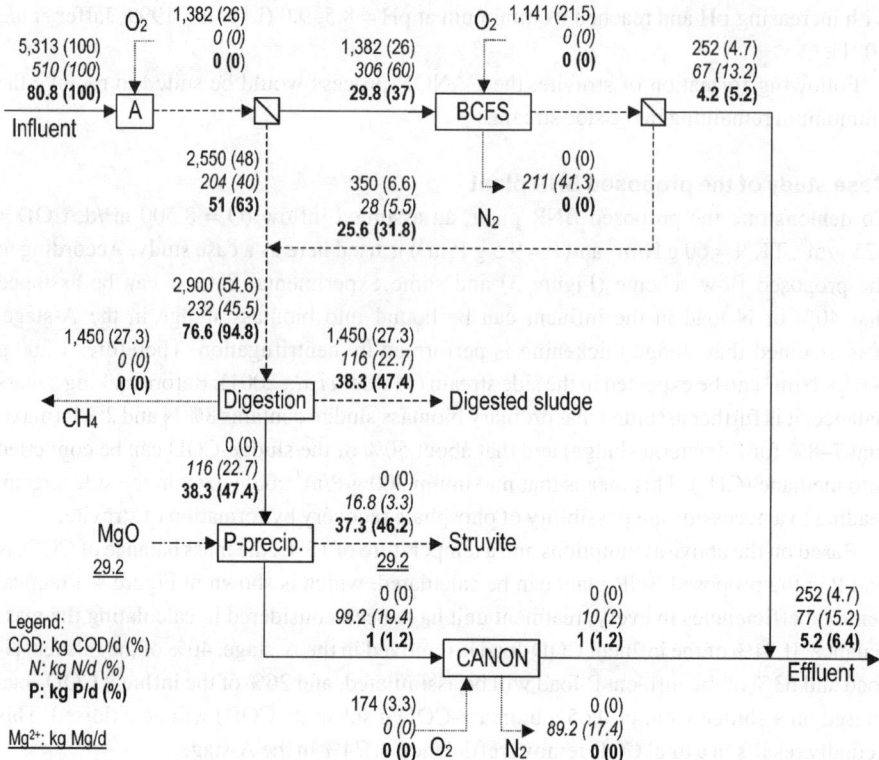

Figure 4 Mass balance in the proposed BNR plant (%: relative to the influent load), indicating what the different line types mean

During digestion, roughly 50% of biomass-COD is converted into methane (CH_4). As a result, the corresponding amounts of nitrogen and phosphorus bound in the sludge are released into the liquid, which will appear in the digester effluent. The other half of biomass-COD remains as excess sludge to be incinerated.

By using magnesium chloride ($MgCl_2$) as a chemical precipitant (Jaffer et al., 2001), 97% of P-removal as struvite can be achieved, which is equivalent to a phosphate recovery of 46.2% of the influent P-load. Under full turbulence and high pH (8.5–9.0), struvite can be formed within 15 min and settle to reach its maximal compact density (2,050 kg/m^3) within only 10 min (Li et al., 1999). This means that a very compact chemical precipitation unit can be expected.

As a final step, the CANON process is proposed to further treat the side stream still containing high ammonium concentration at a temperature of 30°C. Based on the previous studies (Hao et al., 2001c and 2001d), it has been ascertained that at 30°C the CANON process could achieve 90% of N-removal in a 1-mm thick biofilm at an ammonium surface load of 1.5 g NH_4^+ – N/m^2·d associated with a DO concentration level of about 1 g O_2/m^3. If a biofilm has a total surface of 250 m^2 biofilm/m^3·reactor, a biofilm reactor of 265 m^3 is needed.

Although higher concentrations of ammonium and phosphate still remain in the treated side stream after formation of struvite and autotrophic N-removal, this treated stream can be mixed with the effluent of the main stream before being discharged. In this way, the large flow-rate of the main stream with lower concentrations of ammonium and phosphate could dilute the treated side stream. It is also possible to recycle the side stream water back to the influent and to get a slightly better effluent. In this case study, the total effluent concentrations of COD, N and P are calculated to be 30 g COD (inert)/m^3, 9 g N/m^3 and 0.6 g P/m^3, respectively.

Remarks of sustainability for the proposed BRN plant

In the proposed BNR plant, COD oxidation into CO_2 is minimised (26% oxidation in the A-stage), which is needed for the formation of biomass out of soluble COD. 48% of the influent COD-load is converted into biomass sludge in the A-stage. The remaining COD (26%) is used in the second stage for N and P removal in the BCFS® process. Since only 60% of the influent N-load needs to be removed through denitrifying dephosphatation, a corresponding amount of COD saving for denitrification could be expected. Summed over the whole process, 1 mg-N removed needs only 2.6 mg oxygen, which is less than the normal value. As shown in Figure 4, over ¼ (27.3%) of the influent COD load could be converted into CH_4. After formation of struvite, almost half (46.2%) of the influent P-load could be recovered. Compared with a conventional activated sludge WWTP with a low sludge load (0.05 kg BOD_5/kg VSS·d), the main differences are shown in Table 1.

Evaluation of the proposed BNR plant

In the proposed BNR plant, the different technologies such as the A/B process, the BCFS® process, and formation of struvite have been separately applied in practice (Momberg *et al.*, 1992; van Dongen *et al.*, 2001; Brandse and van Loosdrecht, 2001). Digestion is an existing technology for sludge conversion into methane. Although the CANON process is still under research, some co-incidental application cases have been reported (Hippen *et al.*, 1997; Siegrist *et al.*, 1998; Helmer and Kunst, 1998; Helmer *et al.*, 1999). The previous studies of ours (Hao *et al.*, 2001c, 2001d) have given a good qualitative description of this novel concept of autotrophic ammonium removal. This will speed up its deliberate application. Therefore, technological difficulties will not be a problem in application of the proposed BNR plant to practice.

Due to the integration of all the concerned technologies, the infrastructure costs (engineering, construction and materials) of the proposed BNR plant will be relatively high. Attributed to its sustainability, however, the operational costs (mainly consumption of electricity and production of sludge) of the proposed BNR plant will be lower, according to Table 1. Moreover, CO_2 emissions into the atmosphere can be reduced, and recovery of phosphate as struvite can be realised, which could be used as a potential fertiliser in agriculture or horticulture.

In wastewater treatment, increasing sustainability is often associated with decentralized treatment schemes, which are then compared to conventional centralised systems without any recovery. Owing to recovery of COD and phosphate, the proposed centralised plant should be competitive with decentralized treatment schemes.

Table 1 Main differences between the proposed BNR plant and a conventional WWTP

	Proposed plant	Conventional WWTP
Oxygen demand		
kg O_2/kg N removed	2.6	4.65
kg O_2/kg COD removed	0.29	0.6
COD demand		
kg COD/kg N removed	1.4	4–5
Methane production		
kgCH_4-COD/kg COD removed	0.28	0
Sludge production		
kg sludge-COD/kg COD removed	0.29	0.4
Phosphate recovery		
kg P/kg P removed	0.49	0

Conclusions

A sustainable BNR plant for sewage treatment is proposed in this article. Recovery of COD as methane and recovery of phosphate as struvite are emphasised in the proposed BNR plant. Because the BCFS® process for denitrifying dephosphatation and the CANON process for autotrophic ammonium removal can significantly contribute to saving COD in integrated N and P removal, an optimal treatment system is possible with minimal use of energy and resources. The mass balance of the proposed BNR plant indicates that the rest of COD after being mostly regained as biomass is still enough to accomplish denitrifying dephosphatation and that no chemical precipitants are needed for P-removal. A significant amount of P in the sludge water is worthy of being recovered as struvite. Compared to a conventional low loaded activated sludge process, the optimal treatment system is quite attractive for minimisation of energy and resources as well as recovery of phosphate. Technological difficulties will not be a problem in application of the proposed BNR plant to practice.

Acknowledgements

This research is within the framework of a joint research between Dutch and Chinese universities, which was financially supported by the Royal Dutch Academy of Sciences (KNAW) under the projects 99CDP016 and 01CDP003, and by Beijing Municipal Education Committee under the project 01KJ-059.

References

Battistoni, P., Pavan, P., Cecchi, F. and Mata-Alverez, J. (1998). Phosphate removal in real anaerobic supernatants: modelling and performance of a fluidised bed reactor. *Water Sci. Technol.*, **38**(1), 275–283.

Battistoni, P., Angelis, A.D., Pavan, P., Prisciandaro, M. and Cecchi, F. (2001). Phosphorus removal from a real anaerobic supernatant by struvite crystallisation. *Water Res.*, **35**(9), 2167–2178.

Boehnke, B. (1978). Moeglichkeiten Abwasser reinigung durch das "Adsorption-Belebungs-verfahren". GWA 29. Schrifureihe des Instituts fuer Siedlungswasserwirtschaft der RWTH Aachen, Germany.

Brandse, F. and van Loosdrecht, M.C.M. (2002). Sewage treatment and methods for phosphate recovery in the BCFS® process. *Water Sci. and Technol.* (in press).

Dijkman, H. and Strous, M. (1999). Process for ammonia removal from wastewater. Patent PCT/NL99/00446.

Hao, X., Heijnen, J.J., van Loosdrecht, M.C.M. and Qian, Y. (2001a). Contribution of P-bacteria in BNR processes to overall environmental impact of WWTPs. *Water Sci. Technol.*, **44**(1), 67–76.

Hao, X., van Loosdrecht, M.C.M., Meijer, S.C.F., Heijnen, J.J. and Qian, Y. (2001b). Model based evaluation of two BNR processes – UCT and A_2N. *Water Res.*, **35**(12), 2851–2865.

Hao, X., van Loosdrecht, M.C.M., Heijnen, J.J. and Qian, Y. (2001c). Model-Based Evaluation of Kinetic, Biofilm and Process Parameters in a One-Reactor Ammonium Removal (CANON) Process. *Biotechnol. and Bioeng.*, **77**(3), 266–277.

Hao, X., van Loosdrecht, M.C.M., Heijnen, J.J. and Qian, Y. (2001d). Model-Based Evaluation of the Behaviour of the CANON Process with Variable Temperature and Inflow. *Water Res.* **35**, 2851–2860.

Helmer, C. and Kunst, S. (1998). Simultaneous nitrification and denitrification in an aerobic biofilm system. *Water Sci. Technol.*, **37**(4/5), 183–187.

Helmer, C., Kunst, S., Juretschko, S., Schmid, M.C., Schleifer, K.-H. and Wagner, M. (1999). Nitrogen loss in a nitrifying biofilm system. *Water Sci. Technol.*, **39**(7), 13–21.

Hippen, A., Rosenwinkel, K.-H., Baumgarten, G. and Seyfried, C.F. (1997). Aerobic deammonification: a new experience in the treatment of wastewaters. *Water Sci. Technol.*, **35**(10), 111–120.

Jaffer, Y., Clark, T.A., Pearce, P. and Parsons, S.A. (2001). Assessing the potential of full scale phosphorus recovery by struvite formation. *Water Sci. Technol.*

Janssen, L.P.B.M. (1991). *Transport phenomena data companion*. Deltse Uitgeversmaatschaooij, Delft, The Netherlands.

Kerrn-Jespersen, J.P., Henze, M. and Strube, R. (1994). Biological phosphorus release and uptake under alternating anaerobic and anoxic conditions in a fixed-film reactor. *Water Res.*, **28**, 1253–1255.

Kuba, T., van Loosdrecht, M.C.M. and Heijnen, J.J. (1996). Phosphorus and nitrogen removal with minimal COD requirement by integration of nitrifying dephosphatation and nitrification in a two-sludge system. *Wat. Res.*, **30**(7), 1702–1710.

Li, X.-Z., Zhao, Q.-L. and Hao, X. (1999). Ammonium removal from landfill leachate by chemical precipitation. *Waste Mgmt.*, **19**, 409–415.

Momberg, G.A. and Ollermann, R.A. (1992). The removal of phosphate by hydroxyapatite and struvite crystallisation in South Africa. *Water Sci. Technol.*, **26**(3–4), 987–996.

Mulder, A. (1992). Anoxic ammonia oxidation. U.S. Patent documents 427849(5078884). United States patent.

Mulder, J.W., van Loosdrecht, M.C.M., Hellinga, C. and van Kempen, R. (2001). Full scale application of the Sharon process for treatment of reject water of digested sludge dewatering. *Water Sci. Technol.*, **43**(11), 127–134.

Siegrist, H. Reithaar, S., Koch, G. and Lais, P. (1998). Nitrogen removal loss in a nitrifying rotating contactor treating ammonium-rich wastewater without organic carbon. *Water Sci. Technol.*, **38**(8–9), 241–248.

Strous, M. (2000). Anammox and nitrification. In: *Microbiology of Anaerobic Ammonium Oxidation* (PhD thesis) 63–81. ISBN 90-9013621-5, the Netherlands.

Strous, M., van Gerben, E., Kuenen, J.G. and Jetten, M.S.M (1997). Effects of aerobic and micro-aerobic conditions on anaerobic ammonium-oxidising (Anammoxo). *Appl. Envir. Microbiol.*, **63**, 2446–2448.

van der Graaf, J.H.J.M. (2001). What to do after nutrient removal? *Water Sci. Technol.*, **44**(1), 129–135.

van Dongen, U., Jetten, M.S.M. and van Loosdrecht, M.C.M. (2001). The SHARON®-ANAMMOX® process for treatment of ammonium rich wastewater. *Water Sci. Technol.*, **44**(1), 153–160.

van Loosdrecht, M.C.M., Kuba, T., Smolders, G. and Heijnen, J.J. (1992). Biological phosphorus removal under denitrifying conditions (in Dutch). H_2O, **19**, 526–530.

van Loosdrecht, M.C.M., Kuba, T., van Veldhuizen, H.M., Brandse, F.A. and Heijnen, J.J. (1997). Environmental impacts of nutrient removal processes: case study. *J. of Envir. Engrg.*, **123**(1), 33–40.

van Loosdrecht, M.C.M., Brandse, F.A. and de Vries, A.C. (1998). Upgrading of wastewater treatment processes for integrated nutrient removal – the BCFS® process. *Water Sci. Technol.*, **37**(9), 209–217.

Vlekke, G.J.F.M., Comeau, Y. and Oldham, W.K. (1988). Biological phosphorus removal from wastewater with oxygen or nitrate in sequencing batch reactors. *Envir. Technol. Lett.*, **9**, 791–796.

Enhanced biological phosphorus removal process implemented in membrane bioreactors to improve phosphorous recovery and recycling

B. Lesjean*, R. Gnirss**, C. Adam***, M. Kraume*** and F. Luck*

* Anjou Recherche/Vivendi Water, Chemin de la Digue, BP76, 75603 Maisons Laffitte Cedex, France
(E-mail: *boris.lesjean@generale-des-eaux.net*; *francis.luck@generale-des-eaux.net*)
** Berliner Wasserbetriebe, Cicerostr. 24, 10709 Berlin, Germany (E-mail: *regina.gnirss@bwb.de*)
*** Technische Universität Berlin, Institut für Verfahrentechnik, Ackerstr. 71, D-13355 Berlin, Germany
(E-mail: *christian.adam@ivtfg1.tu-berlin.de*; *matthias.kraume@ivtfg1.tu-berlin.de*)

Abstract The enhanced biological phosphorus removal (EBPR) process was adapted to membrane bioreactor (MBR) technology. One bench-scale plant (BSP, 200–250 L) and two pilot plants (PPs, 1,000–3,000 L each) were operated under several configurations, including pre-denitrification and post-denitrification without addition of carbon source, and two solid retention times (SRT) of 15 and 26 d. The trials showed that efficient Bio-P removal can be achieved with MBR systems, in both pre- and post-denitrification configurations. EBPR dynamics could be clearly demonstrated through batch-tests, on-line measurements, profile analyses, P-spiking trials, and mass balances. High P-removal performances were achieved even with high SRT of 26 d, as around 9 mgP/L could be reliably removed. After stabilisation, the sludge exhibited phosphorus contents of around 2.4%TS. When spiked with phosphorus (no P-limitation), P-content could increase up to 6%TS. The sludge is therefore well suited to agricultural reuse with important fertilising values. Theoretical calculations showed that increased sludge age should result in a greater P-content. This could not be clearly demonstrated by the trials. This effect should be all the more significant as the influent is low in suspended solids.
Keywords Enhanced biological phosphorus removal (EBPR); membrane bioreactor (MBR); microfiltration; phosphorus content

Introduction

Stringent water quality standard criteria for phosphorus and nitrogen compounds have been motivated world-wide by the needs to protect surface water bodies against eutrophication phenomena. The current European regulation describes guidelines for total phosphorus and nitrogen in treated effluent to respectively 80 or 70% removal rate, or down to 1–2 mgTP/L and 10–15 mgTN/L (depending on the size of the plant). More stringent regulations on nutrients removal are expected to come soon into force in some countries. Some plants have already to comply with stricter requirements. In Germany for example, the Rödingen WWTP built in 1999 has to comply with a discharge criteria of 0.5 mgP/L in the effluent (Engelhardt and Firk, 2000). Near future water quality standards contemplated in the Netherlands, to match surface water guidelines when the treated effluent flow amounts for a great part of the total river flow, include threshold values of 0.15 mgTP/L and 2.2 mgTN/L. For publicly owned treatment works (POTWs) discharging effluents, the nutrient water quality criteria published in January 2001 by the U.S. Environmental Protection Agency (U.S. EPA), range for total phosphorus from 0.076 mgTP/L and 2.18 mgTN/L for Eco-region VI down to 0.010 mgTP/L and 0.12 mgTN/L for Eco-region II. The application of these latest criteria will require the addition of further treatment steps to the current treatment lanes. Technologies are already available, such as post-denitrification with additional carbon source, when further nitrogen removal is required, flocculation with sand or membrane filtration for phosphorus removal, or ozonation with biological activated carbon (BAC) for final biological polishing.

On the other hand, it is increasingly accepted that natural resources are limited and that the recycling of scarce resources should be encouraged. Phosphate is one of these precious resources for which recovery and recycling is raising increasing interest. For example, Sweden has already proposed a criteria of 75% recovery rate for recycling of phosphates from wastewater and biological wastes. The new technologies will therefore not only need to remove phosphorus very efficiently, but also to allow for significant recovery.

In activated sludge technologies, phosphorous removal results from three main mechanisms. The first one is the physiological phosphorus requirement for cell metabolisms and growth. It amounts to 1 to 2% of the total suspended solid mass in the mixed liquor. When appropriate conditions are insured, such as the presence of anaerobic conditions with carbon source, specific phosphate-accumulating organisms (PAO) can develop, and the mechanism of enhanced biological phosphorus removal (EBPR) can take place. This phenomenon is sometimes referred to as "luxury phosphorus uptake". Jardin and Pöpel (1996) reported that EBPR leads to increased phosphorous concentration up to 7% TS. They operated a system with settled wastewater in the range 1.5–6.5%TS (mean effluent value of 3.3 mgP/L over the whole trials period, but phosphoric acid was spiked at times). Similarly, Mino et al. (1998), quoting Liu (1995), considered that a sludge with very high P removal capacity will have a P-content of 8–12.5%P/VSS (no indication however of P-concentration in effluent or wastewater source). Other authors (Wang et al., 2002) worked with synthetic wastewater in order to develop purer PAO cultures and to identify the organisms responsible for EBPR. They measured phosphorus content up to 10%VSS with glucose and 20%VSS with acetate feed, for an influent P/COD ratio of 20/260 (no indication of P-values in the effluent). Finally, the third mechanism responsible for phosphorus removal in activated sludge relates to physicochemical fixation of phosphate, mainly by precipitation or adsorption. This can occur naturally when the conditions are suitable (appropriate pH, presence of iron or calcium ion, etc.), but it can be also artificially enhanced and controlled when dosing coagulants, generally iron or alum salts, at one stage of the process. The advantages of physicochemical precipitation on EBPR, to achieve enhanced phosphorus removal, are known to be better for control and reliability. However, many drawbacks are associated with physicochemical precipitation, such as up to 25% sludge production increase, additional chemical consumption, salinity increase of the effluent, and potential detrimental impact on the biological nitrification, due to resulting low alkalinity and pH. Moreover, it is now accepted that physicochemical processes hinder recycling of phosphate as fertiliser in comparison to biological synthesis.

The objective of this research project was to adapt and test the EBPR process in membrane bioreactor (MBR) technologies, and to investigate EBPR performances under high solid retention time (SRT). Figure 1 shows theoretical calculations of P-content in sludge depending on SRT and TSS/BOD$_5$ fraction. This was calculated with a P-mass balance and the formula proposed by the German working group ATV (2000) to calculate the sludge concentration under an organic load of 0.05 kgBOD$_5$/kgTS.d and a temperature greater than 10°C. Note that it was considered in this calculation that the P-concentration in the effluent would be always 0.5 mgP/L, whatever the SRT and resulting P-content in the sludge. These results show that the sludge P-content is expected to increase when the SRT increases. This effect is all the more significant as the influent contains low levels of suspended solids. High SRT appears therefore to be favourable when phosphorus recovery is desired. However, recent studies (Wang et al., 2002) indicate that PAOs undergo competitive conditions with glycogen-accumulating organisms (GAOs) at long SRT (>20 days). Also the Berliner Wasserbetriebe observed that phosphorus removal performance decreases significantly when P-content reaches approximately 4.5%P/TS (Peter-Fröhlich, 2001). This is mainly due to particulate release in the secondary clarifier. Another significant degradation occurs when P-content reaches appr. 6%P/TS. This further degradation of performances relates to a release of soluble phosphate in the effluent. These indications tend to show that P-content, and SRT, cannot be increased inconsiderably without impacting significantly on phosphorus removal efficiency. Also it is expected that P-removal and P-recovery can be maximised through EBPR process while identifying an optimum P-content that would be strongly related to an optimum SRT. This optimum P-content and SRT could be greater in the case of membrane bioreactor, where all particles are rejected, than in conventional activated sludge technologies.

Three flexible MBR pilot plants were operated under several Bio-P configurations and operation conditions, with degritted urban wastewater. Two EBPR configurations were adapted from conventional systems to MBR technology, and tested at SRTs of 15 and 26 day. Phosphate spiking trials were carried out in order to determine the full potential of the EBPR processes and to evaluate the maximum P-accumulation capacity of the sludge. This paper sets out the results of this investigation.

Methods

Pilot plants

The three MBR pilot plants were entirely autonomous and flexible with regard to through-flow, process configuration, volume, and operation conditions. One bench-scale plant (BSP, 200 to 250 L), and two pilot plants (PPs, 1,000 to 3,000 L each), were operated on the site of the Berlin-Ruhleben WWTP. They were continuously fed with municipal raw sewage screened through 1 mm punch holes (rotative drum). As seen in Table 2 and Table 3, the daily average COD/N/P ratio of the raw water was around 100/8/1, and the TSS/BOD_5 ratio was around 1.3.

The membrane system of the bench-scale plant was provided by GKSS, Germany, and consisted of a flat-sheet membrane module, directly immersed in one aerated vessel located at the end of the biological reactor (medium pore size approximately 37 nm, and membrane area approximately 1.5 m^2). The filtration system of the two parallel pilot plants was developed by Memcor, Australia. It included one immersed hollow fibre membrane module each, set up in a membrane vessel located in an external sludge recycle as shown in Figure 2

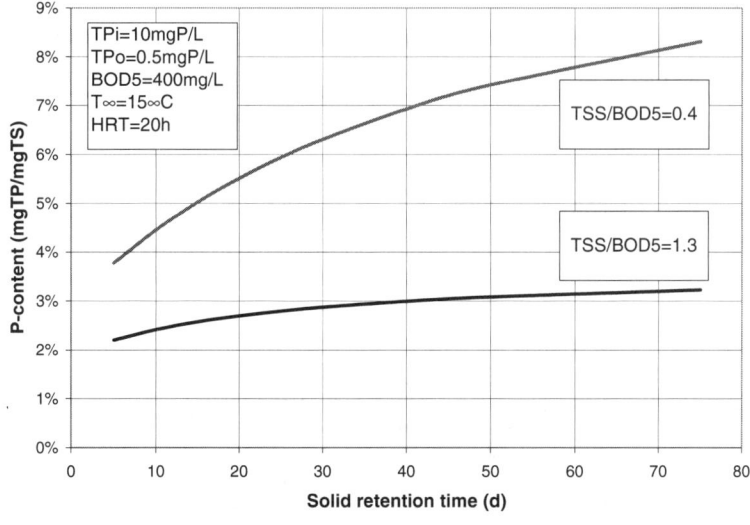

Figure 1 P-content in sludge depending on sludge age and TSS/BOD_5 fraction

(pore size approximately 0.2 μm, and membrane area approximately 10 m² per module). It is noteworthy that the membrane units were operated below their optimum operation ranges. At this stage of the project, the assessment of the EBPR process was the main focus of the study. The optimisation of membrane performances will be undertaken in a later phase.

Process configurations

As seen in Figure 2, two multi-stage EBPR process configurations, adapted from conventional EBPR processes to MBR systems, were selected for the investigations. The main adaptations resulted from the replacement of the sedimentation tank used to separate the effluent from the sludge (operated in anoxic/anaerobic conditions) by the membrane unit (aerobic system). The concentrate sludge returning from the aerated membrane system, nearly saturated in dissolved oxygen (typically 5–8 mg/L), must be therefore directed to the first of the aerobic stages. Similarly, the return sludge to the anaerobic zone, traditionally originating from the clarifier and low in nitrate, must originate from one anoxic stage of the biological reactor.

Configuration 1 is an adaptation of the UCT process (Tchobanoglous and Burton, 1991) where the membrane filtration is inserted in the second sludge recycle. Nitrogen removal is achieved through a conventional pre-denitrification mode. In contrast, Configuration 2 was developed with a post-anoxic zone operated without addition of external carbon source. Justifications and expected benefits of post-denitrification configuration in MBR will be fully detailed elsewhere, as well as a discussion on nitrification and denitrification mechanisms observed in the trials plants, or the impact of post-denitrification on both N-removal and P-removal mechanisms (paper in preparation).

Each configuration included successive anaerobic, anoxic, and aerobic zones. Depending on the configuration tested, the anoxic and aerobic zones consisted of 2 to 5 reactors. The sludge return and recycle ratios, based on average net flow, were set up in the two pilot plants (PPs) at 500% for the membrane loops, 400% for the sludge recycle from the aerobic to the anoxic reactor (configuration 1 only), and 100% from the anoxic to the anaerobic reactor. With the bench scale plant (BSP) configurations, where the membrane module was directly immersed in an additional aerobic reactor, the recycle and return ratios were set up to 400% and 100% (configuration 1).

Operation conditions

The first configuration was initially tested with the bench-scale plant under 15 d SRT. This was to ensure that EBPR was possible with process configurations adapted to MBR tech-

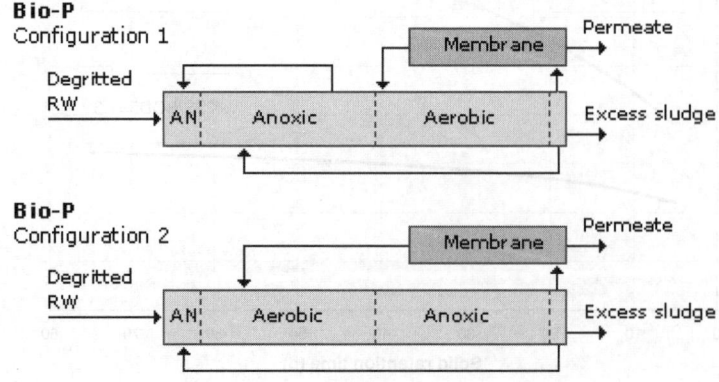

Figure 2 Two Bio-P configurations adapted to MBR technology

nologies under conventional sludge age. The two configurations were then tested in parallel with the two pilot plants under approximately 26 d SRT. Operation conditions of the MBR pilot plants during these tests are given in Table 1.

The differences reported on mass and volume organic load between the two runs resulted from a change of raw water quality. The difference observed in the sludge concentration in the membrane reactors of the two parallel PPs, despite the identical SRT and HRT, resulted from normal concentration and dilution phenomena dependent on sludge recycle configurations and ratios, and commonly observed in MBR. The better mass repartition inherent in configuration 2, due to the sludge recirculation loop from the membrane over the whole reactor volume (Figure 2), leads to lower sludge concentration in contact with the membrane (around 20% less than configuration 1).

Analytical methods

For each run, the pilot units were operated over 2 to 3 sludge ages before conducting extensive analyses for performance assessment. Analytical methods used during the trials are detailed elsewhere (Adam *et al.*, 2001; Gnirss *et al.*, 2002).

Bench-scale plant results at 15 d solid retention time

Table 2 presents the average results of 24 h-sample analyses performed with the bench-scale plant at 15 d SRT. The reactor temperature ranged between 16 and 19°C, and the sludge concentration *in the aerobic zones* was around 6 gMLSS/L (VSS/MLSS~68%).

Phosphorus removal

EBPR was monitored with TP concentrations in the effluent as low as 100 µg/L in configuration 1, and a P/TS ratio of 2.4%. These results coincide with the theoretical calculations presented in Figure 1 (for a TSS/BOD$_5$ ratio of 1.3). Surprisingly, profile measurements could show that most of P-uptake occurred in the anoxic reactors. EBPR dynamics was demonstrated through batch-tests, on-line measurements, profile analyses, and mass balances (Adam *et al.*, 2001).

Nitrogen removal

Daily nitrogen removal efficiency of 86 to 90% was observed. This can be accounted for by nitrogen uptake through cell growth, and denitrification process (theoretically 80% of the remaining nitrogen with a recycle ratio of 400%, and when sufficient biodegradable carbon is available for complete denitrification). These excellent nutrient removal results were certainly favoured by the very high COD/N and COD/P ratios, not typical for municipal

Table 1 Operation parameters of the MBR bench-scale and pilot plants

Parameters	Units	BSP Config. 1	MBR pilot plants Config. 1	2 // PPs Config. 2
Net flow	L/h or m³/h	10 L/h	122 L/h	108 L/h
Biological reactor volume	m³	0.21	2.2	2.0
Hydraulic retention time (HRT)	h	21	18	
Solid retention time (SRT)	d	15	26	
Sludge concentration (recirculation *from membrane zone*)	g$_{MLSS}$/L	~ 7	~ 14	~ 12
Mass organic load (F/M) based on overall mass	kg$_{COD}$/kg$_{MLSS}$·d	0.22	~ 0.10	
Volume organic load based on anoxic and aerobic volume	kg$_{COD}$/m³·d	1.24	1.15	

Table 2 Average influent and effluent concentrations at 15d SRT

	COD (mg/L)	TN (mg/L)	NH$_4$-N (mg/L)	NO$_3$-N (mg/L)	TP (mg/L)
Raw water	998*	69.7	41.3	0.42	10.5
BSP effluent config 1	35.7	9.17	0.49	6.0	0.10
removal (%)	96%	87%	99%	-	99%

* elevated value compared to Table 3 (+35%), probably due to flush out of ferric sulphide accumulated in the pipes of the raw water inlet in the first weeks of the trials

wastewater. Further studies should be undertaken with wastewater showing lower COD/nutrients ratios.

Full EBPR capacity

The very low P-concentration measured in the effluent (<100 µgP/L) and in the excess sludge (low P/TS ratios of 2.4%) tended to indicate that EBPR mechanisms were P-limited and the full EBPR removal capacity was not reached. It was decided to conduct P-spiking trials at the end of the run, without adding any supplementary carbon source, in order to quantify the full EBPR potential of the system. Phosphate was continuously dosed at 30 mgP/L, resulting in influent concentrations of around 40 mgP/L. Ultimate removal efficiency of 18–25 mgP/L could be achieved after a stabilisation phase of one to two SRT. The P-residual in the effluent remained always beyond 4 mgP/L over the spiking period. P/TS ratios stabilised around 6%, and sludge concentration at 8.5 gMLSS/L in the membrane reactor. This demonstrated the high EBPR removal potential of the MBR configuration, and the possibility to achieve high P content in the sludge. Such an MBR system could therefore treat efficiently highly P-loaded effluents, and consequently ensure high P-recovery rates, and produce a sludge of great fertilising value.

Natural P-precipitation phenomena

Batch tests underlined the importance of naturally occurring P-precipitation with ferric ions. This resulted from the presence of ferric sulphide in the raw water, originating from a surface water treatment plant which disposed its sludge in the municipal sewer. This had certainly contributed to stabilising and increasing P-removal rate over the trials, but the exact impact could not be precisely quantified. However it could be demonstrated through stoichiometric mass balances that the additional P-removal observed during the spiking trials was to a great extent related to Bio-P mechanisms. Further research will be focussed on the quantification of P-removal through natural precipitation, and tests will be undertaken with another type of municipal raw water which is not affected by water plant sludge, and which exhibits lower COD/P and TSS/P ratios.

Pilot plants results at 26 d solid retention time

Table 3 presents the average results of 24 h-sample analyses performed during the two parallel runs conducted with 26 d SRT. The reactor temperatures ranged between 12 and 15°C.

Phosphorus removal

The observations made with the BSP at 15 d SRT were confirmed with the two pilot plants at 26 d SRT. High P-removal was observed, with average TP concentrations of 60 µg/L in configuration 1, and 70 µg/L in configuration 2, for P/TS ratios of around 2.3%. This did not correlate precisely with theoretical mass balance calculations, which provide values of around 2.7%. These low P/TS values, despite the high SRT, can be explained by the high TSS$_{influent}$/BOD$_{influent}$ ratio of around 1:3. It demonstrates – against many preconceived ideas – that high SRT conditions do not necessarily impair EBPR mechanisms while leading to high and unstable P/TS ratios. It proves also that under this high TSS/BOD ratio, increasing the sludge age will not improve the P-content significantly. As seen in Figure 1, a sludge age of 50 days would theoretically lead to a P-content of around 3.1%, and therefore similar P-removal performances could be expected. In case of lower TSS/BOD ratio, greater increase of P-content would be expected with higher SRT. Bio-P dynamic could be clearly demonstrated with both configurations through batch-tests, on-line measurements, profile analyses (Gnirss *et al.*, 2002), and mass balances. All these results underpinned that efficient enhanced phosphorus removal can be achieved in MBR systems under high SRT conditions, with both pre-denitrification or post-denitrification process configurations.

Nitrogen removal

The benefit of the post-denitrification configuration was clearly demonstrated, and is highlighted by the nitrogen mass balance. Configuration 2 achieved an average 3.6 mgTN/L in the effluent, i.e. 94%N-removal, whereas configuration 1 was limited by the recycle rate of 300% and fulfilled only 82% N-removal with 11 mgTN/L in the effluent. It is important to note that no additional carbon source was added in configuration 2 to achieve efficient denitrification, unlike other traditional post-denitrification systems, which always require an external carbon source for efficient nitrate removal.

Sludge production

The average MLSS concentrations of the aerobic zone were respectively around 12 and 10 g/L in configuration 1 and 2 (VSS/MLSS approximately 67%). This difference resulted only from normal dilution and concentration phenomena due to the recycle configurations and ratios specific to each process. Mass balances could show that the overall mass present within both systems was proportional to the reactor volume. Therefore no difference in sludge production could be demonstrated. The specific excess sludge production of both configurations was monitored around 0.4 kgMLSS/kgCOD.

P-precipitation phenomena

The 2 pilot plants were operated with the same type of influent as the BSP. They are therefore subject to similar P-precipitation phenomena as described earlier. Further trials undertaken with another type of water should help to understand this competitive mechanism.

Discussion on optimum loading

The profile measurements showed that nutrients removal was already very effective in the middle stage of the biological reactors. It can be therefore extrapolated that shorter HRT, hence higher flow and volume organic load, could achieve also efficient N- and P-removal. For a given sludge concentration, the results of both trials at 15 d and 26 d SRT tend to indicate that the highest organic load – and optimum to sustain a given discharge criteria –

Table 3 Average influent and effluent concentrations at 26d SRT

	OD (mg/L)	TN (mg/L)	NH_4-N (mg/L)	NO_3-N (mg/L)	TP (mg/L)
Raw water	740	61	43	n.d.	9.1
PP effluent config 1	32	11.0	< 0.5	9.0	0.06
removal (%)	96%	82%	> 98%	–	99%
PP effluent config 2	35	3.6	< 0.5	1.0	0.07
removal (%)	95%	94%	> 98%	–	99%

would be achieved at the lowest sludge age. However, lower organic load could be suitable to small decentralised plants that must be highly reliable and robust while coping with significant flow variations. Moreover, longer SRT should ensure lower sludge production. The sludge would be also expected to contain more phosphorus and to be better stabilised; it would therefore best suit agricultural land disposal.

Conclusion

These trials could demonstrate that enhanced biological phosphorus removal is effective in MBR systems, under sludge ages characteristic of traditional wastewater treatment plants, but also under higher sludge ages usually more suitable to MBR systems. Efficient P-removal was achieved with 15 d and 26 d solid retention times. Two process configurations were tested with both sludge age conditions: one "standard" configuration including a pre-denitrification stage, and a configuration characterised by post-denitrification without addition of any carbon source. With both SRTs, both configurations achieved similar and high P-removal performances. Under identical loading conditions, nitrogen removal was greater in the post-denitrification configuration. Average total nitrogen concentration of 3.6 mg/L was monitored in the effluent.

Under all SRT conditions, the sludge exhibited phosphorus contents of around 2.4%TS. When spiked with phosphorus (no P-limitation), P-content could increase up to 6%TS. The sludge is therefore well suited to agricultural reuse with great fertilising values. Theoretical calculations showed that increased sludge age should result in greater P-content. This effect should be all the more significant as the influent is low in suspended solids.

Further tests will be undertaken with another type of raw water coming from a decentralised area with a short retention and separative sewer, and which is not affected by wet industries. The lower COD/P and TSS/P ratios characteristic of the new trials site will enable us to confirm the results of this study.

Acknowledgement

This research project was supported jointly by the Berliner Wasserbetriebe and Vivendi Water in the frame of the KompetenzZentrum Wasser Berlin. The trials undertaken on the bench-scale pilot plant were subcontracted to and conducted by the team of Prof. Kraume (TU Berlin).

References

Adam, C., Gnirss, R., Lesjean, B., Buisson, H. and Kraume, M. (2002). Enhanced biological phosphorus removal in membrane bioreactors. *Wat. Sci. Tech*, **46**(4–5), 281–286.

Arbeitsblatt ATV-DVWK-A 131 (2000). Bemessung von einstufigen Belebungsanlagen, Hennef.

Engelhardt and Firk (2000). 3. Aachener Tagung, February 2000.

Gnirss, R., Lesjean, B., Adam, C. and Buisson, H. (2002). Cost Effective and Advanced Phosphorus Removal in Membrane Bioreactors for a Decentralised Wastewater Technology. *IWA 2nd International World Water Congress*, 7–12 April 2002, Melbourne.

Jardin, N. and Pöpel, H.J. (1996). Behavior of waste activated sludge from enhanced biological phosphorous removal during sludge treatment. *Wat. Env. Res.*, **68**(6), 965–973.

Mino, T., Van Loosdrecht, M.C.M. and Heijnen, J.J. (1998). Review paper. Microbiology and biochemistry of the enhanced biological phsophate removal process. *Wat. Res.*, **32**(11), 3193–3207.

Peter-Fröhlich, A. (2001). Erkenntnisse zur biologischen Phosphatentfernung aus Praxisuntersuchungen. Dissertation 2001 and der TU Dresden.

Tchobanoglous, G. and Burton, F.L. (1991). *Wastewater Engineering, Treatment, Disposal and Reuse*, Metcalf & Eddy, McGraw Hill International Editions, Civil Engineering Series, 3rd Edition, 1991, pp. 733.

Wang, J.C., Park, J.K and Whang, L.M. (2002). Comparison of Fatty Acid Composition and Kinetics of PAO and GAO. *Wat. Env. Res.*, **73**(6), 704–710.

Improved nutrient removal using *in situ* continuous on-line sensors with short response time

P. Ingildsen* and H. Wendelboe**

* Danfoss Analytical, Ellegaardsvej 36, 6400 Soenderborg, Denmark and IEA, Lund University, Box 118, SE-221 00 Lund, Sweden

** Danfoss Analytical, Ellegaardsvej 36, 6400 Soenderborg, Denmark

Abstract Nutrient sensors that can be located directly in the activated sludge processes are gaining in number at wastewater treatment plants. The *in situ* location of the sensors means that they can be located close to the processes that they aim to control and hence are perfectly suited for automatic process control. Compared to the location of automatic analysers in the effluent from the sedimentation reactors the *in situ* location means a large reduction in the response time. The settlers typically work as a first-order delay on the signal with a retention time in the range of 4–12 hours depending on the size of the settlers. Automatic process control of the nitrogen and phosphorus removal processes means that considerable improvements in the performance of aeration, internal recirculation, carbon dosage and phosphate precipitation dosage can be reached by using a simple control structure as well as simple PID controllers. The performance improvements can be seen in decreased energy and chemicals consumption and less variation in effluent concentrations of ammonium, total nitrogen and phosphate. Simple control schemes are demonstrated for the pre-denitrification and the post precipitation system by means of full-scale plant experiments and model simulations.

Keywords Control and automation; full-scale control implementation; *in-situ* nutrient sensors; nutrient removal

Introduction

With the availability of sensors suited for the *in situ* location, i.e. directly in the mixed liquor quite simple control structures and controllers can be used to obtain effective control of WWTPs. This paper describes the requirements for the control of nitrogen removal in pre-denitrification systems and phosphate precipitation in post-precipitation systems, regarding sensors, controllers and automation software. The results are backed up by full-scale experiments at Källby WWTP in Sweden (see description of the plant in Ingildsen (2002)) and model simulations using the benchmark model (Copp, 2002) using the activated sludge model no. 1 (Henze *et al.*, 1987).

Control structure selection

The nitrogen removal process consists of two main processes: the nitrification process that transforms ammonium into nitrate during aerobic conditions and the denitrification process that transforms nitrate into free nitrogen under anoxic conditions by the use of easily degradable organic matter. A typical pre-denitrification plant has five major control handles available for the nitrogen removal process: the airflow rate (and the airflow distribution), the internal recirculation, the sludge outtake, the external carbon dosage and the sludge recirculation. The purpose of control is to manipulate these control handles in order to reach satisfactory performance. The primary outputs from the plant are effluent ammonium, nitrate, organic matter and suspended solids. This means that in principle the system has multiple inputs and multiple outputs, which indicates that a MIMO (multiple input – multiple output) solution seems to be preferable. A MIMO solution implies that the control handles are coordinated according to the couplings in the system. This means that one output variable cannot be controlled by one input variable only but rather depends on the manipulation of more control handles. However the point of view in this paper is that a simple SISO (single input – single output) control structure can perform well when the sensors are applied *in situ*. The *in situ* location so to speak ensures a decoupling of the processes, provided an adequate control structure is applied.

If limiting the area of investigation to the nitrogen removal processes, the control handles affecting nitrification and denitrification in the medium time-scale are aeration, internal recirculation and carbon dosage and the main outputs are ammonium and total inorganic nitrogen (sum of nitrate and ammonium). The control of carbon removal is not dealt with, as the carbon source is often a limiting resource in pre-denitrification systems, which seldom gives rise to problems with the effluent permit provided that the nitrogen removal performs well. The control of suspended solids is also left out of the analysis as this issue primarily pertains to the design and control of the settlers.

By means of a relative gain array analysis (see Shinskey (1979) or Skogestad and Postlethwaite (1996)) it has been shown in Ingildsen (2002) that the control of aeration can be performed independently of the internal recirculation and the carbon dosage. The internal recirculation and the external carbon dosage on the other hand needs to be coordinated as they both mainly affect the denitrification process in the medium time scale. However, as will be shown later this interaction can be dealt with quite easily.

Control of aeration

For the dynamic control of aeration it is of major importance that the ammonium sensor is located in the nitrification process volume rather than after the secondary settlers. For an illustration of the importance of this sensor location, see Figure 1 where a comparison of the signals from a sensor located at the end of the nitrification reactors (effluent from zone 10) and a sensor location after the secondary sedimentation unit is shown. During the period shown in Figure 1 a constant DO setpoint strategy was applied. Besides the delaying effect the settler has on the signal it also has a smoothing effect. This smoothing of the signal means that it is not possible to observe periods of 0 mg/l NH_4-N when looking at the effluent from the settler. Looking at the signal from the end of the nitrification volume on the other hand it can be seen that 30% of the time the ammonium concentration is below 0.1 mg/l NH_4-N. A similar experiment was carried out in Lindau WWTP, Germany which has two pre-denitrification systems in series, a low loaded system followed by a secondary system that is normally loaded (see description of the plant in Ingildsen (2002)). Also in this system there is 0 mg/l NH_4-N during quite a large fraction of the time. These periods of total

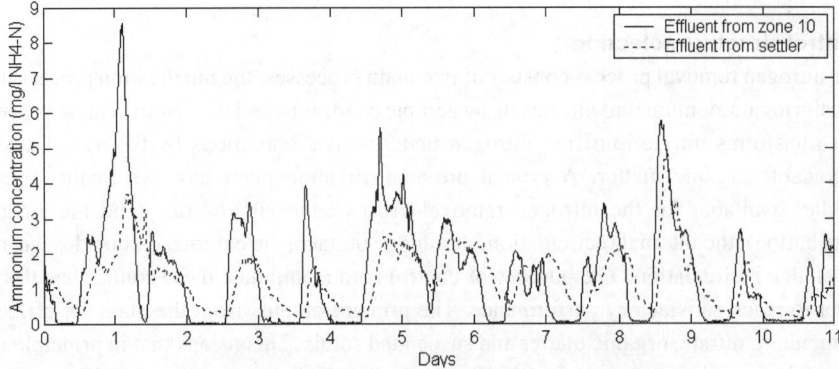

Figure 1 Comparison of the signal from an ammonium sensor located *in situ* (zone 10) and an ammonium sensor located in the effluent of the secondary settler (Källby WWTP, Sweden)

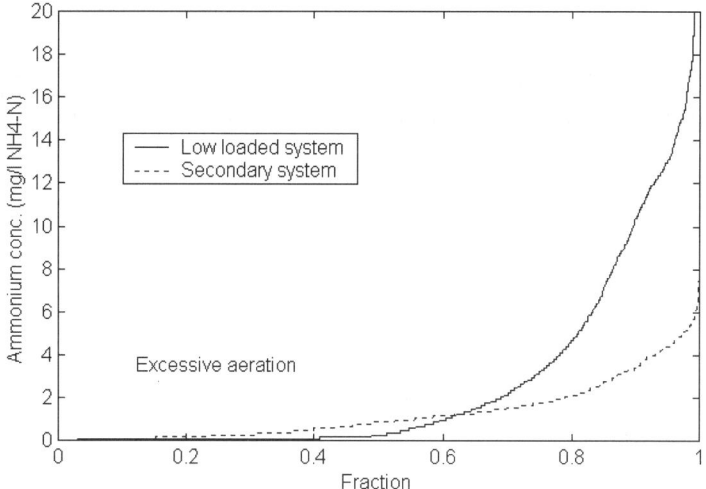

Figure 2 Distribution of the ammonium concentration measurements based on in situ measurements of ammonium in the outlet of the nitrification zone over 37 days (Lindau WWTP, Germany)

ammonium removal indicate excessive aeration and hence that energy consumption for aeration can be reduced by means of better control.

At Källby WWTP four different controllers for the supervisory control of the DO setpoint were tested. The controllers were based on 1) pure feed-forward using a simple hydraulic model, 2) slow integral feedback based on a sensor location in the outlet of the secondary settlers, 3) a combination of controller 1 and 2 and finally 4) an *in situ* controller using the ammonium concentration signal at the end of the aerobic reactor for fast feedback based on a PI controller. An overview of the applied controllers and the sensor locations is provided in Figure 3.

The result of the four tests was that the *in situ* controller (controller 4) performed significantly better than the other controllers in terms of energy savings and the tight control of the concentration of effluent ammonium. At the same time this controller distinguished itself by being very simple (i.e. only working by proportional and integral control) and only being based on one sensor. The performance of the controller during a period where the ammonium setpoint was set at 3 mg/l NH_4-N can be seen in Figure 4. Here it can be seen that the controllers ensures a proper DO setpoint most of the time. At certain periods the maximum DO setpoint does however not provide sufficient nitrification capacity to obtain the required setpoint, i.e. the controller saturates, the same happens at certain periods of time at the minimum DO setpoint. A supervisory ammonium setpoint that uses the average

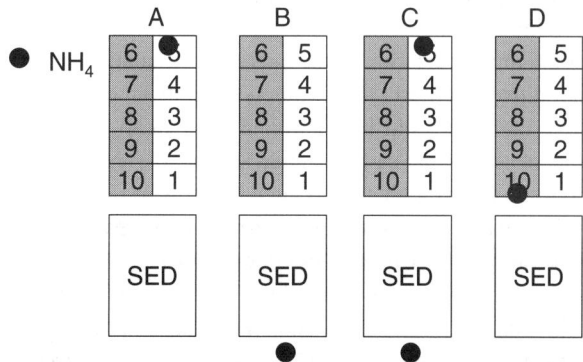

Figure 3 Controller types tested at Källby WWTP

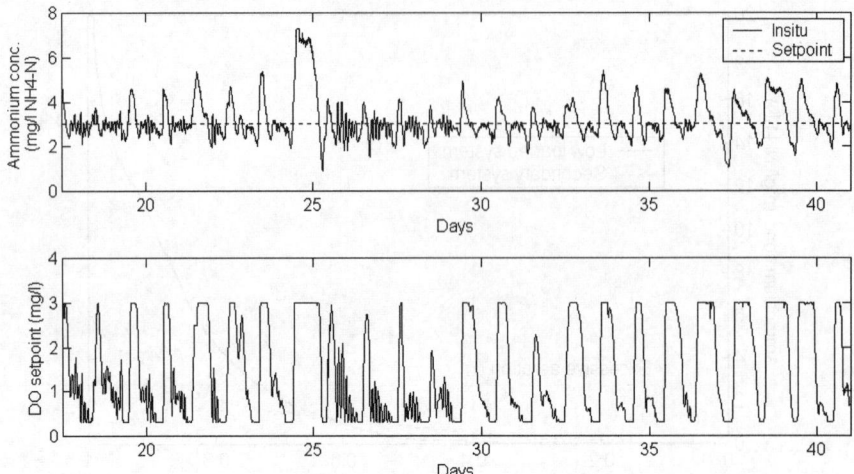

Figure 4 Control of the effluent ammonium by means of the DO setpoint (Källby WWTP)

effluent ammonium concentration over a couple of days as input may correct the ammonium setpoint to take these limitations into consideration over a longer time-scale. Alternatively the control of the number of aerated zones may improve the control method.

The control documented in Figure 4 yielded an energy saving for aeration of 28% compared to a constant DO setpoint aeration strategy (using DO setpoints of 2–2.5 mg/l). This result was at the cost of a slightly increased effluent ammonium concentration. An experiment with an ammonium setpoint of 1.5 mg/l NH_4-N yielded a 12% reduction in the consumed energy for aeration, while at the same time providing a lower effluent ammonium concentration than in the reference line. In principle, this choice of ammonium setpoint avoided situations where the effluent ammonium concentration was 0 mg/l NH_4-N.

Control of internal recirculation and carbon dosage

A controller that controls internal recirculation by means of a nitrate setpoint at the end of the anoxic reactor(s) was proposed by Londong (1992). In Yuan *et al.* (2002) it was shown that this control strategy leads to close to optimal reduction of the effluent total nitrogen. The rationale is that the cost of internal recirculation in terms of energy consumption is negligible compared to other operation costs. Hence, the only objective of such a controller is to minimise effluent total nitrogen. Using a nitrate setpoint in the range of 1–3 mg/l NO_3-N leads to close to maximum inorganic nitrogen removal. Again, a simple PI controller is sufficient for the control of internal recirculation provided the nitrate sensor is located *in situ* at the end of the anoxic reactor and provided the response time of the sensor is short (less than 10–15 minutes).

If the system after the implementation of the internal recirculation controller does still not meet effluent total nitrogen criteria one option is to apply an external carbon source. The dosage of external carbon should be controlled in order to just reach the effluent criteria for total nitrogen, as external carbon source is generally expensive and excessive dosage may also lead to problems with complying with effluent requirements on organic matter. The control of the external carbon source has to be coordinated with the control of the internal recirculation as both control handles affect the anoxic zone in the medium time scale. The internal recirculation flow rate controller is supplemented with a controller that has total inorganic nitrogen (i.e. the sum of ammonium and nitrate) in the outlet of the aerobic reactors as input and the external carbon dosage as output. The structure is depicted in Figure 5.

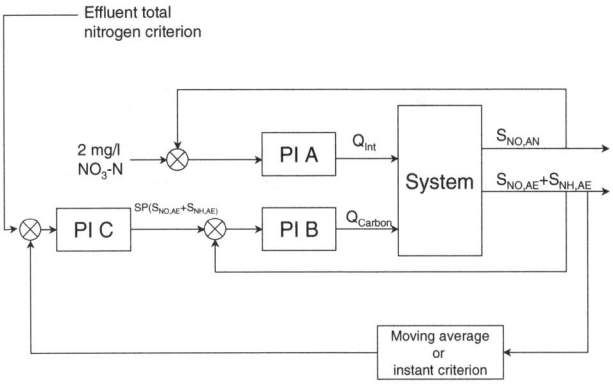

Figure 5 Control structure for the combined control of internal recirculation and external carbon addition

Controller A ensures full utilisation of the anoxic volume and hence of the incoming organic matter regardless if it stems from an external carbon source or the influent. The second controller B controls the carbon dosage so as to obtain an inorganic nitrogen setpoint depending on the effluent criterion. Controller C corrects this effluent total inorganic nitrogen setpoint depending on the type of effluent criterion (i.e. average over a longer time frame or grab sample values). Additionally, some kind of maximum limits should be applied to the dosage of external carbon. For example in the case of strong inhibition due to toxicity, the controller may increase the dosage beyond the reasonable. The controller has been tested in the benchmark simulation platform (Copp, 2002), where it showed good performance, see Figure 6.

A control structure for the control of aeration, internal recirculation and external carbon dosage involves three nutrient sensors located *in situ*: two nitrate sensors and an ammonium sensor. The sensor locations are depicted in Figure 7. This control structure ensures a specified effluent total nitrogen concentration as well as the specified distribution between ammonium and nitrate. The choice of setpoints is a matter of firstly complying with effluent water quality permits and secondly balancing effluent quality, operational cost and robustness of the system.

Control of chemical phosphate removal

At Källby WWTP four different control strategies for the control of chemical precipitation have been compared. The methods are depicted in Figure 8. The control methods are:

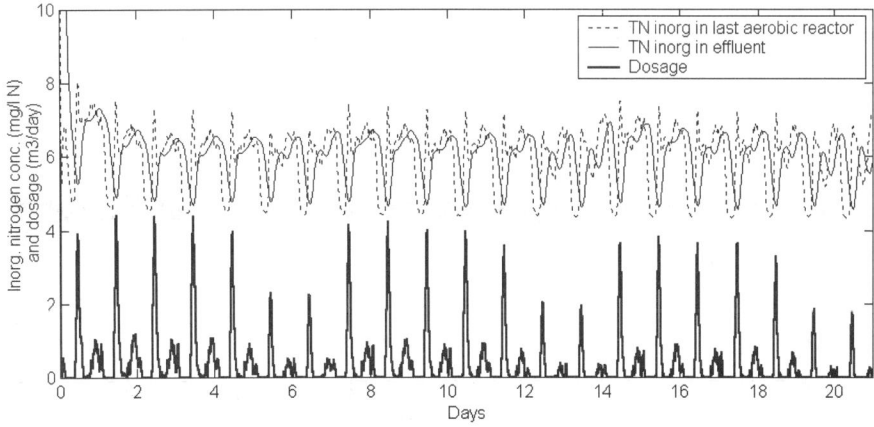

Figure 6 Controller performance with an average effluent total inorganic nitrogen set point of 6 mg/l N

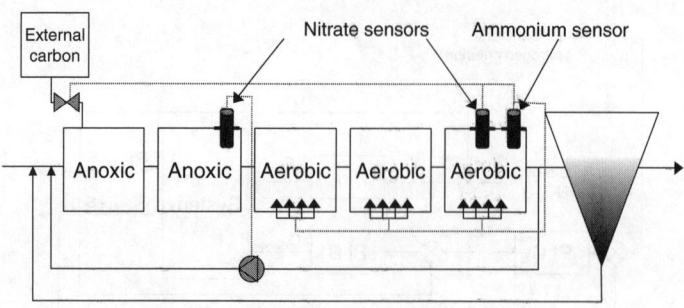

Figure 7 Proposed control structure of nitrogen removal

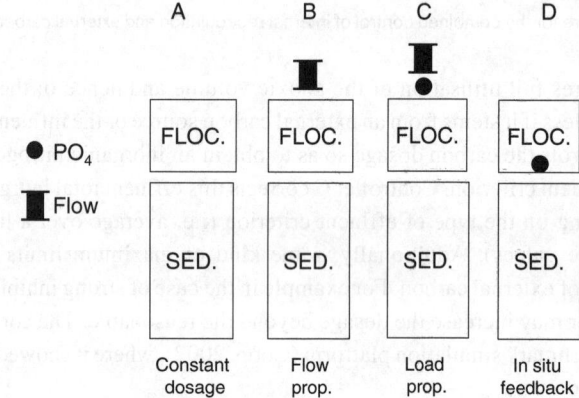

Figure 8 Tested controllers at Källby WWTP

constant dosage (A), flow proportional dosage (B), load proportional dosage (C) and finally an *in situ* feedback control loop (D) where a phosphate sensor is located at the end of the flocculation chamber to provide a feedback signal to the dosage so as to achieve a certain effluent phosphate set point. The control performance of the controllers improves from A towards D, with the *in situ* feed back controller showing the best performance of the four controller types. The result from a seven-day period of control based on this controller with a phosphate setpoint of 0.5 mg/l PO_4-P is shown in Figure 9. The standard deviation of the effluent phosphate concentration is 0.03 mg/l PO_4-P. The *in situ* feedback controller was a simple PI controller with a time constant slightly shorter than the average retention time in the flocculation chamber.

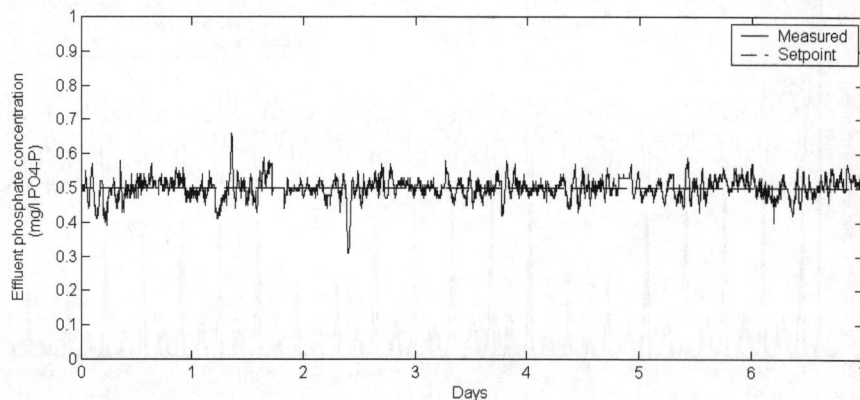

Figure 9 Control of phosphate precipitation

In Ingildsen (2002) it was shown that for the Källby WWTP the total phosphorus concentration correlated linearly with the phosphate concentration. This has also been found in Ammundsen *et al.* (1992). It is believed to be the case for most plants where the clarification of the chemical sludge works well. More suspended solids in the effluent naturally means a higher effluent concentration of total phosphorus. This linear dependency means that it is possible to specify an effluent phosphate setpoint at which compliance to a specific total phosphorus criterion is also ensured. This is rather important as most effluent permits are specified in terms of the concentration of total phosphorus rather than of phosphate.

A comparison between the four strategies outlined in Figure 8 has been carried out. This is done by comparing the dosage that would have to be used to at least 90% of the time dose at least as much as the *in situ* controller suggests. The comparison shows that from A to D the required dosage drops. A full-scale experiment verifies that this type of comparison is reasonable. In the experiment two parallel lines were compared. In one line the *in situ* strategy was applied and in the other the load proportional strategy was applied. The result was a 30% higher dosage in the line with the load proportional strategy than in the line with the *in situ* feedback strategy. This is similar to the 36% proposed in Table 1.

Discussion

This paper shows that it is feasible to control the removal processes of nitrogen and phosphorus by using nutrient sensors and it is the opinion of the authors that nutrient sensors for process control will in the future be regarded as standard equipment at WWTPs. The most important result of the above study is that large improvements can be achieved in pre-denitrification systems by using a reasonably simple SISO control structure based on simple PI controllers. This holds true for the control of biological removal of nitrogen and the chemical removal of phosphate. A pre-requisite for this simple control structure is the *in situ* location of the sensors. The sensors should be located to measure directly in the processes, i.e. directly in the mixed sludge liquor and directly in the flocculation chamber.

The *in situ* location poses quite high demands on the sensors compared to automatic analysers measuring in the effluent. A sensor series manufactured by Danfoss Analytical, described in Lynggaard-Jensen *et al.* (1996) are developed exactly with this purpose in mind. The sensors use an ion filter that ensures that only the ions enter the sensor, while all suspended solids are kept out of the sensors. This removes the need for an ultrafiltration unit, which is requested for most traditional automatic analysers. This is of major importance for reliability as well as for the ease of maintenance. Devisscher *et al.* (2002) estimates the maintenance requirement for an automatic analyser using an ultrafiltration unit as 45 minutes, which has to be carried out daily. For comparison the maintenance requirement for a Danfoss InSitu® sensor is half an hour every four weeks. At these intervals the chemicals and the ion filter are exchanged.

In the timely development of control based on nutrient sensors there are many parallels to the development of DO sensor based control of aeration, which took place in the 1970's. In the beginning quite complex and unreliable sensor systems were used and complex

Table 1 Comparison of phosphate precipitation control strategies

Strategy	Average dosage (l/min)	Relative consumption (to the *in situ* controller)
In situ	0.42	
Constant dosage	0.66	167%
Flow proportional dosage	0.62	156%
Load proportional dosage	0.54	136%

methods for the DO control were proposed. However, today DO control is mostly carried out by the use of simple and reliable DO sensors using simple control structures based on on/off control or PID controllers. Similarly the nutrient sensors are becoming increasingly reliable and simple methods of using the sensors are proposed and have been shown efficient in full-scale implementations.

Conclusions

When sensors are located *in situ* simple control structures can be used for controlling the biological nitrogen removal process and the chemical phosphate precipitation process. For the control of aeration, internal recirculation only two sensors are required, a nitrate and an ammonium sensor. If external carbon is also to be controlled, i.e. if the available organic matter in the influent wastewater flow is not sufficient for a proper reduction of total nitrogen an additional nitrate sensor is requested. All three control loops can be implemented in traditional PLCs or soft PLCs, as simple PI controllers suffice. Similarly a simple PI controller suffices for the control of phosphate dosage provided a sensor is located in the end of the flocculation reactor. The Danfoss InSitu® sensor has been specially developed for this purpose and gives a simple and robust solution to the measurement in the *in situ* position, without the need of ultra-filters and with a low time requirement and frequency of maintenance.

References

Ammundsen, B., Önnerth, T.B. and Nielsen, M.K. (1992). Biological phosphorous removal in traditional wastewater treatment plants (In Danish). *Stads- og Havneingeniøren*, **8**, 79–82.

Copp, J. (2002). *The COST simulation benchmark – description and simulator manual.* COST (European Cooperation in the field of Scientific and Technical Research), Brussels, Belgium.

Devisscher, M., Bogaert, H., Bixio, D., Van de Velde, J. and Thoeye, C. (2002). Feasibility of automatic chemicals dosage control – a full-scale evaluation. *Wat. Sci. Tech.*, **45**(4/5), 445–452.

Henze, M., Grady, Jr. C.P.L., Gujer, W., Marais, G.v.R. and Matsuo, T. (1987). *Activated sludge model No. 1*. IAWPRC scientific and technical reports, London, UK, IAWPRC (IAWQ).

Ingildsen, P. (2002). *Realising full-scale control in wastewater treatment systems using in situ nutrient sensors*. Ph.D. thesis, University of Lund, Sweden, ISBN 91-88934-22-5.

Lant, P. and Steffens, M. (1998). Benchmarking for process control: "Should I invest in improved process control?" *Wat. Sci. Tech.*, **37**(12), 49–54.

Lindberg, C.-F. (1997). *Control and estimation strategies applied to the activated sludge process*. Ph.D. thesis, Uppsala University, Sweden, ISBN 91-506-1202-6.

Londong, J. (1992). Strategies for optimised nitrate reduction with primary denitrification. *Wat. Sci. Tech.*, **26**(5–6), 1087–1096.

Lynggaard-Jensen, A., Eisum, N.H., Rasmussen, I., Svankjaer Jacobsen, H. and Stenstroem, T. (1996). Description and test of a new generation of nutrient sensors. *Wat. Sci. Tech.*, **33**(1), 25–35.

Shinskey, F.G. (1979). *Process control systems*. 2nd edition. McGraw-Hill, New York, USA.

Skogestad, S. and Postlethwaite, I. (1996). *Multivariable feedback control*. John Wiley and Sons, Chichester, UK, ISBN 0-471-94277-4.

Weijers, S. (2000). *Modelling, Identification and Control of Activated Sludge Plants for Nitrogen Removal*. Ph.D. thesis, Technical University of Eindhoven, The Netherlands, ISBN 90-386-0997-3.

Yuan, Z., Oehmen, A. and Ingildsen, P. (2002). Control of nitrate recirculation flow in pre-denitrification systems. *Wat. Sci. Tech.*, **45**(4/5), 29–36.

Impact of separate urine collection on wastewater treatment systems

J. Wilsenach and M. van Loosdrecht

Department of Biochemical Engineering, Delft University of technology, Julianalaan 67, 2628 BC, The Netherlands (E-mail: *J.A.Wilsenach@tnw.tudelft.nl*)

Abstract Wastewater treatment should not only be concerned with urban hygiene and environmental protection, but development of a sustainable society must also be considered. This implies a minimisation of the energy demand and potential recovery of finite minerals. Urine contains 80% of the nitrogen (N) and 45% of the phosphorus (P) in wastewater. Separate collection and treatment would improve effluent quality and save energy in centralised biological nutrient removal (BNR). BNR processes are not optimal to treat water with very low N concentration resulting from separate urine collection. Relying on nutrient removal through sludge production, methanation of the sludge, subsequent nutrient removal from the digestion effluent results in optimised and more sustainable wastewater treatment. This paper quantitatively evaluates this option and discusses the potential.

Keywords Energy; nutrients; Sharon/Anammox; struvite; urine; wastewater

Introduction

Wastewater treatment has slowly changed over the last century from a system for prevention of diseases in urban society towards a system for protecting the natural environment. In recent years, society started to demand that technical systems contribute to the "sustainability" of the same society. Although "sustainability" can not be quantified easily, it is generally accepted that recovery of resources from waste contributes to an increased sustainability.

Nutrients (ammonium and phosphate) and organic carbon can potentially be recovered form municipal wastewater. However, the dilute nature of wastewater makes recovery economically and energetically expensive. Several studies have shown that up to 80% of the total N load and around 45% of the total P load in municipal wastewater originate from urine (Larsen and Gujer, 1996; Hanæus et al., 1997). Experience from ecological villages in Sweden showed that separate urine collection can be done efficiently (Hanæus et al., 1997; Jönsson et al., 1997). Separate urine collection largely improves the potential for nutrient recovery, because the concentrations of N and P are a hundred times higher than in wastewater. On the other hand, in e.g. The Netherlands, the nutrients in human waste amount to less than 20% of the amount in animal manure. From a societal point of view, nutrient recovery from animal manure should get prime attention, because manure can be collected more easily and can be treated locally on a large scale.

The importance of urine separation is recognised but the effect on central treatment processes has not yet been quantified. Separate urine collection would not be worthwhile if it only had a marginal impact on central wastewater treatment systems. Improvement of the overall wastewater management system should be a stronger driving force than nutrient recovery alone. If it can be shown that advanced treatment processes would benefit from separate urine collection, then wastewater treatment in general would benefit form urine separation. Most advanced wastewater treatment works operate according to variants of the modified UCT process. For this study we chose the BCFS® process (Biological/Chemical Phosphorus and Nitrogen removal) as a conventional reference process. The treatment plant at Hardenberg in The Netherlands is taken as an example, because a calibrated model was available (Meijer et al., 2001). The system as such is already fully optimised for biological N and P removal and produces good effluent quality (current annual average: 3.5 gN_{tot}/m^3, 0.5 gNH_4-N/m^3, 0.15 gP/m^3).

The N content of bacteria is between 9% and 12% and the P content is between 1% and 3% of the volatile suspended solids (Ekama and Marais, 1984). This means that for wastewater with an influent COD:N:P ratio of approximately 100:5:1 almost all the N and P will be used for cell growth (with the yield factor Y_H = 0.63 gCOD/gCOD). If urine were collected separately, the influent nutrient load could then be reduced to match this cell growth requirement. The N and P normally removed through nitrification/denitrification and dephosphatation, e.g. in a BCFS® process, could then be removed by collecting a large percentage of urine separately. The N and P remaining in the influent could be removed through waste activated sludge production at short solid retention times. A major decrease in the influent N and P load would allow for simple treatment processes that will not be effective if all urine remains diluted in wastewater. This paper evaluates quantitatively the advantages of separate urine collection for the design, operation and sustainability aspects of a centralised wastewater treatment system. Hereto we made use of sub-systems already applied at large scale.

Methodology

Influent characteristics

Average Dutch influent flow rates and concentrations were used for this study (CBS, 2000; STOWA, 2002). Influent concentrations for chemical oxygen demand, total nitrogen, ammonium and total phosphorus were COD_{tot} = 537 gCOD/m^3, N_{tot} = 50 gN/m^3, NH_4^+ = 40 gN/m^3 and P_{tot} = 8 gP_{tot}/m^3. The COD, N and P fractions in influent wastewater were divided into different model components, similar to Hao et al. (2001). The current wastewater influent at Hardenberg (8,500 m^3/d) is still less than the design flow rate. At the current flow rate, the total N effluent concentration is much lower than the effluent standard of 10 gN/m^3. The flow rate was maximised to the extent that the system just complied with the effluent demand and was determined by iteration to be 13,500 m^3/d. Effects of changing influent nutrient loads were compared to this reference (zero scenario).

Modelling urine separation

Urine contributes only a small volume to the total wastewater volume. However, the water currently used to flush urine is a significant fraction of the total. Urine flush water was assumed to be 35 l/p.d (Jönsson et al., 1997). Therefore, if 100% urine separation could be achieved, wastewater discharge to the treatment works could be reduced by 36.25 l/p.d (including 1.25l/p.d urine). Modern source separation toilets use a small amount of flush water. The production of urine (including flush water) was assumed 2 l/p.d.

From the influent nitrogen concentration, 40 gN/m^3 can be attributed to urine. Nitrogen in urine is mainly present as urea, $CO(NH_2)_2$. This soluble compound rapidly hydrolyses to NH_4^+ and HCO_3^-. The N-load contributed by urine was assumed to consist of soluble ammonium (S_NH4) only (Helström et al., 1999 and Hellström and Johansson, 1999). At 100% urine separation, the total N influent concentration would drop from 50 gN/m^3 to 11.4 gN/m^3 (including the effect of the decreased flow rate). The ammonium concentration at different urine separation efficiencies was determined according to Eq. (1):

$$NH_{4\_25} = (NH_{4\_00} \times V_{00} - 0.25 \times NH_{4\_urine} \times I)/V_{25} \qquad (1)$$

Where, NH_{4_25} = ammonium concentration in wastewater with 25% urine collected separately (gN/m^3), NH$_{4\_00}$ = ammonium concentration in wastewater without urine separation (gN/m^3), V$_{00}$ = wastewater flow rate without urine separation (m^3), NH$_{4\_urine}$ = ammonium load in urine (12 gN/p.d), I = number of individuals connected to treatment plant (45,000, at flow rate = 13,500m^3/d) and V$_{25}$ = wastewater flow rate with 25% urine collected separately.

The phosphorus and COD concentration due to separate urine collection was determined in the same way. The inflow concentration of P$_{tot}$ was 8 gP/m^3 in wastewater without urine separation. The phosphorus load from urine was assumed 1 gP/p.d. At 100% urine separation, the total influent phosphorus concentration would then be 5.3 gP/m^3. Separate urine collection will lead to a small decrease in the COD load discharged to wastewater treatment works (12 gCOD/p.d @ 100% urine separation).

Modelling the BCFS® process

Hao et al. (2001) modelled the BCFS® process at Hardenberg. We used the computer software package AQUASIM 2.0 (Reichert, 1998) to implement the dynamic simulation of the BCFS® process model, which is schematically represented in Figure 1.

The total volume of the five compartments is 10,000 m^3. A secondary settling tank (2,800 m^3) is downstream of the final aeration basin. Mixed liquor is returned to different reactors and the settled sludge return rate was equal to the inflow. Waste activated sludge was withdrawn from the clarifier's sludge compartment. In all simulations of the BCFS® process the total suspended solids concentration (TSS) was kept constant at 5,000 g/m^3. In practice, solids retention time (SRT) controls the TSS. In the simulations a TSS equal to 5,000 g/m^3 was maintained by adjusting the SRT. The sludge volume index of existing BCFS® processes is below 120 ml/g. It was assumed that the sludge will separate and settle well in the secondary settling tank. A conservative temperature, common for the colder half of the year in north-west Europe, of 12°C was used for all simulations.

Treatment optimisation and model integration

A second set of simulations was done to evaluate a proposed system for treatment of separately collected urine and wastewater. Figure 2 presents a flow diagram for the integration of existing processes. Effluent concentrations and removal efficiencies for different process units were based on literature information. The aerobic reactor was simulated as described above. The sum of influent wastewater (Q$_1$), the pre-thickener overflow (Q$_4$) and effluent from the Sharon/Anammox process (Q$_{10}$) gives the influent flow rate and concentrations (Q$_3$) as shown in Figure 2. The aerobic reactor had a volume of 1000 m^3 (hydraulic retention time of two hours). The volume of the clarifier's sludge compartment was assumed 10% of the volume of the aerobic reactor.

Based on the substrate ratio required for bacterial growth (COD:N:P = 100:5:1) it is clear that nutrients are present in excess of the requirement; N$_{exc}$ = 24 g/m^3 and P$_{exc}$ = 1.5 g/m^3. According to these figures, nutrients remaining in wastewater after 60% urine

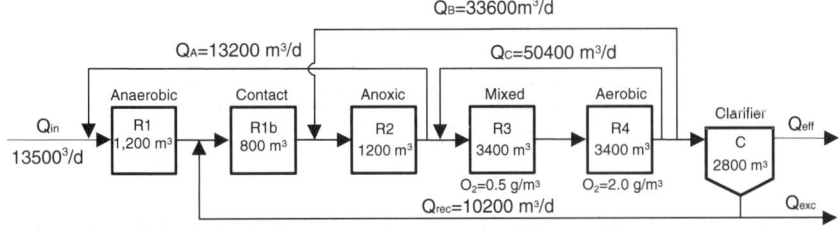

Figure 1 Schematic process diagram of the Hardenberg wastewater treatment works

separation could be removed with waste sludge. Waste sludge should be gravity thickened (TSS = 25 kg/m^3, Tchobanoglous, 1991) before entering anaerobic digestion (Q_5). We assumed that 100% of slowly biodegradable COD (X_S) and 90% of the COD of bio-mass (X_H and X_I) can be transformed into bio-gas (Q_{12}). The CH_4:CO_2 ratio was assumed 65:35 for this study (Malina and Pohland, 1992). Digested sludge (Q_6) can be thickened (80 kg/m^3) and dewatered (200 kg/m^3) before incineration. Supernatant and centrate (Q_7) are mixed with the separately collected urine stream. The COD concentration in anaerobic digester supernatant is normally around 1,200 g/m^3 (Stowa, 2000, 25). It could be assumed that no gaseous nitrogen escapes the digester and that at steady state, all nitrogen entering the digester leaves as either NH_4^+, or nitrogen in dewatered sludge (Q_{13}).

Struvite ($MgNH_4PO_4.6H_2O$) precipitates naturally in urine or in anaerobic sludge digesters and downstream piping. Relatively low phosphate concentrations can be expected form anaerobic digester supernatant; P_{tot} = 50 g/m^3. In this case, supernatant and filtrate is combined with urine; P_{tot} = 800 gP/m^3. Addition of MgO would increase pH sufficiently to precipitate struvite from the combined stream (Q_8) and yield a constant effluent concentration of 18 gP/m^3 (Schuiling, 1999). Recovered struvite could be used as fertiliser (Q_{15}).

The mixture of urine (9,000 gNH_4^+-N/m^3) and supernatant (1,200 gNH_4^+-N/m^3) leaving the sruvite crystalliser (Q_9) would still contain around 2,400 gNH_4^+-N/m^3. Recent development of the Sharon/Anammox technology made removal of highly concentrated NH_4^+ from wastewater more sustainable. Up to 95% of the influent ammonium is removed as nitrogen gas (Q_{14}) and only 5% of the influent total nitrogen leaves the combined process (STOWA, 2000–25). The effluent from these processes would be returned to the aerobic reactor (Q_{10}). A combination of computer simulation and mass balance calculations was used to evaluate the performance of this integrated process.

Assessment of energy demand

The energy consumption of integrated separate urine and wastewater treatment was evaluated. The energy consumption of a BCFS® process without urine separation or pre-settling was used as a reference. Energy consumption of the two processes was determined theoret-

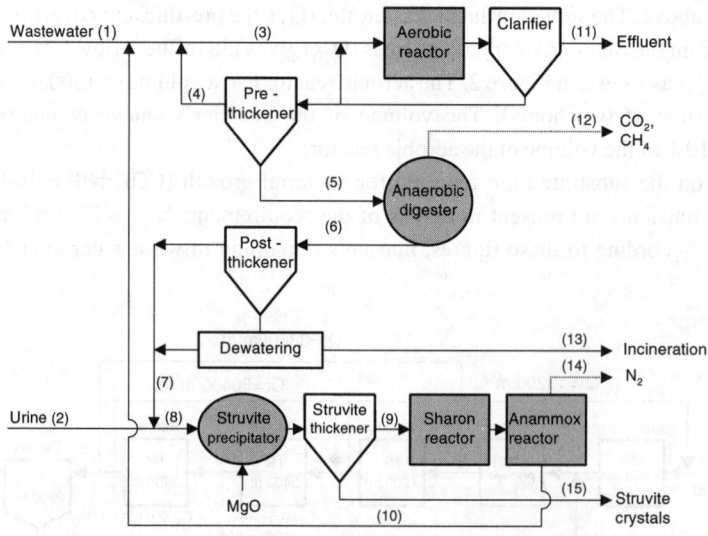

Figure 2 Proposed process flow diagram to treat wastewater and urine separately (Flow numbers refer to subscripts in the text, Q_1, Q_2, Q_3, etc.)

ically, based on aeration (E_{aer}), sludge dewatering (E_{dew}), sludge incineration (E_{inci}), pumping (E_{pump}), mixing of tank reactors (E_{mix}) and methane gas (E_{CH_4}) produced by the anaerobic sludge digestion. The total net energy requirement (E_{net}) can be expressed as:

$$E_{net} = E_{aer} + E_{dew} + E_{inci} + E_{pump} + E_{mix} + E_{heat} - E_{CH_4} \qquad (2)$$

The total oxygen requirement is the sum of oxygen required for sludge production in the aerobic reactor and the oxygen required in the Sharon process. The oxygen requirement for the aerobic reactor was simulated with a full activated sludge model in AQUASIM. Oxygen requirement for the Sharon process was based on the stoichiometric ratio, 1.82 g O_2/gNH_4-N influent (STOWA, 2000).

The calculation of energy involved in incineration, dewatering and methane combustion was described by van Loosdrecht *et al.* (1997). Energy required for pumping and mixing is usually neglected in estimations. The total recirculation flow of the BCFS® process is almost tenfold the simple aerobic reactor and becomes significant when processes are compared. Pump rates were based on average daily flows. The mixing of anaerobic and anoxic compartments in the BCFS® process has to be compared to mixing in the struvite precipitator and Sharon/Anammox reactor. Furthermore, the anaerobic digester requires a considerable amount of mixing. The power requirement of mixing was based on an average figure of 10 W/m³ mixed volume (Grady and Lim, 1980).

Results and discussion
Effect of urine separation on nutrient removal in a BCFS process

The effects of separate urine collection on an existing BCFS process with raw wastewater were simulated. The main results are shown in Figure 3.

Due to a decrease in the number of autotrophic bacteria (nitrifiers), effluent ammonium concentration ($NH_4^+{}_{eff}$) increases slightly with increasing urine separation. Effluent nitrate concentration (NO_{3_eff}) decreases with increasing urine separation. The COD/TKN ratio increases with increasing urine separation and therefore the denitrification potential increases. Less nitrate is produced (less ammonium oxidation) while the capacity to reduce nitrate increases. While NO_{3_eff} decreases non-linearly, the amount of nitrogen gas produced decreases linearly with increasing urine separation. Total N in the effluent (N_{tot_eff}) is the sum of ammonium, nitrate and nitrogen contained in suspended solids (not settled in the clarifier). The model predicts N_{tot_eff} = 3.2 g/m³ at 50% urine separation, which is the current effluent concentration at Hardenberg, at influent flow rate of 8,500 m³/d compared

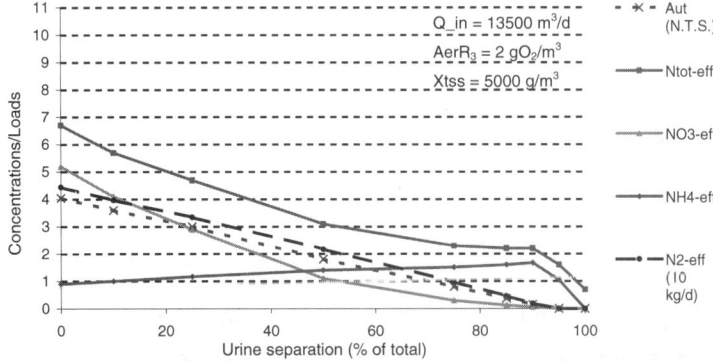

Figure 3 Effects of urine separation on N removal in advanced UCT-type (Figure 1) wastewater treatment processes

to 13,500 m^3/d in the model simulations. The N removal capacity of a BCFS process can be increased by 60% with 50% urine separation. One observes a substantial decrease in N$_{tot\_eff}$ with urine separation up to 50%. The decrease is less obvious above 50% urine separation.

Virtually all phosphate had already been removed in the case of zero urine separation and urine separation therefore had little effect on the P-removal efficiency.

Effect of urine separation on nutrient removal through increased sludge production

In the second set of simulations, the effect of urine separation on nutrient removal through sludge production was evaluated. Hereto a short SRT is preferable. At too high sludge withdrawal rates (short SRTs), sludge growth rate limits the process. The relation between SRT and sludge production in an aerobic zone integrated with other processes (Figure 2), with 75% of urine collected separately, is shown in Figure 4(a). The figure shows a drastic increase in effluent nutrient concentrations (little sludge produced, but much COD in effluent) at SRT < 0.5 days. The oxygen consumption is also an indication of the sludge activity at low SRT. Higher SRT and resulting lower waste sludge removal results in higher effluent nutrient concentrations. The optimum SRT is around one day. Nitrification starts at SRT $\cong$ 5 d for a temperature of 12°C.

Temperature of wastewater also determines the growth rate of bacteria and sludge formation. Figure 4(b) shows that the effluent concentrations of N and phosphorus vary relatively little because of changing temperature. Temperature variations are not too important at SRT = 0.8 d. The figure also shows that working with 12°C gives a conservative estimate of the sludge production and related nutrient removal.

Figure 5 illustrates the effect of increased urine separation efficiency. This leads to a lower nutrient load in the final effluent (Q$_{11}$, Figure 2). Higher urine separation also results in more nitrogen gas (Q$_{14}$, Figure 2) and struvite production (Q$_{15}$, Figure 2). However, at urine separation efficiencies > 75%, the unavailability of ammonium limits sludge production, as can be seen from the increased COD load in the effluent. Limiting sludge production also leads to higher effluent N$_{tot}$ and P$_{tot}$ concentrations.

Energy balance

Table 1 gives a brief summary of energy requirements for different scenarios. The total energy requirement of separate urine and wastewater treatment is shown for urine separation efficiencies of 50%, 65%, 75% and 85%. The energy demand of the reference system is also shown.

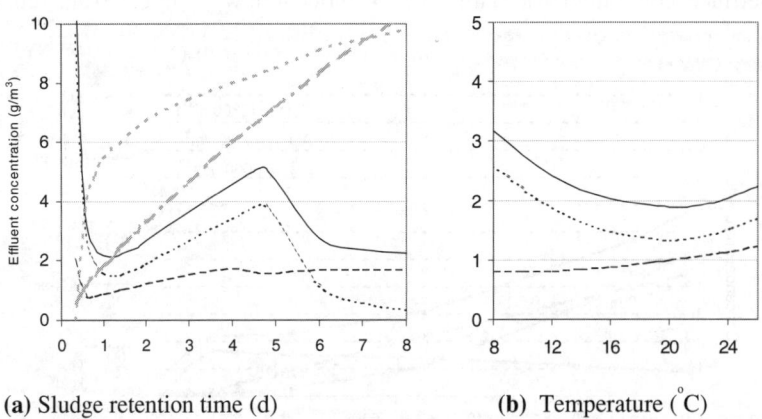

(a) Sludge retention time (d)   (b) Temperature (°C)

Figure 4 Effects of (a) Sludge retention time, with T = 12°C, and (b) Temperature, with SRT = 0.8 d, on nutrient removal at 75% urine separation N$_{tot}$ —— NH$_4$ - - - - P$_{tot}$ - - - - X_TSS (1:2500) —·—·— O$_2$_net (1:20) ········

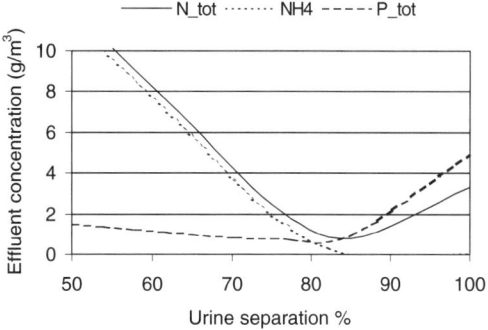

Figure 5 Effluent concentrations as a function of urine separation efficiencies for the flow scheme of Figure 2, with T = 12°C and SRT = 0.8 d

Table 1 Summary of energy requirements for urine separation system and reference systems. (Negative number for total energy indicate net production)

Urine separation	0	50	65	75	85	%
Digested sludge mass	2,111	1,917	1,888	1,881	1,760	kg/d
Urine mass	0	45,000	58,500	67,500	76,500	kg/d
Total energy	15,302	−6,204	−5,671	−5,467	−4,666	MJ/d
	6.25	−1.60	−1.46	−1.41	−1.20	W/pers

The energy requirement for sludge handling (in the integrated processes) is still less than the reference system (where less of the produced sludge is transformed to methane, due to a higher sludge age and lower degradable fraction). The amount of energy generated via methane combustion in the integrated processes is more than three times the potential of the reference system. The energy requirement for aeration in the reference system is four times as high as the combined energy requirement for aeration of urine and activated sludge production. The model shows that a net energy production is possible with separate treatment of urine and wastewater. The continuous power demand for treatment of normal wastewater (including dilute urine) in the reference process is around 6 W/p. The potential net power generation (resulting from methane combustion and low aeration) is more than 1 W/p.

Conclusions

1. Advanced biological nutrient removal processes would benefit from separate collection and treatment of urine. Total nitrogen effluent concentrations could be reduced from 7.5–2.5 gN/m³ at around 60% urine separation. Separation efficiencies over 60% show little further improvement, because the process is not optimal for low ammonium influent concentrations.
2. Existing processes can be integrated and optimised to treat urine and wastewater on central scale, with more than 60–70% urine separation. Effluent with very low ammonium, nitrate and phosphate concentrations can be produced (all less than 1g/m³).
3. The actual nutrient content of particulate influent COD and nutrient content of sludge strongly influence the nutrient removal efficiency. Default values of N and P content in sludge suggest that complete nutrient removal is possible with 75% urine separation.
4. Urine separation decreases the energy requirement for wastewater treatment radically. Where advanced BNR processes require around 6 W/p, an integrated process to treat urine and wastewater separately could produce more than 1 W/p. The energy available for separate collection and transport of urine may therefore not exceed 7 W/p.

References

Ekama, G.A. and Marais, G.v.R. (1984). *Theory, design and operation of nutrient removal activated sludge processes*, South-African Water Research Commission report.

Grady, C.P.L. and Lim, H.C. (1980). *Biological Wastewater Treatment – Theory and Applications*, Marcel Dekker Inc. ISBN 0-8247-1000-2.

Hanæus, J., Hellström, D. and Johansson, E. (1997). A study of a urine separation system in an ecological village in northern Sweden, *Water Science and Technology*, **35**(9), 153–160.

Hao, X. Heijnen, J.J., Qian, Y. and van Loosdrecht, M.C.M. (2001). Contribution of P-bacteria in BNR processes to overall effects on the environment, *Water Science and Technology*, **44**(1), 67–76.

Hellström, D. and Johansson, E. (1999). Swedish experience with urine separating systems, *Wasser & Boden*, **51**(12), 26–29.

Hellström, D., Johansson, E. and Grennberg, K. (1999). Storage of human urine: acidification as a method to inhibit decomposition of urea, *Ecological Engineering*, **12**, 253–269.

Jönsson, H., Stenström, T.-A., Svensson, J. and Sundin, A. (1997). Source separated urine-nutrient and heavy metal content, water savings and faecal contamination, *Water Science and Technology*, **35**(9), 145–152.

Larsen, T.A. and Gujer, W. (1996). Separate management of anthropogenic nutrient solutions (human urine), *Water Science and Technology*, **34**(3–4), 87–94.

Malina, J.F. and Pohland, F.G. (1992). Design of anaerobic processes for the treatment of industrial and municipal wastes, *Water Quality Management Library*, Volume 7, Technomic Publishing Company Inc.

Meijer, S.C.F., van Loosdrecht, M.C.M. and Heijnen, J.J. (2001). Metabolic modelling of a full-scale enhanced biological phosphorus removing WWTP, *Water Research*, **35**(11), 2711–2723.

Reichert, P. (1998). AQUASIM 2.0 – *Computer Program for the Identification and Simulation of Aquatic Systems*, EAWAG, ISBN 3 906484-16-5.

STOWA (2000). *Het gecombinmeerde Sharon/Anammoxproces: Een duurzame methode voor N-verwijdering uit slibgistingswater*, STOWA report 2000–25.

Van Loosdrecht, M.C.M., Kuba, T., Van Veldhuizen, H.M., Brandse, F.A. and Heijnen, J.J (1997). Environmental impacts of nutrient removal processes: Case study, *Journal of Environmental Engineering*, **123**(1), 33–40.

Investigation of the effectiveness of source control sanitation concepts including pre-treatment with Rottebehaelter

D.R. Gajurel, Z. Li and R. Otterpohl

Technical University Hamburg-Harburg, Institute of Municipal and Industrial Wastewater Management, Eissendorfer Strasse 42, D-21073 Hamburg, Germany (E-mail: *gajurel@tuhh.de, www.tuhh.de/aww*)

Abstract High levels of nutrients recovery can be achieved with source control sanitation – technologies are already available. Separation toilets for example separate urine that can be used in agriculture with some crop restrictions as a fertiliser after about 6 months of storage. The grey water has very low loads of nitrogen and can be treated in different combinations of biological and physical treatment and reused. Faecal matter with flush water from the separation toilet can be discharged into Rottebehaelter (an underground pre-composting tank) that retains solid material and drains liquid to a certain extent. Investigation of Rottebehaelter in the different sites and laboratory experiments showed that retained faecal material still contained a high percentage of water. However, odour was not noticed in those Rottebehaelters that have been examined. One of the major advantages of this system over other forms of pre-treatment as the septic tanks is that it does not deprive agriculture of the valuable nutrients and soil conditioner from human excreta. It has to be stated that maintenance is a crucial factor. As an intermediate result of the intensive research of Rottebehaelter it seems that these systems are rather a way of solids retaining, de-watering and long-term storage before the contents are further treated.

Keywords Brown water; grey water; nutrients; Rottebehaelter; source control sanitation; yellow water

Introduction

The linear flow of nutrients from farming land to water bodies due to flush sanitation results in high demands of mineral fertiliser. Production of fertiliser is energy intensive and causes environmental problems. Generating nitrogen from air requires a considerable amount of energy; and mining and refining the raw materials for phosphate production generates huge amounts of hazardous wastes. Reserves of phosphate (P), potassium (K) and also sulphur (S) are definitely limited on a time scale of a couple of human generations especially with regard to economic constraints. Therefore, nutrients present in wastewater should rather be reused instead of discharging them into the water body in order to minimise the production of mineral fertiliser.

It is well recognised that transfer of nutrients from terrestrial to aquatic ecosystem causes on one hand, eutrophication in water bodies and on the other hand nutrient deficiency in agricultural land. These problems are increasing greatly particularly in the developing world, where there is hardly any provision to interrupt the nutrients discharge into the water body and, in return, loss of nutrients from agricultural land is compensated only with hardly affordable commercial fertiliser in order to feed a rapidly growing population. In the developed world, nutrient transfer to water bodies through wastewater has been relatively controlled by costly high technology for the larger treatment plants. However, even with high investment, which is not affordable for most of the developing countries, more than 20% of nitrogen, more than 5% of phosphorus and more than 90% of potassium are still emitted from wastewater treatment plants into the aquatic environment (Otterpohl *et al.*, 1997). Those nutrients, which are captured in sludge are often contaminated with heavy metals such as cadmium (Cd) and organic compounds such as PCB (polychlorinated biphenyl), which pose potential toxic risks to plants, animals and humans (Metcalf and Eddy, 1991). Therefore, large amounts of sewage sludge are disposed of in landfills or incinerated. A very small part is applied to the agricultural land.

In order to recover and reuse nutrients ecologically and economically, household wastewater should be treated separately. High levels of nutrients recovery are possible with the concept of source control sanitation (Henze *et al.*, 1997; Esrey *et al.*, 1998; Jönsson *et al.*, 1999; Larsen and Udert, 1999; Otterpohl *et al.*, 1999a; Otterpohl, 2001). This paper does show the effectiveness and limitations of source control sanitation concepts with pre-treatment by pre-composting tanks (Rottebehaelter) to recover the particulate fraction of nutrients from household wastewater.

Source control sanitation

A vision of source control sanitation for household wastewater is based on the fact of very different characteristics of grey water (washing water from kitchen, shower, washbasin and laundry), yellow water (urine with or without flush water) and brown water (faeces, toilet paper and flush water). The typical characteristics of the streams of household wastewater, shown in Table 1, clearly reveal that urine contains most of the soluble nutrients, whereas grey water, despite its very large volume compared to urine, contains only a small amount of nutrients. Furthermore, faeces, which is about 10 times smaller in volume than urine, contains nutrients, high organic load and the largest part of pathogens.

If urine is separated and reused in agriculture, not only nutrients will be reused fully, but also a high level water protection will be reached. Unlike wastewater containing urine and faeces, grey water can be treated with simple and low cost processes and reused. There are many cost efficient biological treatments and membrane technology that can produce high quality water. Even with the combination of ground infiltration grey water can be processed to tap water. If faeces is separated and kept in a small volume with non or low-flush toilet, this will be a favourable condition for high hygienisation and production of soil conditioner. Therefore, separated treatment of different flows according to their characteristics

Table 1 Typical characteristics of household wastewater components (Compiled from: Geigy, Wissenschaftliche Tabellen, Basel 1981, Vol. 1, Larsen and Gujer, 1996 and Fitschen and Hahn, 1998)

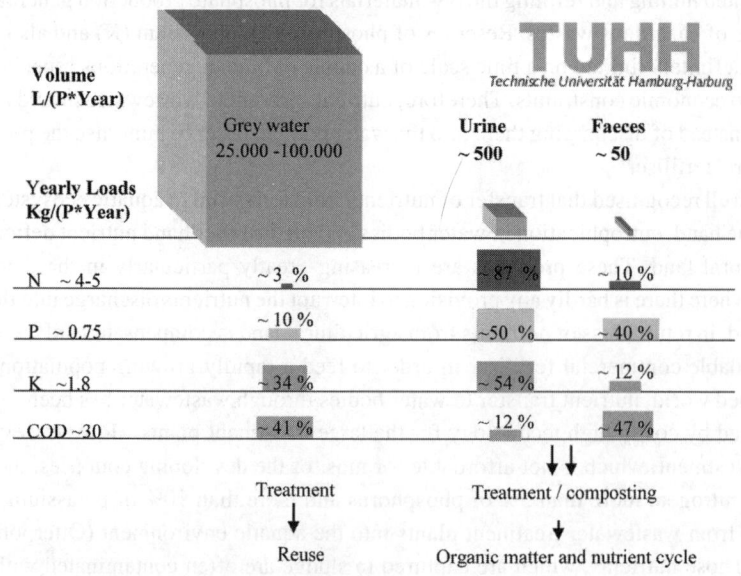

can lead to full reuse of resources and a high hygienic standard. In order to realise the source control, technologies have already been developed in Europe (Esrey *et al.*, 1998, Otterpohl *et al.*, 1999b and 2001). A separation toilet which has been increasingly used in Sweden, Denmark and Germany, for example, is a suitable technology to separate the urine and faeces at source. The toilet has two bowls, the front one for urine and rear one for faeces.

Treatment of separated flows
Yellow water treatment
Urine is relatively sterile and can be reused without further treatment (Wolgast, 1993). However, due to faecal contamination, pathogens have been found in yellow water; but in low concentration, which will pose low hygienic risk of using yellow water as a fertiliser, if it is stored at least for 6 months before being used in agriculture (Jönsson *et al.*, 1999; Hellström and Johansson, 1999). Since the practising of separated collection of yellow water, farmers in Sweden have been collecting it in the underground storage tank for applying to their agricultural land (Esrey *et al.*, 1998). This practice is good for a region where the farmland is near to the housing area; otherwise, transportation of a large amount of urine solution for longer distance has many negative environmental impacts (Hellström and Johansson, 1999).

Recently, many methods for treatment and reduction of volume of collected yellow water have been studied. One method is dewatering by evaporation with and without nitrification and freeze concentration (Gulyas, 2000). By freezing, it is possible to concentrate 80% of nitrogen and phosphorus in 25% of the original volume (Ban *et al.*, 1999). By nitrification in combination with drying, it is possible to concentrate over 70% of the nitrogen in 10% of the original volume (Hellström and Johansson, 1999). These methods could be beneficial, when a larger volume of urine solution has to be transported a long way to the agriculture farm. These methods have been investigated only in small scale so far.

Grey water treatment
Since grey water contains very low nitrogen concentrations (Table 1), nitrification and de-nitrification processes are not necessary for grey water treatment. Therefore, unlike total domestic wastewater, grey water is relatively easy to treat although concentrations are not necessarily lower than in end-of-pipe systems due to a far lower dilution. Choice of household chemicals that can be mineralised (degradability is no useful indicator) is helpful for achieving good quality. Grey water treatment with vertical constructed wetlands with sizes of 2 m^2 per inhabitant in the Flintenbreite settlement has shown good performance (Otterpohl, 2001). Constructed wetlands are cheap in construction and operation. However, because of space scarcity, it is not always appropriate for the densely settled urban areas. For these areas, among other methods, SBR can be suitable. Treatment of grey water from the Flintenbreit settlement in small scale SBR was studied. Details are given in Zifu Li *et al.*, 2001. The results showed that, SBR can greatly reduce organic matter, nutrients and turbidity. For reuse purposes, tertiary treatment is needed. Treatment with a slow sand filter can meet quality requirements for groundwater recharge. *E. coli* was greatly eliminated to acceptable levels for recharge. Therefore, grey water treated with the combination of SBR and slow sand filter is suitable for ground water recharge. Treatment with membrane-bioreactors will probably be the choice of the future especially if reuse is intended.

Brown water treatment
A relatively new technology called Rottebehaelter or pre-composting tank, which is usually an underground concrete tank having two filter beds at its bottom or two filter bags

that are hung side by side and used alternately in an interval of 6–12 month, has been increasingly applied in Austria, Switzerland and Germany for domestic wastewater pre-treatment (Figure 1). Those investigated in Germany have demonstrated to be beneficial and can be combined with concept of source separation (Gajurel et al., 2001). One of the major advantages of this system over other forms of pre-treatment systems is that it does not deprive agriculture of the valuable nutrients and soil conditioner from human excreta. The retained materials that have already been de-watered and pre-composted in the Rottebehaelter for 8–12 months can be further composted with biological kitchen and/or garden waste in a local composter for at least a year and used in agriculture. This avoids expensive tanker-trucks which are extensively used in conventional systems to transport sludge. Moreover, compared to septic tanks methane emission can be very low as the outer parts of the retained material maintains the aerobic conditions. However, handling of the bags or the material is not a simple task and should be improved for the future. There are concepts in Austria in areas with strong gradients in the ground where the tanks are accessible with agricultural fork-lifters and can be removed and emptied in a simple way.

Investigation of existing Rottebehaelter in Lambertsmuehle pilot project
Background
The source control sanitation system has been installed in the pilot project in Lambertsmuehle near Cologne city in Germany since the summer of 2000. Details can be found in Otterpohl et al. (2001a). The yellow water, by the means of a separating toilet, is collected separately in an underground tank, where it is kept till it is ready for use in agriculture. The brown water is discharged into the Rottebehaelter, which is madesup of concrete monolithically and constructed underground outside the building (Figure 2). It is covered with a prefabricated concrete slab and provided with ventilation. A shutter of a concrete slab for changing the filter bag, inspection and cleaning has been provided on the covering of the Rottebehaelter. Inside the Rottebehaelter, two filter bags are hung side by side in such a way that when one is full, the influent is manually diverted with the help of a diversion pipe into the next empty bag. The filtrate, due to urine separation, is nutrient poor, and is mixed with grey water and treated in a constructed wetland.

Performance of the Rottebehaelter
In September 2001, samples from both active and inactive filter bags were analysed. The results are shown in Table 2. In both filter bags – active as well as inactive-moisture content was higher than the optimal range (40–60% for composting). Moisture content above 70% leads to anaerobic conditions (Bidlingmaier, 1983). Thus, anaerobic conditions must have

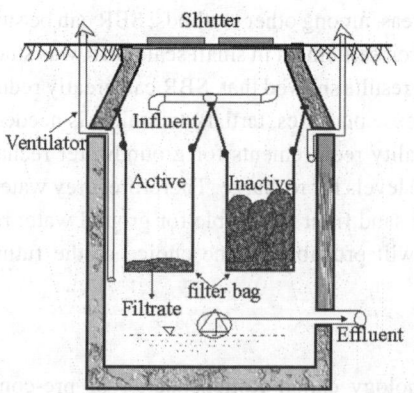

Figure 1 Rottebehaelter (pre-composting tank)

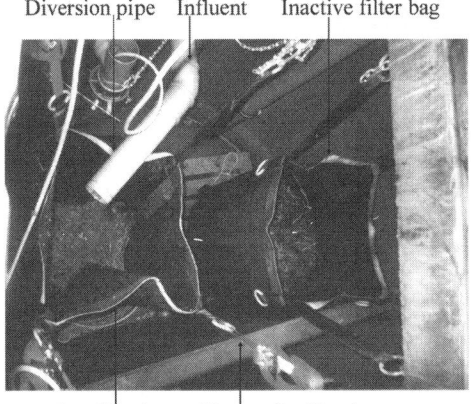 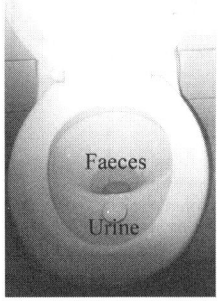

Diversion pipe Influent Inactive filter bag

Active filter bag Hanger for filter bag

Figure 2 Lambertsmuehle project: Rottebehaelter with filter bag and separation toilet

Table 2 Characteristics of retained materials (FM: Fresh matter, DM: Dry matter)

Parameter sample from	Water content (% FM)	Loss on ignition (% DM)	C (% DM)	N (% DM)	C/N	P (% DM)	K (% DM)	pH	Temp °C
Active filter bag	87.72	95.48	46.60	6.74	6.91	0.69	1.07	7.21	18
Inactive filter bag	83.14	93.26	50.30	7.16	7.025	0.61	1.61	6.30	20

taken place in both bags. However, no odour was noticed during the sampling. Also people living in the house have not complained about odour problem so far. Low temperature and low reduction of volatile solids suggest that a slow decomposition process took place in both filter bags. It might have caused slow and low emission of odour, which was not detected by the human nose in the open air.

Low C:N ratio showed that structural bulking agent was not added sufficiently in order to maintain a C:N ratio of composting material in the filter bags of between 20:1 and 30:1. Struciaral material also helps to reduce water content and increase air circulation inside the material. In the inactive filter bag, the pH was lower than the optimal range, 7–8. It was most probably due to formation of volatile organic acids. Phosphorus and potassium concentrations in both retained materials were lower than concentration in faeces. This was mostly due to loss through the filter bag in filtrate.

Investigation of the performance of Rottebehaelter in controlled conditions
Methods and materials

The aim of this investigation was to find out the effectiveness of a filter bag in separating and pre-composting solid stuff from brown water. For this, two small scale experiments, one with straw as structural additive and another without straw, were carried out in a small container of volume 30 l with filter bag having a pore size of 1 mm (Figure 3).

For the experiments, faeces was collected directly in the plastic bag at the time of excretion and weighted prior to filling into a small container where toilet paper (10 g) and flush water (6 l) were also added. After that, the mixture, called brown water, was filled manually into the container having a filter bag inside once a day. The filtrate that fell at the bottom of the container was pumped out everyday. Temperatures in the filter bag and surrounding were measured. The experiment was carried out in two phases – the filling phase for 30 days and the silence phase for 45 days. After the silence phase, the pre-composting

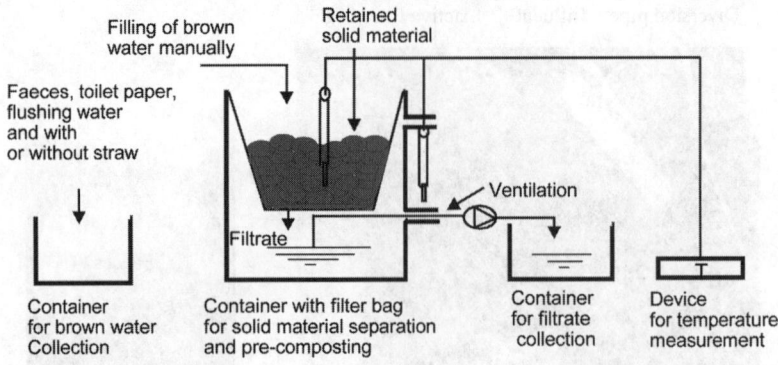

Figure 3 Experimental set up

materials were mixed thoroughly and analysed. In experiment 1, only brown water was filled; whereas in experiment 2, 2.5 g of straw as structural additive was added at every filling.

Results and discussions
Effectiveness of Rottebehaelter in pre-composting

Table 3 shows the characteristics of retained materials. Water content of the material was too high in both the experiments and the water content in the lower part was higher than that in the upper part. Therefore, anaerobic decomposition must have occurred at the inner of the lower part. Slight odour near the container during the experiment also supports this idea.

pH value was relatively lower in upper part of the experiment 1 – without straw, whereas it was higher in the upper part of the experiment 2 – with straw. Loss of organic substance was more in experiment 2 – with straw than experiment 1 without straw. Therefore, addition of sufficient amount of dry structural bulking agent e.g. straw, bark etc. is required for the composting. It does provide optimal C/N ratio, water content and air circulation of the pre-composting material, which are prerequisite parameters for the composting. In both experiments, these parameters were not optimal. In experiment 2, straw was added, but too little. Therefore, further investigation with addition of sufficient amount of dry structural bulking agent is needed.

Effectiveness of Rottebehaelter in retaining carbon and nutrients

Figure 4 shows the percentage of carbon and nutrients retained in the filter bag. Among nutrients, nitrogen was retained above 60% of the total influent in both experiments. Despite the loss due to anaerobic conditions, nitrogen retained in the filter bag was consid-

Table 3 Characteristics of retained materials (FM: Fresh matter, DM: Dry matter)

Parameter sample from	Water content (%FM)	Loss on ignition (% DM)	C (% DM)	N (% DM)	C/N	pH
Experiment 1: withoutstraw						
• Upper Part	66.81	85.95				7.15
• Lower Part	81.67	85.87				7.85
• Mix of both parts			44.2	7.15	6.12	
Experiment 2:with straw						
• Upper Part	74.14	83.29				8.61
• Lower Part	82.12	85.57				8.51
• Mix of both parts			38.5	6.3	6.11	

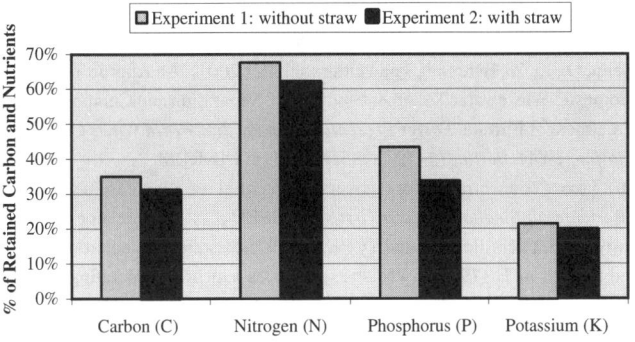

Figure 4 Carbon and nutrients retained in the filter bag

erably higher than phosphorus and potassium. However, in experiment 1, phosphorus retained was above 40% of the influent. In general, the performance of the filter bag in retaining carbon and nutrient was satisfactory.

Conclusions

Technologies for source separation and treatment have been developing rapidly. So far, performance of the small scale experiments and pilot projects have given a positive impression. Investigation of Rottebehaelter has shown its potential for solids-liquid separation and pre-treatment (de-watering to a certain extent) and collection/storage of solid stuff in household wastewater. One of the major advantages of this system over other systems is that it does not deprive agriculture of the valuable nutrients and soil conditioner from human excreta and does not require expensive tanker-trucks. However, water content of the retained material was very high, which must be lowered to an optimal level. Moreover, measures such as prevention of heat loss from the Rottebehaelter and good ventilation are important for effective performance. Last but not least, addition of sufficient amount of structural bulking agents such as bark, straw etc. into the retained material are important factors for maintaining optimal C:N ratio, moisture and oxygen required for composting. Combination of the urine source separation concept with Rottebehaelter was demonstrated to be effective to recover nutrients. It can be the most appropriate system for application in the regions where there is a demand for local reuse of the end product. It has to be stated that maintenance is a crucial factor, removal and handling of the pre-composted material has to be improved. In addition, proper procedures of further composting and usage will have to be established. Compared to septic tanks, there are a couple of advantages that make further development worthwhile.

Acknowledgement

We are thankful to Heinrich Böll Foundation, Germany for providing scholarship to carry out the research.

References

Ban, Z, Byden, S. and Lind, B.-B. (1999). Concentration, Crystallosation and mineral adsorption for nutrient recovery and reuse from human urine. Presented at the *4th International Conference on Ecological Engineering for Wastewater Treatment*, Norway.

Bidlingmaier, W. (1983). Das Wessen der Kompostierung von Siedlungsabfällen. In: *Muell Handbuch* (Bd.4, KZ 5305), Kösel, G. , Bilitewski, W., Schburer, H. (Hrsg.), Erich Schmidt Verlag, Berlin.

Esrey, S.A., Gough, J., Rapaport, D., Sawyer, R., Simpson-Hébert, M., Vargas, J. and Winblad, U. (1998). *Ecological Sanitation*. SIDA, Stockholm, Sweden.

Fitschen, I. and Hahn, H.H. (1998). Characterisation of municipal wastewater part of human urine and a preliminary comparision with liquid cattle excretion. *Wat. Sci. Tech.*, **38**(6), 9–16.

Gajurel, D.R., Benn, O., Li, Z., Behrendt, J and Otterpohl, R. (2001). An Appropriate Ecological Sanitation concept for domestic Wastewater Treatment and Reuse: Separation and Composting of Solids and Biological Treatment of Filtrate. *Poster Presentation in the 2nd World Water Congress of International Water Association (IWA), Berlin, 15–19 October, 2001.* In CD-ROM.

Gulyas, H. (2000). Freeze concentration for recovery of potential recyclable from yellow water. In: *Hamburger Berichte zur Siedlungswasserwirtschaft*, GFEU e.V. TU Hamburg-Harburg (In German).

Henze, M., Somolyody, L., Schilling, W. and Tyson, J (1997). Sustainable sanitation. *Wat. Sci. Tech.*, **35**(9).

Hellström, D. and Johansson, E. (1999). Swedish experiences with urine separating systems. *Wasser & Boden* **51**(11), 21–25, Blackwell Wissenschafts-Verlag, Berlin (In German).

Larsen, T.A. and Gujer, W. (1996). Separate management of anthropogenic nutrient solution (Human urine). *Wat. Sci. Tech.*, **34**(3–4), 87–94.

Larsen, T.A. and Udert, K.M. (1999). Urine separation – a way of closing the nutrient cycles. *Wasser & Boden* **51**(11), 21–25, Blackwell Wissenschafts-Verlag, Berlin (In German).

Jönsson, H., Vinnerås, B., Hoeglund, C. and Stenstroem, T.-A. (1999). Source separation of urine. *Wasser & Boden* **51**(11), 21–25, Blackwell Wissenschafts-Verlag, Berlin (In German).

Metcalf and Eddy (1991). *Wastewater Engineering: Treatment Disposal Reuse*. McGraw-Hill, Inc.

Otterpohl, R., Grottker, M. and Lange, J. (1997). Sustainable water and waste management in urban areas. *Wat. Sci. Tech.*, **35**(9), 121–133.

Otterpohl, R. (2001). Design of highly efficient source control sanitation and practical experiences. In: *Decentralised Sanitation and Reuse* (Lens, P.; Zeemann, G. and G. Lettinga, Eds). IWA Publ.

Otterpohl, R., Oldenburg, M. and Zimmermann, J. (1999a). Integrated Wastewater Disposal for Rural Settlements. *Wasser & Boden*, **51**/11,1999, 10–13, Blackwell Wissenschafts-Verlag, Berlin (In German).

Otterpohl, R., Albold, A. and Olderburg, M. (1999b). Source Control in Urban Sanitation and Waste Management: 10 Systems with Reuse of Resource. *Wat. Sci. Tech.*, **39**(5), 153–160.

Otterpohl, R., Bastian, A., Londong, J. and Oldenburg, M. (2001a). The Lambrtsmühle case: Ecologically and economically efficient decentral wastewater and resources mangement for rural and peri-urban area. *Paper Presentation in the 2nd World Water Congress of International Water Association (IWA), Berlin, 15–19 October, 2001.* In CD-ROM.

Wolgast, M. (1993). *WM-Ekologen ab*, PO Box 11162, S-10061 Stockholm, Sweden.

Zifu Li, Deepak Raj Gajurel and Ralf Otterpohl (2001). Appropriate Technologies for Greywater Recycling in Urban Areas. *Poster Presentation in the 2nd World Water Congress of International Water Association (IWA)*, Berlin, 15–19 October, 2001. In CD-ROM.

Nitrification and autotrophic denitrification of source-separated urine

K.M. Udert, C. Fux, M. Münster, T.A. Larsen, H. Siegrist and W. Gujer

Swiss Federal Institute for Aquatic Science and Technology (EAWAG) and Swiss Federal Institute of Technology (ETH), 8600 Dübendorf, Switzerland (E-mail: *larsen@eawag.ch*)

Abstract In laboratory experiments, source-separated urine was stabilised with nitrification and denitrified via nitritation and anaerobic ammonium oxidation. The highest total ammonia concentration in the influent was 7,300 gN/m^3, the maximum pH 9.2. In a moving bed biofilm reactor (MBBR) with Kaldnes® biofilm carriers, we stabilised urine as a 1:1 ammonium nitrate solution. The maximum nitrification rate was 380 gN/m^3/d corresponding to 1.7 gN/m$^2_{biofilm}$/d. Nitrite ammonium solutions were produced in a continuous flow stirred tank reactor (CSTR) with 4.8 days sludge retention time (SRT) at 30°C and in a sequencing batch reactor (SBR) with more than 30 days SRT. Nitrate build-up was negligible in both reactors. Nitritation rates were 780 gN/m^3/d in the CSTR and 280 gN/m^3/d in the SBR, respectively. However, shortening the cycles would increase nitritation in the SBR. High concentrations of nitrous acid, salts, and presumably hydroxylamine suppressed nitrite oxidation in the nitritation reactors. In all three nitrification reactors, maximally 50% of the influent total ammonia was oxidised without pH control. None of the common inhibition or limitation approaches could explain why ammonia oxidation always stopped at pH values around 6. In a batch experiment, we showed that source-separated urine can be denitrified autotrophically by anammox bacteria.

Keywords Anammox; biofilm carrier; nitrification; nitritation; urine separation

Introduction

Separate treatment of urine is a promising alternative to the present urban wastewater systems, because it may better conserve our natural resources (Larsen and Gujer, 1997; Larsen *et al.*, 2001). However, source-separated urine is a highly concentrated and unstable solution. After collection in separating toilets (NoMix toilets) or waterless urinals, urine is usually stored for several days or even weeks. During storage, bacterial urease hydrolyses urea to ammonia and bicarbonate, causing a pH increase. The high pH triggers precipitation of calcium phosphate, struvite, and calcite. As a result, 90% of total nitrogen is present as ammonia or ammonium, the pH is close to 9, and at least 30% of phosphorus is precipitated (Udert *et al.*, 2003a).

After storage, urine contains a large amount of ammonia NH$_3$, which can volatilise in case the urine solution is agitated during transport or application as fertiliser (Hellström and Johansson, 1999). In general, there are two objectives for nitrogen treatment of source-separated urine: either stabilise and recover nitrogen for later use (e.g. as fertiliser), or remove nitrogen as dinitrogen gas. Whenever possible, the first should be favoured, because recovery and conservation of resources is a primary goal of urine separation. However, if efficient reuse is not possible, nitrogen removal from urine may be a good method for saving energy and resources (Maurer and Larsen, 2003).

In this study, we investigated whether biological nitrification can be used to stabilise nitrogen in urine. Nitrification not only lowers the pH but also converts ammonia to nitrite and nitrate. Both effects conserve nitrogen in the solution. Our aim was to produce an ammonium nitrate solution, which may be used as fertiliser in terms of nitrogen composition. Ammonium nitrate is the main fertiliser in Europe (UNIDO, 1998). Additionally, we examined nitritation and autotrophic denitrification as methods for nitrogen removal.

Nitrification and denitrification of urine

Stable biological nitrate production requires that nitrite oxidation is as fast as ammonia oxidation. Changes in environmental conditions affect ammonia and nitrite oxidisers differently and may destabilise the overall process. The most important factors that influence nitrification are

- oxygen (nitrite oxidisers have a lower affinity to oxygen than ammonia oxidisers)
- temperature (the growth rate of nitrite oxidisers increases more slowly with temperature than the growth rate of ammonia oxidisers)
- substrate inhibition and limitation (ammonia NH_3 and nitrous acid HNO_2 are the substrates for ammonia and nitrite oxidisers, respectively)
- product inhibition (ammonia oxidisers are strongly inhibited by nitrous acid, though nitrate inhibition of nitrite oxidisers is usually negligible)
- inhibition by intermediates (nitrite oxidisers are inhibited by hydroxylamine NH_2OH, which is an intermediate of ammonia oxidation)
- growth limitation by inorganic carbon or phosphate
- alkalinity (high alkalinity prevents strong pH changes)

pH is a key parameter. It determines the acid-base equilibria of ammonia, nitrite, and hydroxylamine. Furthermore, the proton concentration influences the metabolism of nitrite and ammonia oxidisers. Since the factors that control nitrification are still not fully understood, the proposed kinetic constants and threshold values vary widely. Comprehensive overviews and discussions are given by Nowak (1996) and Dombrowski (1991). Recently, several metabolic pathways of ammonia oxidation and denitrification have been detected (Schmidt et al., 2002). The new knowledge may enable us to better control nitrification processes.

Nitrification of source-separated urine has already been investigated by Johansson and Hellström (1999). They treated stored urine with a total nitrogen concentration of 4.6 ± 0.4 kg/m^3. Without pH control 50% of the influent total ammonia was oxidised to nitrite and nitrate. Complete nitrification to nitrate was not achieved. With $NaHCO_3$ addition, ammonia oxidation could be increased.

In the last years, several new denitrification processes have been investigated and developed (Verstraete and Philips, 1998). One promising process is the combination of partial nitrification with anaerobic ammonium oxidation by anammox bacteria (Strous et al., 1998). Since anammox bacteria are autotrophic, no organic substrate is necessary. Additionally, sludge production is very small.

Materials and methods

Urine solutions

We used stored urine from two collection systems. After complete urea hydrolysis, total ammonia concentrations were $1,800 \pm 400$ gN/m^3 and $3,800 \pm 100$ gN/m^3, respectively. The pH values varied between 8.9 and 9.1. The ratio of total COD to total ammonia was 0.88 ± 0.14 gCOD/gN. Phosphate precipitated as struvite and hydroxyapatite during storage. The concentration of dissolved phosphate was at least 80 gP/m^3 after precipitation (Udert et al., 2003a). Diluted urine solutions were prepared with distilled water. To increase the total ammonia concentration to values up to 7,300 gN/m^3, we spiked the urine with a 3 mol l^{-1} ammonia solution (1.5 mol l^{-1} NH_4HCO_3 and 1.5 mol l^{-1} NH_3).

Moving bed biofilm reactor (MBBR) for nitrate production

To produce an ammonium nitrate solution, we used a MBBR with Kaldnes® biofilm carriers (Kaldnes Miljøteknologi AS, Tønsberg, Norway). The Kaldnes® biofilm carriers were polyethylene tubes with a diameter of 10 mm and a length of 8 mm. The volumetric surface available for bacterial growth was assumed to be 460 m^2/m^3 $_{filled\ reactor}$ (Maurer *et al.*, 2001). The filling ratio of the reactor (total volume 2.8 l.) was 50%. The biofilm carriers had already been used in a nitritation reactor (Fux *et al.*, 2003) and were stored for 9 months at 4°C. The temperature in the reactor was 25.3 ± 0.5°C. The oxygen concentration was kept between 3 and 5.2 gO$_2$/m^3. We continuously logged pH and oxygen concentrations. Respiration rates were calculated online. Before sampling started (day 0, see Figure 1), the reactor had been running for eleven days with pH control at 7.9 (1 mol l^{-1} HCl). From day 0 to day 4, the pH was controlled to values between 7.0 and 7.8, until the nitrification was strong enough to decrease the pH below 7.0. Thereafter, pH was only controlled, when a pH increase had occurred due to an aeration failure (day 16).

Continuous flow stirred tank reactor (CSTR) for nitritation

We inoculated the CSTR with activated sludge from another nitritation experiment (Fux *et al.*, 2002). Before feeding urine, digester supernatant with increased ammonia concentrations was used as influent (total ammonia 4,500 gN/m^3). The reactor had a volume of 2.8 l. pH was not controlled. Temperature was fixed to 30.0 ± 0.4°C. The oxygen concentration was kept between 2.5 and 4 gO$_2$/m^3. pH and oxygen were continuously monitored and logged and respiration rates were calculated online.

Sequencing batch reactor (SBR) for nitritation

The reactor volume was 7.5 l., the total cycle duration 2 days. For each cycle, 30 min. were used for sedimentation, 7 min. for drain and max. 3 min. for filling. Half of the reactor volume was exchanged, which resulted in a hydraulic retention time HRT of 4 d. Except for sedimentation and drain, the reactor was aerobic with oxygen concentrations between 2 and 4.5 gO$_2$/m^3. Neither pH nor temperature were controlled. The temperature was 24.5 ± 0.5°C. pH values dropped from maximally 8.8 to a minimum of 6.

Batch reactor for anaerobic ammonium oxidation

To investigate autotrophic denitrification we used anammox sludge from a pilot plant, which was running with digester supernatant (Fux *et al.*, 2002). The reactor had a volume of 2.2 l. The temperature was kept constant at 30.0 ± 0.1°C and the pH was controlled at 8.01 ± 0.02 by addition of 1 mol l^{-1} HCl. Before adding the substrate, the sludge was sparged with nitrogen gas for half an hour to establish anaerobic conditions.

Analytical methods

Due to the high concentrations in source-separated urine, samples had to be strongly diluted before measurement. The estimated variation coefficients for all analyses were generally less than 5% except for inorganic carbon (80% at low concentrations). We used the following analytical methods: for COD test tubes from Hach and Dr. Lange; for total ammonia flow injection analysis (reaction with bromocresol purple); for nitrite and nitrate ion chromatography, flow injection analysis (reduction of nitrate to nitrite with cadmium, reaction of nitrite with sulphanilamide), and Dr. Lange test tubes; for phosphate flow injection analysis (ammonium molybdate method). pH, temperature and oxygen were measured with WTW measurement devices. The carbonate alkalinity was determined with titration subtracting the alkalinity of nitrite, ammonia and phosphate species. Total carbonate concentrations were calculated form the carbonate alkalinity. COD measurements were corrected for nitrite concentrations.

Calculations

To calculate the concentration of ammonia and nitrous acid, we used the pK$_a$ values 9.25

and 3.29, respectively (T = 25°C, I = 0 M, Stumm and Morgan, 1996). Ion activities were calculated with the Davies approach (Stumm and Morgan, 1996). We assumed that the ionic strength is proportional to the total ammonia concentration. After urea hydrolysis, undiluted urine has a total ammonia concentration of 8,200 gN/m³ and an ionic strength of 0.55 eq/l (Udert et al., 2003a). Growth rates and necessary SRT of nitrite oxidisers were calculated with the kinetic model of Hellinga et al. (1999). This model includes limitation and inhibition by nitrous acid and limitation by oxygen. Since decay processes are not considered, we supplemented the kinetics with a decay coefficient of $0.05 \cdot \mu_{max}$. The kinetic data from Hellinga et al. (1999) were also used for calculating the inhibition of ammonia oxidation by nitrous acid.

Nitrate production in the MBBR

In this experiment, we investigated whether bacterial nitrification is suitable for converting stored urine into a 1:1 ammonium nitrate solution. We aimed at a nitrification rate of 400 gN/m³/d. In order to acclimatise the bacteria to highly concentrated urine, we started with diluted urine (total ammonia 1,300 gN/m³) and slowly increased the inflow rate until we nearly reached the maximum total ammonia load (Figure 1). Then, we gradually raised the influent urine concentration and concomitantly decreased the inflow rate. The final total ammonia load was 750 ± 50 gN/m³/d at an influent concentration of 7,100 ± 100 gN/m³. This is 87% of the expected total ammonia concentration in undiluted stored urine (8,200 gN/m³, Udert et al., 2003b). The pH in the influent was 9.2. On day 73, the reactor was kept at steady state until the end of the experiment (day 79). 11% of the inflow was lost due to evaporation during periods with high aeration (starting on day 20). Nitrogen loads in the influent and effluent differed only by 4%, which is in the range of measurement variations. Thus, nitrogen loss by denitrification was negligible.

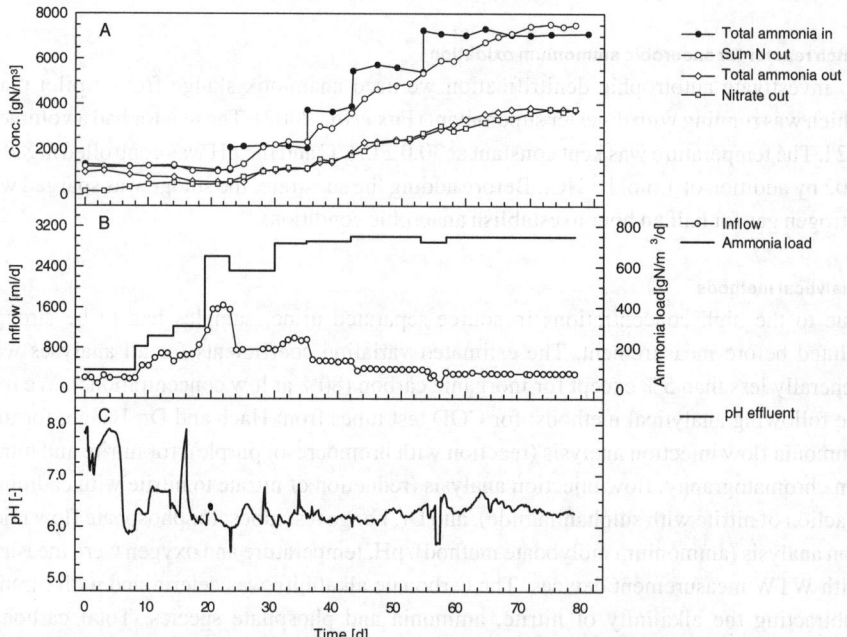

Figure 1 Measurements in the MBBR. A: nitrogen compounds in the influent and effluent. Sum N is the sum of nitrite, nitrate and total ammonia. B: inflow rate and total ammonia load in the influent. C: pH in the outflow. Day 0 marks the beginning of regular sampling. After day 0, nitrite in the effluent was always below 1% of the total measured nitrogen compounds. T = 25.3 ± 0.5°C. No pH control after day 4 except for a pH correction on day 16 (aeration failure)

Nitrification capacity

About 50% of the total ammonia in the influent was oxidised to nitrate. This fraction was reached on day 23 and maintained for the rest of the experiment. Nitrite in the effluent was less than 1% of the influent total ammonia. The maximum nitrification rate was 380 ± 30 gN/m^3/d (day 42 to 79), which corresponds to a specific nitrification rate per biofilm surface of 1.7 ± 0.1 gN/m^2/d. Rusten *et al.* (1995) reported similar maximum nitrification rates (1.4 to 1.6 gN/m^2/d) in a wastewater treatment plant with post-denitrification (8.0–15.6°C). Rostron *et al.* (2001) reached 4.0 gN/m^2/d at 25°C in a laboratory reactor with a synthetic influent (500 gNH$_4$-N/m^3/d). Consequently, higher nitrification rates may also be possible in the treatment of source-separated urine.

Limitation and inhibition effects

None of the common limitation and inhibition effects can explain why only 50% of total ammonia was oxidised. Usually, MBBRs are diffusion-limited by either oxygen or ammonia (Hem *et al.*, 1994), but in this reactor nitrification rates did not correlate with total ammonia or oxygen concentrations at all. Inhibition by heterotrophic growth (Hem *et al.*, 1994) is unlikely, since nitrification rates were not sensitive to changes in the influent organic substrate load (data not shown). As we could show in a SBR experiment (see nitritation in the SBR), limitation by inorganic carbon is not the reason why only 50% of total ammonia is oxidised.

Ammonia and phosphate did not limit the process at anytime. While the ammonia concentration rose from 0.34 gN/m^3 to 2.7 gN/m^3 (day 21 to 77), the nitrification rate was unaffected. Phosphate concentrations were persistently high, at least 80 gP/m^3.

However, another unknown effect related to the pH value seems to impede the ammonia oxidisers. In all experiments (see nitritation in the CSTR and SBR), ammonia oxidation did not occur at pH values far below 6, which corresponds to experiences with municipal wastewater treatment (Painter, 1986). Since ammonia oxidation itself produces protons, the alkalinity of the solution determines the extent of ammonia oxidation.

Finally, ammonia oxidisers are inhibited by their substrate ammonia (Dombrowski, 1991; Nowak, 1996). Ammonia inhibition was negligible when nitrification was stable, but it had to be prevented by acid addition during the initial phase of the experiment.

Long-term application

Whereas nitrification of source-separated urine seems to be stable when inflow and influent concentrations are varying slightly, high pH fluctuations can cause a breakdown of the process. Nitrification stopped when the pH decreased after a failure of the inflow pump (day 23 and 55). However, when the pump was switched on again, nitrification recovered immediately. More severe problems are caused by pH increases (see nitritation in the CSTR). High ammonia concentrations persisting over a long time period strongly affect ammonia oxidisers. Therefore, stable nitrate production requires that the influent load of total ammonia and the pH in the MBBR vary only in a small range. This can be achieved by controlling the pH via the inflow rate.

Nitritation in the CSTR

In the CSTR, we produced an ammonium nitrite solution for autotrophic denitrification. At the given temperature (30 ± 0.4°C), ammonia oxidisers generally have a higher maximum specific growth rate than nitrite oxidisers, so that nitrite oxidisers can be selectively washed out (Hellinga *et al.*, 1999). Before we used urine as influent, the reactor had been running with ammonia enriched digester supernatant (total ammonia $4{,}830 \pm 280$ gN/m^3). On day 0 (Figure 2), we started to feed enriched urine (total ammonia $7{,}300 \pm 250$ gN/m^3). The pH in the influent was 9.2. We slowly increased the load of total ammonia from 450 gN/m^3/d to 1,580 gN/m^3/d. From day 21 onward the reactor was in steady state until the end of the experiment (day 35). The steady state concentration of particulate COD was 830 ± 50 gCOD/m^3. However, since the reactor walls were not cleaned, biofilm growth occurred. In steady state, 9% of the influent water evaporated from the reactor. The nitrogen load in the effluent was 3.8% higher than in the influent indicating that denitrification was negligible.

Nitritation performance
At steady state the effluent nitrite to ammonium ratio was 1.0 ± 0.05 and the nitritation rate 790 ± 20 gN/m^3/d. Nitrite to ammonium ratios of 1 are also common for nitritation of digester supernatant (van Dongen *et al.*, 2001). The maximum nitritation rate was somewhat higher than in the nitritation reactors described by van Dongen *et al.* (2001) (influent: 1,180 gN/m^3 total ammonia, nitritation rate 630 gN/m^3/d). During the whole experiment, nitrate was less than 1.5% of the total ammonia in the influent.

Inhibition of nitrite oxidisers
The apparent lack of nitrite oxidisers is largely due to nitrous acid inhibition. Based on the model of Hellinga *et al.* (1999), which includes nitrous acid as sole inhibitor, we calculated the minimum SRT for the growth of nitrite oxidisers (model parameters 0.71 gHNO$_2$-N/m^3 and 3.3 gO$_2$/m^3). The calculated minimum SRT (4.9 d) was equal to the actual SRT at steady state (4.8 ± 0.3 d). However, nitrite oxidisers were probably also inhibited by hydroxylamine (Stüven *et al.*, 1992). Hydroxylamine concentrations are particularly high during nitritation of high-strength ammonia solutions. Stüven *et al.* (1992) also showed that hydroxylamine but not ammonia inhibited the nitrite oxidisers. Furthermore, high salt concentrations reduce the growth rates of nitrite oxidisers and ammonia oxidisers as well (Hunik *et al.*, 1993). Apparently, inhibition of nitrite oxidisers by nitrous acid, salt, and hydroxylamine was sufficient to prevent significant nitrate build-up in the CSTR.

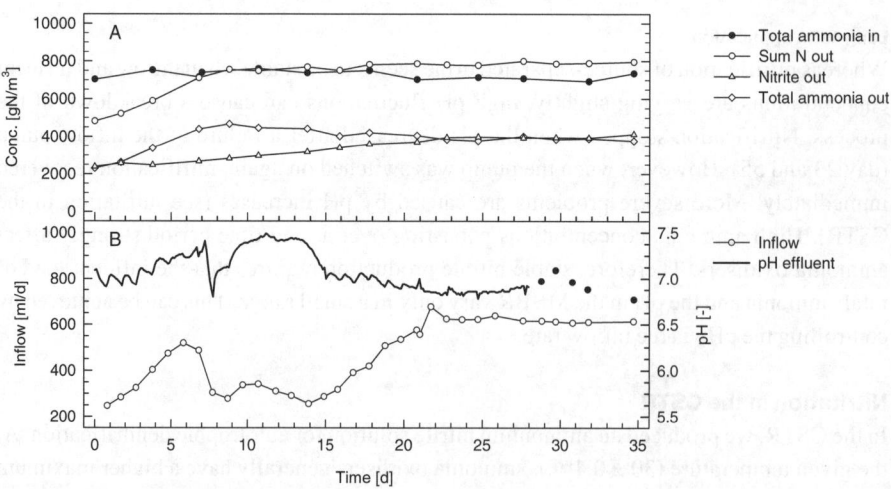

Figure 2 Measurements in the nitritation CSTR. A: nitrogen compounds in the influent and in the reactor. Sum N is the sum of nitrite, nitrate and total ammonia. B: inflow rate and pH in the reactor (after day 29 only occasional pH measurements). New influent on day 7. Period of stable nitritation (steady state phase) starts on day 21. HRT = 4.8 d. T = 30.0 ± 0.4°C

Inhibition of ammonia oxidisers

As we have postulated in the MBBR section, ammonia oxidisers are inhibited by an unknown effect around pH 6. In nitritation reactors, however, ammonia oxidation is also inhibited by nitrous acid. Using the kinetic data compiled by Hellinga et al. (1999) (non-competitive inhibition, $K_i = 0.21$ gN/m^3), we calculated that nitrous acid reduced the growth rate by nearly 80% at steady state. Once again, substrate inhibition of ammonia oxidisers was negligible because of low pH values.

Long-term application

CSTRs are more sensitive to extreme pH values than MBBRs, because bacteria can be washed out. Sudden increases of the inflow rate or lack of oxygen are particularly harmful. On the one hand higher inflow rates decrease the SRT, on the other hand high ammonia concentrations reduce the growth rate of the ammonia oxidisers. Consequently, the bacteria cannot rapidly adapt to the high influent rates, so that they may be washed out. To sum up, ammonia inhibition has a positive feed back and destabilises ammonia oxidation. Corresponding to the MBBR, we suggest controlling the pH via the inflow rate to prevent high pH fluctuations.

Nitritation in the SBR

In the SBR, we produced an ammonium nitrite solution for experiments on chemical nitrite oxidation (Udert et al., 2003c). The reactor was inoculated with activated sludge from a denitrifying wastewater treatment plant. After an initial phase of some weeks, the reactor was steadily running for more than 3.5 years with total ammonia concentrations in the influent between 700 and 1,300 gN/m^3. From January to March 2002, after 3.5 years of continuous operation, we carried out additional experiments with higher influent concentrations (see Table 1). In all experiments, pH values in the reactor ranged from 8.8 at the beginning and 6.0 at the end of the cycle (see Figure 3). According to three mass balances 4 to 12% of nitrogen was lost during a cycle. However, the uncertainty of mass balances was high, because the influent flow rate and reactor volume were estimated based on previous measurements. If nitrogen was lost, it was mainly due to denitrification. Ammonia volatilisation was negligible, as we had determined in a separate experiment. The SRT based on the concentration of suspended solids was between 30 and 100 days, but large amounts of biomass were attached to the reactor walls. The concentration of suspended solids ranged from 2,000 to 3,600 gCOD/m^3 (see Table 1).

Nitrification performance

Nitritation in the SBR proved to be very stable with total ammonia concentrations in the influent as high as 2,240 gN/m^3 (Table 1). The nitrite to ammonium ratio was 1.0 ± 0.1 on average. Nitrate in the effluent accounted for less than 5% of the influent total ammonia. Figure 3 shows that ammonia was oxidised only during a short time of the cycle. During the rest of the cycle, nitrite and total ammonia concentrations did not change significantly. Nitrate build-up occurred similarly: in the beginning when pH values were high, nitrate increased faster than in the idle phase with pH 6.0. The overall nitritation was low, maximally 280 gN/m^3/d (cycle 16.3.02), but during the phase of ammonia oxidation, the nitritation rate reached values of 1,300 gN/m^3/d (cycle 18.2.02).

Inhibition of nitrite oxidisers

Generally, nitrite oxidation was slow and even decreased after ammonia oxidation had stopped (Figure 3). At pH 6, nitrite oxidisers were strongly inhibited by their substrate nitrous acid. According to the kinetic model of Hellinga et al. (1999), nitrous acid inhibition reduced the growth rate by 80%. During ammonia oxidation, however, pH values were higher and nitrous acid did not inhibit strongly. Nevertheless, nitrate build-up was also low. In this phase, the nitrite oxidisers were probably inhibited by hydroxylamine (see nitritation in the CSTR).

Inhibition and limitation of ammonia oxidisers

Independent of the influent concentrations, ammonia oxidation stopped at pH 6 (Table 1). As we have discussed above, none of the common inhibition and limitation approaches can explain our observation. Ammonia limitation can be excluded once again, because the ammonia concentrations at which nitritation stopped differed by a factor 3.7 (Table 1). Nitrous acid inhibition slowed down ammonia oxidation (see nitritation in the CSTR), but model estimations showed that nitrous acid inhibition was insufficient to completely inhibit ammonia oxidation.

Limitation by inorganic carbon has often been mentioned as the reason for bad nitrification performance (e.g. Wett et al., 1998). At the end of the cycle the total carbonate concentration was as low as 3 mmol l^{-1}. Due to the high measurement variations the actual concentrations may have been even lower. We tested, therefore, if lack of carbon dioxide was the reason why ammonia oxidation stopped at pH 6. In a parallel experiment (data not shown) one SBR was aerated with synthetic air (20% O_2, 80% N_2) and another with a mixture of 10% CO_2, 18% O_2 and 72% N_2. The results showed that stabilising the levels of carbon dioxide had no effect on the overall nitritation. The nitrite to ammonium ratio in the effluent was always 1 in both reactors.

In the experiments presented in this paper, low availability of carbon dioxide may have

Table 1 Concentrations in the SBR at various influent concentrations

Date of cycle start			15.12.98	10.02.02	18.02.02	16.03.02
Influent	Total ammonia	[gN/m^3]	720	1,380	1,840	2,240
	pH	[–]	8.9	8.8	8.9	8.8
End of cycle	Total ammonia	[gN/m^3]	410	680	860	1,100
	Nitrite	[gN/m^3]	410	720	750	1,180
	Nitrate	[gN/m^3]	8	50	60	30
	Nitrite/ammonium	[gN/gN]	1.00	1.06	0.87	1.07
End of ammonia oxidation	pH	[–]	6.0	6.0	6.0	6.1
	NH_3	[gN/m^3]	0.18	0.30	0.39	0.64
	HNO_2	[gN/m^3]	0.73	1.08	1.09	1.28
Biomass in reactor	Suspended solids	[gCOD/m^3]	2,000	3,100	3,000	3,400

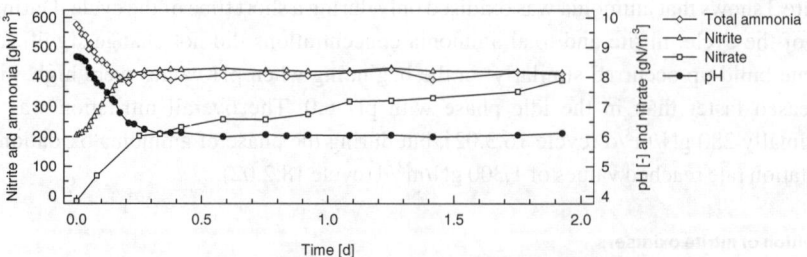

Figure 3 Total ammonia, nitrite, nitrate and pH in the SBR. Total ammonia concentration in the influent 720 gN/m^3. 2-day cycle. 50% of reactor volume exchanged per cycle. HRT = 4 days. SRT ≥ 30 d. T = 24.5 ± 0.5°C

slowed down nitrification at pH values near 6, but it was not the reason why nitrification stopped.

Long-term application

The SBR was running steadily for 3.5 years, but the overall nitrification rates were low. However, the overall nitrification performance could be strongly enhanced by shortening the idle phase of the process. Concomitantly, the effluent nitrate concentration would slightly increase because the cycle period with nitrous acid inhibition was shortened. A further increase of the total ammonia concentration in the influent seems to be possible. However, the volume of added influent solution must be reduced to prevent substrate inhibition of ammonia oxidisers.

Overview of nitrification performance

In Table 2 we compiled the data from the nitrification reactors. MBBR and CSTR values are given for steady state. SBR values show the reactor performance at the highest total ammonia concentration in the influent.

Elimination of organic substances

The concentration of organic substances decreased strongly in all nitrification reactors. On average, 82 ± 5% of the dissolved COD was eliminated. In the influent, about 88 ± 5% of the total COD was dissolved. The ratio of COD to total ammonia in stored urine was on average 0.88 ± 0.14 gCOD/gN.

Anaerobic ammonium oxidation

We conducted batch experiments with anammox sludge to determine whether anaerobic ammonium oxidation is suited for nitrogen removal from partially nitrified urine. Firstly, we carried out a reference experiment, in which we added synthetic $NaNO_2$ to the anaerobic batch reactor. The initial nitrite concentration was 50 gN/m^3. The sludge had excess ammonium, so that additional spiking was not necessary. Anaerobic ammonium oxidation took place immediately (Figure 4). After all the nitrite had been consumed, partially oxidised urine from the CSTR reactor (26.5 ml; total ammonia 4,100 gN/m^3; nitrite 3,400 gN/m^3) was added to the batch reactor. The initial concentration of total COD at the very beginning was 11,400 $gCOD/m^3$ and did not change in the course of the experiments.

Table 2 Data on nitrification performance in the steady state phase (MBBR and CSTR) or at maximum loading (SBR), respectively. SD: standard deviation. *Overall ammonia oxidation in one cycle, **maximum ammonia oxidation

Reactor		MBBR	SD	CSTR	SD	SBR	SD
Influent ammonia	gN/m^3	7,100	100	7,300	250	2,240	100
Influent ammonia load	$gN/m^3/d$	750	50	1,580	30	560	30
SRT	d	–		4.8	0.3	>30	
Reactor temperature	°C	25.3	0.5	30.0	0.4	24.5	0.5
Maximum ammonia oxidation rate	$gN/m^3/d$	380	30	790	20	280*	20
						1,300**	100
	$gN/m^2_{biofilm}/d$	1.7	0.1	–		–	
Suspended solids	$gCOD/m^3$	–		830	50	3,640	100
Effluent total ammonia	gN/m^3	3,750	20	3,880	110	1,100	50
Effluent nitrite	gN/m^3	75	2	3,930	130	1,180	50
Effluent nitrate	gN/m^3	3,730	60	60	30	30	2
Effluent pH	–	6.3	0.05	6.9	0.05	6.1	0.05

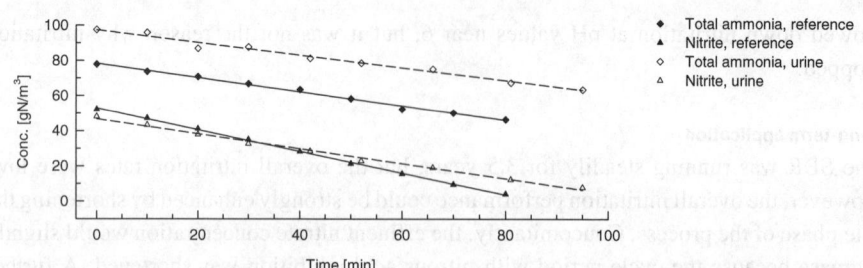

Figure 4 Total ammonia and nitrite degradation caused by anaerobic ammonium oxidation. Comparison of a reference experiment with synthetic NaNO$_2$ and partially nitrified urine as nitrogen source. T = 30°C

The total nitrogen elimination rate was 1,300 gN/m^3/d in the reference experiment and 1,000 gN/m^3/d with partially nitrified urine. In both tests, nitrate was produced at a rate of 100 gN/m^3/d. The ratios of total ammonia elimination to nitrite elimination to nitrate production were 1:1.53 ± 0.05:0.25 ± 0.02 for the reference and 1:1.18 ± 0.07:0.19 ± 0.03 for urine, respectively. Although these ratios vary, they are in the same range as literature values. Van de Graff et al. (1996), for example, reported the ratios 1:1.31 ± 0.06:0.22 ± 0.02.

According to these results, nitrogen removal from source-separated urine is possible with anaerobic ammonium oxidation. However, the salt concentration in the reactors was lower than in source-separated urine. In long-term application, the salt concentrations in the reactors will increase and may diminish the activity of the anammox bacteria. Furthermore, to completely eliminate ammonium, the alkalinity in the preceding nitrification reactor must be increased, so that an ammonium nitrite ratio of 1:1.3 will be obtained.

Conclusions

Source-separated urine can be stabilised with biological nitrification. In a moving bed biofilm reactor (MBBR), enriched urine (total ammonia 7,100 gN/m^3) was oxidised to a 1:1 ammonium nitrate solution. At steady state, the rates were 380 gN/m^3/d and 1.7 gN/m$^2_{biofilm}$/d, respectively (25.3°C, filling ratio 50%). In terms of nitrogen composition, the effluent is a suitable fertiliser.

Furthermore, we showed that nitritation combined with autotrophic denitrification is suited to remove nitrogen from urine. In this study, we particularly focussed on the first step of the process, nitritation. Ammonium nitrite solutions were produced in two reactors, a continuous flow stirred tank reactor (CSTR) and a sequencing batch reactor (SBR). In either reactor, maximally 50% of total ammonia could be oxidised without pH control. The nitritation rate in the CSTR was 790 gN/m^3/d at steady state with a total ammonia concentration in the influent of 7,300 gN/m^3 (T = 30°C, SRT = 4.8 d). The maximum nitritation rate in the SBR was 1,300 gN/m^3/d, but due to long idle phases without ammonia oxidation, the overall nitritation rate was only 280 gN/m^3/d at maximum (influent total ammonia 2,200 gN/m^3, T = 24.5°C, SRT > 30 d). However, shortening the cycles could substantially increase the overall nitritation rates.

In a batch experiment with anammox sludge, we showed that source-separated urine can be denitrified with anaerobic ammonium oxidation. However, further experiments are necessary to investigate the long-term performance. For complete ammonium removal, nitritation must be slightly enhanced by increasing the alkalinity.

Owing to the high concentrations in source-separated urine, inhibition strongly affects nitrification. Ammonia and an unknown effect around pH 6 are the governing inhibitors of ammonium oxidation. The latter determines that maximally 50% of total ammonia can be oxidised without pH control. Nitrous acid and high salt concentrations contribute to the inhibition of ammonia oxidisers. Limitation by inorganic carbon may slow down ammonia oxidation but is not responsible for the nitritation stop at pH values around 6. Limitation by ammonia and phosphate is negligible.

Nitrite oxidisers are mostly affected by high concentrations of nitrous acid and presumably hydroxylamine. Both inhibitors suppress nitrite oxidation in nitritation reactors. Furthermore, high salt concentrations seem to reduce the growth rate as well.

For the long-term operation of MBBRs and CSTRs, we suggest controlling the pH via the inflow to prevent high pH fluctuations. In SBRs, the exchanged volume per cycle must be so low that the initial ammonia concentration does not prevent ammonia oxidation.

Acknowledgements

We thank our laboratory staff for their great support and endurance during the experimental work. Special thanks go to Sarah Thuring, Norbert Baumgärtel and Martin Biebow, who have contributed to this study as part of their diploma thesis or practical training.

References

Dombrowski, T. (1991). *Kinetik der Nitrifikation und Reaktionstechnik der Stickstoffeliminierung aus hochbelasteten Abwässern. (Nitrification. Kinetics and Techniques of Nitrogen Elimination from High-Strength Wastewaters)*. VDI Verlag, Düsseldorf, Germany. *In German*.

Fux, C., Böhler, M., Huber, P., Brunner, I. and Siegrist, H. (2002). Biological treatment of ammonium-rich wastewater by partial nitritation and subsequent anaerobic ammonium oxidation (anammox) in a pilot plant. *Journal of Biotechnology*, **99**(3), 295–306.

Fux, C., Egli, K., Böhler, M., Monti, A., Huang, D., Gujer, W. and Siegrist, H. (2003). Stable nitritation as an essential prerequisite for anaerobic ammonium oxidation (anammox). In preparation for *Environmental Science and Technology*.

Hellinga, C., van Loosdrecht, M.C.M. and Heijnen, J.J. (1999). Model based design of a novel process for nitrogen removal from concentrated flows. *Mathematical and Computer Modelling of Dynamical Systems*, **5**(4), 351–371.

Hellström, D. and Johansson, E. (1999). Swedish experiences with urine separating systems. *Wasser & Boden*, **51**(11), 26–29.

Hem, L.J., Rusten, B. and Ødegaard, H. (1994). Nitrification in a moving bed biofilm reactor. *Water Research*, **28**(6), 1425–1433.

Hunik, J.H., Meijer, H.J.G. and Tramper, J. (1993). Kinetics of Nitrobacter agilis at extreme substrate, product and salt concentrations. *Applied Microbiology and Biotechnology*, **40**(2–3), 442–448.

Johansson, E. and Hellström, D. (1999). Nitrification in combination with drying as a method for treatment and volume reduction of stored human urine. In: Johansson E. (1999). *Urine Separating Wastewater Systems: Design Experiences and Nitrogen Conservation*. Licentiate Thesis, Lulea University of Technology, Lulea, Sweden.

Larsen, T.A. and Gujer, W. (1997). The concept of sustainable urban water management. *Water Science and Technology*, **35**(9), 3–10.

Larsen, T.A., Peters, I., Alder, A., Eggen, R., Maurer, M. and Muncke, J. (2001). Re-engineering the toilet for sustainable wastewater management. *Environmental Science and Technology*, **35**(9), 193A–197A.

Maurer, M., Fux, C., Graff, M. and Siegrist, H. (2001). Moving-bed biological treatment (MBBT) of municipal wastewater: denitrification. *Water Science and Technology*, **43**(11), 337–344.

Maurer, M., Schwegler, P. and Larsen, T.A. (2003). Nutrients in urine: energetical aspects of removal and recovery. *Water Science and Technology*, **48**(1) 37–46 (this issue).

Nowak, O. (1996). *Nitrifikation im Belebungsverfahren bei massgebendem Industrieabwassereinfluss. (Nitrification in the activated sludge process under considerable influence of industrial wastewater)*. Wiener Mitteilungen: Wasser – Abwasser – Gewässer, Volume 135, Technical University of Vienna, Vienna, Austria. *In German*.

Painter, H.A. (1986). Nitrification in the treatment of sewage and waste-waters. In: Prosser, J.I. (ed.) *Nitrification*. Special Publications of the Society for General Microbiology, vol. 20, IRL Press Limited, Oxford, UK.

Rostron, W.M., Stuckey, D.C. and Young, A.A. (2001). Nitrification of high strength ammonia wastewaters: comparative study of immobilisation media. *Water Research*, **35**(5), 1169–1178.

Rusten, B., Hem, L.J. and Ødegaard, H. (1995). Nitrification of municipal wastewater in moving-bed biofilm reactors. *Water Environment Research*, **67**(1), 75–86.

Schmidt, I., Sliekers, O., Schmid, M., Cirpus, I., Strous, M., Bock, E., Kuenen, J.G. and Jetten, M.S.M. (2002). Aerobic and anaerobic ammonia oxidizing bacteria – competitors or natural partners? *FEMS Microbiology Ecology*, **39**(3), 175–181.

Strous, M., Heijnen, J.J., Kuenen, J.G. and Jetten, M.S.M (1998). The sequencing batch rector as a powerful tool for the study of slowly growing anaerobic ammonium-oxidizing microorganisms. *Applied Microbiology and Biotechnology*, **50**(5), 589–596.

Stumm, W. and Morgan, J.J. (1996). *Aquatic Chemistry*, 3rd ed., John Wiley & Sons Inc. New York, USA.

Stüven, R., Vollmer, M. and Bock, E. (1992). The impact of organic matter on nitric oxide formation by Nitrosomonas europaea. *Archives of Microbiology*, **158**, 439–443.

Udert, K.M., Larsen, T.A. and Gujer, W. (2003a). Estimating the precipitation potential in urine-collecting systems. Submitted to *Water Research*.

Udert, K.M., Larsen, T.A. and Gujer, W. (2003b). Precipitation dynamics in urine-collecting systems. Submitted to *Water Research*.

Udert, K.M., Larsen, T.A. and Gujer, W. (2003c). Microbially mediated chemical nitrite oxidation of high-strength ammonium-nitrite solutions. *In preparation*.

UNIDO (1998). *Fertilizer manual*, 3rd ed, United Nations Industrial Development Organization, International Fertilizer Development Center, Kluwer Academic Publishers, Dordrecht, The Netherlands.

van de Graaf, A.A., De Bruijn, P., Robertson, L.A., Jetten, M.S.M. and Kuenen, J.G. (1996). Autotrophic growth of anaerobic ammonium-oxidizing micro-organisms in a fluidized bed reactor. *Microbiology-UK*, **142**, 2187–2196.

van Dongen, U., Jetten, M.S.M. and van Loosdrecht, M.C.M. (2001). The SHARON®-Anammox® process for treatment of ammonium rich wastewater. *Water Science and Technology*, **44**(1), 153–160.

Verstraete, W. and Philips, S. (1998). Nitrification-denitrification processes and technologies in new contexts. *Environmental Pollution*, **102**(suppl.), 717–726.

Wett, B., Rostek, R., Rauch, W. and Ingerle, K. (1998). pH-controlled reject-water-treatment. *Water Science and Technology*, **37**(12), 165–172.

Treatment of domestic sewage in a combined UASB/RBC system. Process optimization for irrigation purposes

A. Tawfik*, G. Zeeman**, A. Klapwijk**, W. Sanders**, F. El-Gohary* and G. Lettinga**

* National Research Centre, Water Pollution Control Department, El-tahrir St., Dokki, Giza, Egypt
(E-mail: Tawfik8@yahoo.com; Tawfik8@hoymail.com)
** Wageningen University and Research centre, Agrotechnology and Food Sciences Department, Sub-department of Environmental Technology, P.O. Box 8129, 6700EV, Wageningen, The Netherlands

Abstract A Rotating Biological Contactor (RBC) was fed with raw domestic wastewater or anaerobic effluents. The experiments were conducted at increasing operational temperatures viz. 11, 20 and 30°C to assess the potential increase in removal efficiencies for the different COD fractions (COD_{total}, $COD_{suspended}$, $COD_{colloidal}$ and $COD_{soluble}$), E. coli and in the nitrification rate. The results clearly show that, the RBC at HRT of 2.5 h and OLR of 47 $gCOD/m^2 \cdot d$ provided a very high residual COD_{total} value of 228 mg/l when treating domestic wastewater. This was not the case as compared to the results obtained for the system when operated at the same HRT but at lower OLR's of 27, 20 and 14.5 g $COD/m^2 \cdot d$ with a UASB effluent at operational temperatures of 11, 20 and 30°C respectively. The residual COD_{total} values amounted to 100, 85 and 72 mg/l in the final effluents. Moreover, a high removal of ammonia and low residual values of E. coli was found for the RBC when treating a UASB effluent at operational temperature of 30°C as compared to the situation for treatment of domestic wastewater and UASB effluent at lower temperatures of 11 and 20°C. It can be concluded that an efficient pre-treatment of sewage implies a substantial reduction of OLR applied to the RBC and consequently improves the residual of COD_{total}, ammonia and E. coli in the final effluent. Therefore, this study supports using a combined system UASB/RBC for treatment of domestic wastewater for reuse in irrigation.
Keywords COD; E. coli; nitrification; post-treatment; RBC; sewage; UASB

Introduction

In view of the rapidly growing shortage of renewable water resources in many parts of the world, there is a growing interest in the reuse of effluent from wastewater treatment plants. Besides possible industrial and urban reuse, treated wastewater can be reused in agriculture depending on local effluent requirements. For instance, Egyptian Environmental legal (law 48/82) required the following (BOD = 60 mg/l, COD = 80 mg/l, TSS = 50 mg/l and NO_3-N = 50 mg/l) for wastewater reuse in restricted irrigation. The often applied complete nitrification/denitrification process, which re-circulates potentially useful nitrogen via atmospheric N_2. This process proceeds at the expense of energy and financial input. This approach appears very inefficient from both an energy and resource utilization point of view. The nutrient rich effluent can be directly used for irrigation purposes especially in developing countries, where agriculture often is limited by water and nutrients. The crop choice and irrigation system to be applied will determine the necessity and extent of nutrient and pathogen removal. Several studies have shown that the crop yield is higher with treated wastewater irrigation as compared to freshwater irrigation (Gijzen and Veenstra, 2001). A major dilemma in this context is how to choose the most appropriate treatment technology to achieve optimal reuse of water and nutrients at minimal energy expense. Anaerobic treatment doesn't require oxygen and therefore energy input and will yield energy in the form of CH_4 gas. Anaerobic treatment plants have limited space requirement and therefore can be planned at locations within or just outside the city. Besides the generation of energy, anaerobic treatment of wastewater may have additional benefits for re-use oriented treatment schemes. The anaerobic conditions in the reactor may reduce the level of (toxic) metal compounds in the effluent due to the formation of insoluble metal-sulfides, which precipitate into the anaerobic sludge. As a result the potential danger of accumulating metals into the food chain via agricultural applications of the effluent may be significantly reduced.

Combined with a proper post-treatment, anaerobic treatment provides a sustainable and appropriate method for providing a good quality effluent from domestic sewage, not only for developing countries but also for advanced countries. It is being used successfully in tropical countries (Goncalves et al., 1999), and there are some encouraging results from subtropical and temperate regions (El-Gohary and Naser, 1999).

Effluents from anaerobic reactors cannot be used for restricted irrigation without proper post-treatment. Therefore post-treatment alternatives were researched. Recently, the use of many different biofilm systems for post-treatment have been reported, such as trickling filters (Augusto et al., 2000), fixed media submerged biofilters (Goncalves et al., 1998), granular media biofilters (Goncalves et al., 1999), fluidised bed reactors (Collivignarelli et al., 1990). Each system has its advantages and disadvantages. Based on the results of previous work carried out by Tawfik et al. (2002a) the RBC has been selected for this study. It is a relatively compact system with a sufficiently long biomass retention time, allowing the application of higher volumetric loading rates at low energy cost. Moreover the system is easy to operate at high process stability. Despite their enormous potentials, RBC so far is rarely used especially in developing countries for post-treatment of anaerobically pre-treated wastewater. This paper gives an evaluation of the applicability of a Rotating Biological Contactor (RBC) for post-treatment (polishing) of different anaerobic effluent qualities with emphasis on COD, ammonia and *E. coli* removal. Optimisation of the process conditions in the two subsequent reactors (UASB/RBC) for complying different reuse standards was evaluated.

Materials and methods

Five experiments were carried out by using a RBC system (Figure 1). The system was fed with different wastewaters e.g. raw sewage and effluents of a 6 m^3 UASB pilot-plant at different operational temperatures of 11, 20 and 30°C, previously investigated by Grin et al. (1985) for treatment of domestic wastewater.

UASB effluents
The main characteristics of the domestic wastewater and UASB reactor effluents, i.e. the feed of the RBC system, are given in Table 1a.

Pilot-plant
The schematic diagram of the pilot-plant is shown in Figure 1. The RBC system with a

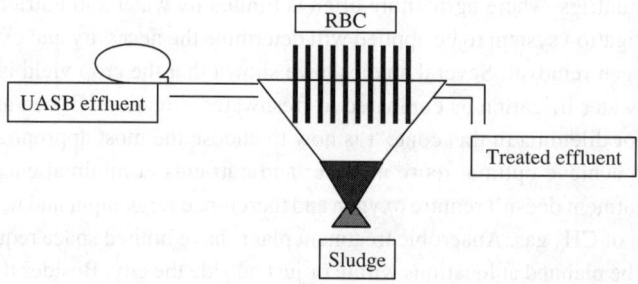

Figure 1 Schematic representation of the RBC system treating different qualities of UASB reactor effluent

Table 1a Mean characteristics of the raw sewage and UASB reactor effluents subsequently used for treatment in a RBC, during the 1st, 2nd, 3rd and 4th experiment

Parameter Samples	Experiment No.	Time of the year Month no.	COD (mg/l)				NH_4-N mg/l	TKN mg/l	E. coli /100ml
			Total	Sus.	Colloidal	Soluble			
Raw sewage	1	4	527 ± 160	287 ± 147	91 ± 58	149 ± 22	49 ± 8	58 ± 6	8.5×10^6
UASB eff. (11°C)	2	6	276 ± 38	117 ± 16	58 ± 34	107 ± 59	52 ± 6	54 ± 6	3.9×10^6
UASB eff. (20°C)	3	4	225 ± 29	77 ± 27	74 ± 28	74 ± 15.3	49 ± 3.5	56 ± 4	1.9×10^6
UASB eff. (30°C)	4	4	164 ± 21	57 ± 11	28 ± 11	82 ± 22.0	59 ± 12	64 ± 11	1.2×10^6

working capacity of 60 l is equipped with 10 polystyrene foam disks with a total effective surface area of 6.5 m² was used. The disk diameter was 0.6 m with a thickness of 0.02 m and the discs were spaced at 0.02 m distance and operated at 5 rpm. The submerged surface amounted to 40%. The disks were mounted on a steel shaft.

Operational conditions

The RBC system has been operated for 16 month under different conditions as shown in Table 1b. The first 30 days of operation were considered as a start-up period while the periods of 20 days were considered as acclimatisation periods to the new experiment.

Sampling and analytical methods

The performance of the reactor was monitored by analysing 48 hr composite samples (twice per week) of the influent and the effluent. The samples were collected in a fridge at 4°C. Dissolved oxygen; pH and temperature were measured regularly *in situ*. The COD was analysed using the micro-method as described by Jirka and Carter (1975). Raw samples were used for COD_{total}, 4.4 μm folded paper filtered (Schleicher and Schuell 595 1/2) samples for $COD_{filtrate}$ and 0.45 μm membrane filtered (Schleicher and Schuell ME 25) samples for dissolved COD ($COD_{soluble}$). The $COD_{suspended}$ and $COD_{colloidal}$ were calculated by the difference between COD_{total} and $COD_{filtered}$, $COD_{filtered}$ and $COD_{soluble}$, respectively. Ammonia, nitrite and nitrate were determined on a stationary auto-analyser (SKALAR SA-9000) total Kjeldahl nitrogen according to the Dutch Standard Normalised Methods (1969) and *E. coli* according to the method described by Havelaar et al. (1988).

Results and discussion
Performance of an RBC at an OLR of 47 gCOD/m².d at a temperature of 17°C
(Reference experiment no. 1)

Treating raw domestic wastewater in an RBC, at an HRT of 2.5 h and at an OLR of 47 g $COD m^{-2}.d^{-1}$, provides low removal efficiencies, resulting in a relatively high residual COD_{total}, $COD_{suspended}$, $COD_{colloidal}$ and $COD_{soluble}$ (Table 2), not complying with any reuse standards. Similar results were achieved for the removal of ammonia and TKN viz. 8

Table 1b Operational conditions of a RBC system

Exp. No.	Type of feeding wastewater	Operational period	T °C	HRT (h)	Flow rate (m³/d)	OLR (gCOD/m².d)
*1st	Raw sewage	4 month	17	2.5	0.576	47
2nd	UASB effl. at T = 11°C.	3 month	13	2.5	0.576	27
3rd	UASB effl. at T = 11°C.	3 month	13	5	0.288	13
4th	UASB effl. at T = 20°C.	3 month	21	2.5	0.576	20
5th	UASB effl. at T = 30°C.	3 month	30	2.5	0.576	14.5

* The 1st experiment with raw sewage was considered as a reference

Table 2 Effluent characteristics of the RBC treating raw sewage and operated at an OLR of 47 gCOD/m^2.d and HRT of 2.5 h

Parameter	COD (mg/l)				NH$_4$	Nitrogen (mg/l)		E. coli /100 ml
	Total	Sus.	Coll.	Soluble		NO$_3$	TKN	
Influent	527 ± 160	287 ± 147	91 ± 58	149 ± 22	49 ± 8		58 ± 6	8.5 × 10^6 ± 5.1 × 10^6
RBC eff.	228 ± 46	84 ± 34	60 ± 41	85 ± 30	45 ± 7	0.1 ± 0.09	49 ± 5	2.6 × 10^6 ± 1.4 × 10^6

and 16% respectively, mainly for assimilation of biomass. A minor nitrate concentration was produced. Also as expected 69% removal was achieved for *E. coli*. This low removal of *E. coli* is likely mainly due to the low removal of COD$_{suspended}$ and COD$_{colloidal}$ at the applied high loading and short HRT.

Performance of the RBC at an OLR of 27 and 13 gCOD/m^2.d and operational temperature of 13°C (2nd and 3rd experiment)

The RBC system treating 11°C-UASB effluent was investigated at an HRT of 2.5 and 5 h and an OLR of 27 and 13 g COD/m^2.d. The results in Table 3 show effluent qualities with respect to different COD fractions, at two different loading rates. Decreasing the HRT from 2.5 to 5 hours created the difference in OLR and resulted in a significant decrease in the effluent total COD value. The colloidal COD value hardly changed as a result of the increased HRT, but was low under both conditions.

The results presented also in Table 3 show that by decreasing the OLR from 27 to 13 g COD/m^2.d and increasing the HRT from 2.5 h to 5 h, the ammonia removal increased from 20 to 30%. However, the nitrification rate was very low, viz. respectively 0.15 and 0.18 g NO$_3$-N/m^2.d at the imposed loading rates. The produced nitrite plus nitrate were compared with the ammonia removal. About 31 and 38% of ammonia was nitrified, respectively.

The *E. coli* count was substantially reduced by 1.4 × 10^5 /100 ml in an RBC by increasing the HRT from 2.5 to 5 h (Table 3). These results demonstrate clearly that the HRT affect seriously the removal of *E. coli* (Polprasert *et al.*, 1983).

Performance of the RBC at an OLR of 20 gCOD/m^2.d and operational temperature of 21°C (4th experiment)

The RBC treating 20°C-UASB effluent was investigated at an HRT of 2.5 h and an OLR of 20 g COD/m^2.d. The results presented in Table 4 show that the system provided residual COD$_{total}$, COD$_{suspended}$ and COD$_{colloidal}$ values of 85, 14 and 17 mg/l respectively. As the major part of the biodegradable COD$_{soluble}$ was already removed in the UASB reactor at a process temperature of 20°C, as expected, little COD$_{soluble}$ was removed in the RBC reactor.

The results presented in Table 4 show that the RBC achieved a removal efficiency of 18 (± 2.5)% for ammonia and 25% (± 3.9) for TKN. The nitrate produced amounted to only 1.5 mg/l (± 0.5). The achieved nitrification efficiency was very low at imposed OLR of

Table 3 Effluent characteristics of the RBC system treating UASB effluent and operated at OLR of 27 and 13 gCOD/m^2.d and HRT of 2.5 and 5 h

Sample	HRT (h)	OLR	COD (mg/l)				Nitrogen (mg/l)				E. coli /100ml
			Total	Sus.	Coll.	Soluble	NH$_4$	NO$_2$	NO$_3$	TKN	
Influent			309 ± 46	85 ± 13	81 ± 14	148 ± 27	45 ± 13			50 ± 13	5.5 × 10^6
RBC eff.	2.5	27	100 ± 7	23 ± 8	14 ± 4	62 ± 8	36 ± 12	1.1 ± 1	1.7 ± 1	38 ± 11	7.9 × 10^5
Influent			291 ± 39	64 ± 22	79 ± 18	148 ± 43	48 ± 12			53 ± 7	7.4 × 10^6
RBC eff.	5	13	76 ± 21	16 ± 14	10 ± 5	50 ± 15	34 ± 17	1.5 ± 1	5.2 ± 5	37 ± 15	6.5 × 10^5

20 gCOD/m².d, because of the inhibitory effect of heterotrophic cells. The growth of heterotrophs decreases the density of nitrifiers in the aerobic part of the biofilm at high OLR. From the results of the study carried out by Boller *et al.* (1987), using RBC's operated at different OLRs, it was concluded that nitrification starts at an organic load of 15 g COD/m².d and was fully developed at about 8 g COD/m².d. According to Boongorsrang *et al.* (1982) for nitrification in a rotating disc contactor, the COD loading rate should even be less than 2.54 g COD/m².d. Based on these observations no nitrification could be expected for the applied conditions. The latter is confirmed by the results, as shown in Table 4.

The removal efficiency of *E. coli* varied from 58 to 97% with an average value of 81% (± 17.6) as shown in Table 4. The *E. coli* count in the final effluent amounted to 2.5×10^5/100 ml which is complying with standards for reuse in restricted irrigation. According to the results of the research work of Tawfik *et al.* (2002b), a 2nd and 3rd stage RBC is undoubtedly needed for complete removal of *E. coli*. In that case unrestricted agricultural reuse is allowed.

Performance of the RBC at an OLR of 14.5 gCOD/m².d and operational temperature of 30°C (5th experiment)

The RBC system treating 30°C-UASB effluent was investigated at an HRT of 2.5 h and an OLR of 14.5g COD/m².d: The COD_{total}, $COD_{suspended}$, $COD_{colloidal}$ and $COD_{soluble}$ removal at these conditions, are presented in Table 5. The results clearly reveal that the system achieved a substantial reduction of COD_{total} resulting in an average effluent concentration of only 72 mg/l. The high removal efficiency of the UASB reactor at the high temperature implemented (30°C) positively affected the performance of the RBC. The system provided an almost complete removal of $COD_{colloidal}$ with only 5 ± 2.7 mg/l remaining in the final effluent (Table 5). This excellent performance towards the removal of colloidal matter can be attributed to entrapment or/and adsorption, followed by hydrolysis and degradation on the biofilm. The system achieved only 29% removal for $COD_{soluble}$, as almost all biodegradable $COD_{soluble}$ was already eliminated in the UASB reactor.

The RBC system exhibited not only excellent COD removal but also a considerably ammonia removal at temperatures of 26°C. The results presented in Table 5 reveal that about 43% ammonia was eliminated at an HRT of 2.5 h or OLR of 14.5 g COD_{total}/m².d and temperature of 30°C. Nitrate and nitrite data reveal that 71% of the ammonia removed occurred through nitrification. The remaining portion of ammonia removed (5 mg/l) probably occurred as a result of adsorption. According to Temmink *et al.* (2001) ammonia

Table 4 Effluent characteristics of the RBC treating UASB effluent (20°C) and operated at OLR of 20 gCOD/m².d and HRT of 2.5 h

Parameter	COD (mg/l)				Nitrogen (mg/l)				E. coli
	Total	Sus.	Coll.	Soluble	NH_4	NO_2	NO_3	TKN	/100ml
Influent	225 ± 29	77 ± 27	74 ± 28	74 ± 15	49 ± 3	–		56 ± 4	$1.9 \times 10^6 \pm 9.9 \times 10^5$
RBC eff.	85 ± 21	14 ± 14	17 ± 23	57 ± 12	40 ± 3	0.7 ± 0.4	1.5	42 ± 4	$2.5 \times 10^5 \pm 2.1 \times 10^5$

Table 5 Effluent characteristics of the RBC treating UASB effluent (30°C) and operated at OLR of 14.5 gCOD/m².d and HRT of 2.5 h

Parameter	COD (mg/l)				Nitrogen (mg/l)			E. coli
	Total	Sus.	Coll.	Sol.	NH_4	NO_2	NO_3	/100 ml
Influent	164 ± 20	50 ± 13	43 ± 11	72 ± 9	42 ± 7			$1.6 \times 10^6 \pm 4.7 \times 10^5$
RBC eff.	72 ± 17	16 ± 16	5 ± 3	51 ± 3	24 ± 7	1.6 ± 0.3	11.5 ± 2	$1.2 \times 10^5 \pm 5.6 \times 10^4$

can be adsorbed in the biofilm. It is known that the biofilm consists mainly of bacterial cells and extracellular polymeric substances (EPS) which all have a negative surface charge. Consequently various cations, of mono-, di-, and trivalent can be bound, including ammonium.

The results presented in Table 5 show that the RBC achieved a substantial removal of *E. coli*, with values ranging from 89 to 96% (average value of 93.4 ± 2.6%). The *E. coli* count in the final effluent amounted to 1.2×10^5/100 ml, which means that according to prevailing WHO-standards (1989) the effluent only can be reused for restricted irrigation purposes. Further treatment is required for reuse in unrestricted irrigation purposes. According to the results of Tawfik et al. (2002b) a three stage RBC at an HRT of 10 h provided a residual value of 9.8×10^2 for *E. coli* which is complying with WHO (1989) standards (1,000 *E. coli*/100 ml) for reuse in unrestricted irrigation.

General discussion

For restricted irrigation in countries suffering from water shortage, like the case in the Middle East and North African region, the COD concentration in the treated wastewater to be reused is an important parameter. Additionally, the crop choice and irrigation system to be applied will determine the necessity and extent of nitrogen and pathogen removal.

The results of the investigation revealed that the RBC system at an HRT of 2.5 h and OLR of 47 gCOD/m^2.d provide a very high residual COD value of 228 mg/l when treating domestic wastewater. This was not the case as compared to the results obtained for the RBC system when operated at the same HRT but at lower OLR's of 27, 20 and 14.5 gCOD/m^2.d with a UASB effluent at operational temperatures of 11, 20 and 30°C respectively. At OLR of 14.5 g COD/m^2.d, the RBC system provided a residual value of 72 mg/l for COD$_{total}$ which didn't significantly increase by increasing the OLR (20 gCOD/m^2.d) applied to the RBC system treating UASB effluent (T = 20°C). Whereas, the residual value of COD$_{total}$ was significantly increased to 100 mg/l by increasing the OLR to 27 gCOD/m^2.d, when treating UASB effluent at low operational temperature of 11°C. Such a COD value in the latter case which is not complying for effluent reuse in restricted irrigation (COD<80 mg/l). In order to improve the effluent quality, the HRT was prolonged to 5 h and OLR was decreased to 13 gCOD/m^2.d. The system achieved a residual value of 76 mg/l for COD$_{total}$. This indicates that the effect of the anaerobic pre-treatment step, therefore, is quite substantial, i.e. the required HRT of the RBC system treating UASB reactor effluent at operational temperatures of 30 and 20°C is 50% less than that of the system treating UASB reactor effluent at a temperature of 11°C. However, the results obtained by El-mitwalli et al. (2000) achieved a high effluent quality in an anaerobic filter followed by anaerobic hybrid at low temperature (13°C). The volume of the RBC as post-treatment can be declined.

The next question is whether or not it would be beneficial to introduce an efficient UASB prior RBC system in case nitrification is needed and high *E. coli* removal efficiency needs to be accomplished. The results obtained revealed that the nitrification rate in the RBC system treating UASB effluent (T = 30 C) at an OLR of 14.5 gCOD/m^2.d is significantly higher than that of the system treating domestic wastewater, anaerobic effluents at operational temperatures of 20 and 11 at OLR of 47, 27 and 20 gCOD/m^2.d, respectively. The nitrification rate amounted to 0.97 gNO$_3$-N/m^2.d, while it was only 0.2 in the latter cases. This can be attributed to a higher COD particulate concentration (COD$_{suspended}$ and COD$_{colloidal}$) in the domestic wastewater and UASB effluent (t = 11, 20°C) by 40 and 76% respectively. This negatively affects the nitrification rate due to its entrapment in the biofilm, diluting the fraction of nitrifying organisms in the biofilm and consuming part of the oxygen which otherwise would have been available for the nitrifiers (Temmink et al., 2001). Despite the RBC system treating UASB effluent (t = 11°C) being operated at HRT of

5 h and OLR of 13 gCOD/m^2.d, the nitrification rate was still lower than that obtained at OLR of 14.5 gCOD/m^2.d and HRT of 2.5 h by 0.8 gNO$_3$-N/m^2.d. This can be due to the higher operational temperatures of the RBC treating UASB effluent (t = 30°C). Boller *et al.* (1987) found that the nitrification rate increases by about 4.5% per 1°C in the RBC system which can also be shown theoretically from biofilm kinetics (Nowak *et al.*, 1998).

It is not surprising to achieve a high *E. coli* removal efficiency of 94% in the RBC system treating UASB effluent (30°C) as compared to that obtained for treatment of domestic sewage, and anaerobic effluents at operational temperatures of 20°C and 11°C which amounted to 69, 86 and 81% respectively. Once again this is due to a higher removal of COD particulate in the preceding UASB reactor at high operational temperature.

Conclusions

The use of a combined UASB-RBC system for treatment of domestic wastewater is suitable to achieve the required effluent quality complying for reuse in irrigation for different climate areas i.e.

- For tropical areas, with average sewage temperatures (22–30°C), anaerobic pre-treatment proceeds efficiently, provided the system is well designed. The results obtained in this study with high quality anaerobic effluent using the RBC-system revealed that an RBC is recommended for achieving a low COD$_{total}$ (72 mg/l) and for partial ammonia and partial *E. coli* removal at an imposed OLR of 14.5 gCOD/m^2.d and HRT of 2.5 h.
- In subtropical areas (average winter time wastewater temperature > 10°C and summer time wastewater temperatures > 20°C with an UASB effluent COD of 200 mg/l, the required HRT for the RBC system should be >2.5 h and the OLR should not exceed 20 g COD$_{total}$/m^2.d in order to achieve a residual COD value of 85 mg/l.
- In moderate climate areas (average sewage temperature can occasionally (a few days) drop down to values as low as 4°C in winter while they can raise to 20°C in the summer, The performance of conventional UASB reactors will be rather poor during wintertime. The results of our investigations with a poor quality anaerobic effluent revealed that, the required HRT for RBC should be prolonged to 5 h and the imposed OLR reduced to values below 13 gCOD/m^2.d in order to achieve a residual effluent COD value of 76 mg/l. The RBC can be operated in different ways, i.e. during summer time with temperatures around 20°C and better UASB effluent quality, the RBC can be operated at higher OLR and lower HRT.

Acknowledgement

The authors would like to express their gratitude to EU for financial support of this research (CORETECH PROJECT ICA3- (T-1999-000/2).

References

Augusto, C., Chernicharo, L. and Nascimento, M.C.P (2000). A new configuration of trickling filter applied to the post-treatment of effluents from UASB reactors. *VI Latin-American Workshop and Seminar on Anaerobic Digestion*, 5–9 November, Brazil.

Boller, M., Eugster, L., Weber, A. and Gujer, W. (1987). *Nitrifikation in nachgeschalteten rotierenden Tauchkorpern*. EAWAG-report.

Boongorsrang, A., Suga, K. and Maeda, Y. (1982). Nitrification of wastewaters containing carbon and inorganic nitrogen by rotating disc contactor. *J. Ferment. Technol.*, **60**, 357–362.

Collivignarelli, C., Urbini, G., Farneti, A., Bassettin, A. and Barbaresi, U. (1990). Anaerobic-aerobic treatment of municipal wastewaters with full-scale up-flow anaerobic sludge blanket and attached biofilm reactors. *Wat. Sci. Tech.*, **22**(1/2), 475–482.

Dutch Standard Normalised Methods (1969). The Netherlands Normalisation Institute, Delft, The Netherlands.

El-Gohary, F. and Naser, F.A. (1999). Cost-effective pre-treatment of wastewater. *Wat. Sci. Tech.*, **39**(5), 97–103.

El-mitwalli, A.T. (2000). Anaerobic treatment of domestic sewage at low temperature. Ph-D thesis, Sub-department of Environmental Technology, Wageningen University, The Netherlands.

Gijzen, H.J. and Veenstra, S. (2001). Duckweed based wastewater treatment for rational resource recovery and reuse. In. Olguin, E. *et al.* (eds) *Environmental Biotechnology and Cleaner Bioprocesses*, Taylor and Francis, in press.

Goncalves, R., Araujo, V. and Chernicharo, C. (1998). Association of a UASB reactor and a submerged aerated biofilter for domestic sewage treatment. *Wat. Sci. Tech.*, **38**(8–9), 189–195.

Goncalves, R., Araujo, V. and Bof, V.S. (1999). Combining up-flow anaerobic sludge blanket (UASB) reactors and submerged aerated biofilters for secondary domestic wastewater treatment. *Wat. Sci. Tech.*, **40**(8), 71–79.

Grin, P.C., Roersma, R.E. and Lettinga, G. (1985). Anaerobic treatment of raw sewage in UASB reactors at temperatures from 9–20°C In: Proceedings of the seminar/workshop anaerobic treatment of sewage, Amherst, 109–124.

Havelaar, A.H and M. During on behalf of a working group (1988). Evaluation of the Anderson Baird – Parker direct plating method for enumerating *Escherichia Coli* in water. *Journal of Applied Bacteriology*, **64**, 89–98.

Jirka, A. and Carter (1975). Micro-semi – automated analysis of surface and wastewaters for chemical oxygen demand. *Analytical Chemistry*, **47**, 1397–1401.

Nowak, O., Schweighofer, P. and Nikolvcic, B. (1998). *Aspekte zweistufiger Verfahren and von biofilmverfahren.*

Polprasert, C., Dissanayake, M.G. and Thanh, N.C. (1983). Bacterial die-off kinetics in waste stabilization ponds. *J. Wat. Pollut. Control. Fed.*, **55**, 285–296.

Tawfik, A., Klapwijk, A., El-Gohary, F. and Lettinga, G. (2002a). Treatment of anaerobically pre-treated domestic sewage by a Rotating Biological Contactor. *Wat. Res.*, **36**(1), 147–155.

Tawfik, A., Klapwijk, A., El-Gohary, F. and Lettinga, G. (2002b). Post-treatment of anaerobically pre-treated domestic sewage in an RBC system. *Wat. Sci. Tech.*

Temmink, H., Klapwijk, A. and de Korte, K.F. (2001). Feasibility of the BIOFIX-process for treatment of municipal wastewater. *Wat. Sci. Tech.*, **43**(1), 241–249.

WHO (1989). Health guidelines for the use of wastewater in agriculture and aquaculture. Technical report series No. 778. Geneva. World Health Organization.

Optimization of phosphorus precipitation from swine manure slurries to enhance recovery

R.T. Burns, L.B. Moody, I. Celen and J.R. Buchanan

The University of Tennessee, Biosystems Engineering and Environmental Science, 2506 E.J. Chapman Drive, Knoxville, Tennessee, 37996-4531 USA
(E-mail: rburns@utk.edu; lbeal@utk.edu; icelen@utk.edu; jbuchan7@utk.edu)

Abstract Laboratory experiments were conducted using magnesium chloride ($MgCl_2 \cdot 6H_2O$, 64% solution) to force the precipitation of phosphorus and reduce the concentration of soluble phosphorus (PO_4^{3-}) in two swine wastes. One of the swine wastes tested contained a high concentration of PO_4^{3-} (initially ~1,000 mg/L), and the other swine waste tested contained a low concentration of PO_4^{3-} (initially ~230 mg/L). The precipitation reactions were performed to determine the required reaction time, pH, magnesium addition rate and seed material for future precipitate recovery work. For the high and low concentration waste, a 10-minute reaction time at a pH of 8.6 was sufficient to remove 98 and 96% of the PO_4^{3-} from solution. A molar ratio of $Mg^{2+}:PO_4^{3-}$ of 1.6:1 was determined to be effective for PO_4^{3-} removal from both the low and high strength wastes. At a molar ratio of 1.6:1, the PO_4^{3-} in the high concentration waste was reduced from 590 to 12 mg/L. In the low concentration waste, the PO_4^{3-} concentration was reduced from 157 to 15 mg/L. Seeding the reaction did not significantly enhance the recovery process.
Keywords Phosphorus; precipitation; struvite; swine

Introduction

The over-application of phosphorus, as manure by animal feeding operations, is a threat to surface water quality. Over application of manures to cropland resulting from the concentration of animal feeding operations has led to a build-up of phosphorus on many farms (Greaves et al., 1999). Relative to crop needs, manure slurries contain higher levels of phosphorus than nitrogen. When manure is applied to meet crop nitrogen needs, phosphorus is over applied. Laboratory studies show that phosphorus content in swine manure can be reduced by recovering a portion of the phosphorus as a crystalline precipitate containing struvite (magnesium ammonium phosphate hexahydrate, $MgNH_4PO_4 \cdot 6H_2O$) (Wrigley et al., 1992; Beal et al., 1999; Nelson et al., 2000; Burns et al., 2001; Kalyuzhnyi et al., 2001). Precipitation of phosphorus prior to land application of manure offers the potential to recover excess phosphorus from animal manures and move it to cropping areas that require phosphorus fertilizer inputs. The cost effective relocation of excess phosphorus would allow existing animal feeding operations to successfully implement phosphorus-based nutrient management plans on their current land base. Proposed USEPA regulations regarding concentrated animal feeding operations must be finalized by December 2002. These regulations will likely limit land application of manure to phosphorus-based rates. Comparisons of nitrogen and phosphorus-based nutrient management plans indicate that some poultry-broiler producers, swine producers, and diary producers may require as much as ten, eight, and four times more land, respectively, if required to shift to a phosphorus-based plan (Burns et al., 1998). Producers who do not have an adequate land base will be faced with transporting manure nutrients off-site. Recovery of phosphorus as precipitated struvite has the potential to substantially reduce transportation costs by isolating the excess phosphorus and converting it into a crystalline form that can be cost-effectively transported to a cropping system that requires phosphorus input.

While investigators have examined phosphorus precipitation in swine wastes on a laboratory scale, little work has been done to develop this process for field scale application (Nelson et al., 2000). Burns et al. (2001) has shown a 90% reduction in soluble phosphorus via struvite precipitation in a 140,000 L swine slurry holding pond under field conditions. The next step in the development of this technology for farm-scale recovery of phosphorus is the optimization of the recovery process and an economic assessment of the cost effectiveness of the method as a manure management option. In Europe and Japan, large municipal sewage-handling facilities have already embraced phosphorus recovery technology (Battistoni et al., 2001; Gaterell et al., 2001; Kumashiro et al., 2001; Liberti et al., 2001; Mitani et al., 2001; Piekema and Giesen, 2001; Ueno and Fujii, 2001). Livestock producers have yet to benefit from these practices because farm-scale applications have not been developed.

Because the limiting ion for struvite formation in animal manure slurries is usually magnesium, manure slurries are typically amended with magnesium to force the precipitation of struvite. Possible magnesium amendments include magnesium hydroxide, magnesium oxide, and magnesium chloride. Miles and Ellis (2001) initially used a 50% magnesium hydroxide slurry and phosphate fertilizer to reduce ammonia through struvite precipitation. However, they incurred insolubility problems with the magnesium hydroxide and changed to the use of magnesium oxide. Beal et al. (1999) used magnesium oxide (MgO) in bench scale reactions during initial struvite experiments. Phosphorus reductions of greater than 90% (1,256 to 105 mg P L^{-1} and 1591 to 81 mg P L^{-1}) were achieved following the addition of MgO. Magnesium oxide had the additional benefit of increasing pH to aid the struvite reaction. However, because of the insolubility of the material, reaction time was long (20 min) and residual MgO existed after the reaction. Further bench scale experiments showed that magnesium chloride ($MgCl_2 \cdot 6H_2O$) was a good source of Mg^{2+} for struvite formation (Burns et al., 2001). Because of its solubility, magnesium chloride was easier to handle and it reduced the reaction time that was required to bring Mg^{2+} into solution. However, because $MgCl_2 \cdot 6H_2O$ is slightly acidic (pH of 5), it does not increase pH as MgO does. In laboratory experiments where magnesium chloride was added and the pH was not adjusted, there was a 76% reduction in soluble phosphorus (572 to 135 mg P L^{-1}). When pH was adjusted to 9, using sodium hydroxide, 91% of the soluble phosphorus was removed (572 to 50 mg P L^{-1}). In this study Burns et al. (2001) added magnesium chloride at a rate calculated to provide a 1.6:1 magnesium:total phosphorus molar ratio.

Recovery of the precipitated material could be enhanced by increasing the particle size of the precipitate. Particle formation is referred to as nucleation or induction. Homogenous nucleation occurs when the phosphorus precipitate is the nucleus. If other suitable nuclei are present, for example sand grains, the nucleation process is heterogeneous (Parsons, 2001). Amending the precipitation process with nuclei is referred to as seeding the reaction. While facilities in Europe and Japan are seeding to increase particle size and enhance recovery (Battistoni et al., 2001; Kumashiro et al., 2001; Mitani et al., 2001; Ueno and Fuji, 2001), experiments to use this technology on animal manures have not yet been performed. Stratful et al. (2001) performed batch experiments using de-ionized water dosed with Mg^{2+}, NH_4^+, and PO_4^{3-}. From these reactions it was determined that struvite particle size increases with reaction time. As reaction time increased from 1 to 180 min, struvite particle size increased from 0.1 to 3.0 mm.

Methods

The wastewater used in this work was obtained from two swine facilities. Supernatant was collected from a pull-plug pit under a swine farrowing unit (referred to henceforth as high concentration waste) and from a holding pond at a feeder pig unit operating as a recycle flush system (referred to henceforth as low concentration waste). The high concentration waste contained ~1,000 mg/L PO_4^{3-} and 51,000 mg/L COD. The low concentration waste contained ~230 mg/L of PO_4^{3-} and 410 mg/L COD. The initial Mg^{2+}:NH_4^+:PO_4^{3-} ratio (Mg^{2+} was measured as soluble magnesium) in the high and low concentration wastes were 0.26:18:1 and 0.58:12:1, respectively. Wastewater was collected in 19-L containers and refrigerated at 4°C. Experiments were carried out over a two-month period following collection of the waste.

All of the precipitation experiments were carried out as batch reactions in 500-mL beakers with a waste volume of 200 or 400 mL. Reactions were mixed using a magnetic stirrer. For each reaction, a representative sample was retrieved from the collected wastewater. All reactions took place at room temperature (approximately 23°C). The magnesium source for the reactions was $MgCl_2 \cdot 6H_2O$ in 64% solution. For all tests where pH was adjusted, NaOH was used to raise the waste pH.

Samples were analyzed for soluble phosphorus, soluble magnesium, ammonia and COD. Soluble phosphorus (dissolved reactive phosphorus, PO_4^{3-}) was analyzed using *QuickChem* Method 12-115-01-1-H (*Lachat Insturments*, Milwaukee, Wisconsin, USA). Soluble magnesium was analyzed with atomic absorption spectrophotometery using a *Perkin Elmer* method for Analysis of Exchangeable Cations (AY-2) (*Perkin Elmer*, Norwalk, Connecticut). Ammonia was analyzed using Standard Method 4500-NH_3 B and C for distillation and titration (Standard Methods for the Examination of Water and Wastewater, 1998). Chemical oxygen demand was measured using Standard Method 5220, a colorimetric, reactor digestion method (Standard Methods for the Examination of Water and Wastewater, 1998).

Reaction time and pH tests

To determine the effect of reaction time and pH on the precipitation reactions, the high and low concentration swine wastes were reacted with $MgCl_2 \cdot 6H_2O$ for up to 40-minutes with and without pH adjustment. Magnesium was added at a Mg^{2+}:PO_4^{3-} rate of 1.6:1. The reaction volume for the experiment was 400 mL. Samples were extracted from the continuously mixed reactions at 5, 10, 20, 30 and 40 minutes; samples were analyzed for soluble phosphorus and soluble magnesium.

Molar ratio tests

Previous research (Burns *et al.*, 2001) used molar ratios based on Mg^{2+}:P (P being total phosphorus) to determine the amount of magnesium to be amended for a swine waste precipitation reaction. We have found however, that swine wastes with similar total phosphorus concentrations can have an order of magnitude difference in soluble phosphorus levels. In this research, Mg^{2+} addition was based on the molar ratios of Mg^{2+}:PO_4^{3-} (PO_4^{3-} being dissolved reactive phosphorus, referred to here as soluble phosphorus). For the molar ratio tests, the pH of the waste was increased to 8.5 using NaOH and the reaction time for the tests was 10 min. Each waste was tested at five different molar Mg^{2+}:PO_4^{3-} ratios ranging from 1.6:1 to 3.5:1. The reaction volume for the experiment was 200 mL. Samples were analyzed for ammonia, soluble phosphorus and soluble magnesium.

Seeding tests

Studies were carried out to determine the effect of adding a seed material to the reaction to enhance precipitation. Three seeding treatments were used; 1) a control, no seed material, 2) a struvite seeded reaction, and 3) a sand seeded reaction. The reaction volume was 200 mL, and the reaction time was 120 min. The reactions were carried out at a pH of 8.5. Samples were collected at 1, 60, and 120 min. Supernatant from the reactions was analyzed for ammonia, soluble phosphorus and soluble magnesium. Precipitate from the reaction was analyzed for size using a dissecting microscope with an ocular micrometer.

Results and discussion

Over the two-month period that the precipitation reactions were carried out, the soluble phosphorus concentration in the collected swine waste decreased with time. As a result, the experiments discussed in this paper were carried out at variable initial PO_4^{3-} concentrations. The complexity of the waste stream makes a singular explanation of this observation difficult. However, we believe that reductions in soluble phosphorus with time were primarily a result of the formation of calcium phosphate in the stored waste.

Reaction time and pH tests

Tests were performed on the two swine wastes to determine optimum precipitation reaction times to be used in the molar ratio experiments. While performing these tests, the effect of increasing wastewater pH in the high and low concentration waste to 8.5 from 7.4 and 7.5, respectively, was also analyzed.

The PO_4^{3-} concentration in the high concentration waste at the time of the reaction and pH experiments was 1,057 mg/L. Test results showed an increase in PO_4^{3-} removal as reaction time increased from 0 to 40 min (Figure 1A and 1B). However, the resulting PO_4^{3-} concentrations were only 3.4 and 1.3% higher after 10 min into the reaction than after 40 min into the reaction. As the pH was increased from 7.4 to 8.5, there was an increase in PO_4^{3-} removal. The resulting PO_4^{3-} concentrations at a pH of 7.4 and 8.5 were 95 and 21 mg/L after 10 min and 59 and 7 mg/L after 40 min, respectively. As a result, the molar ratio tests on the high concentration waste were carried out using a 10 min reaction time at a pH of 8.5.

The initial PO_4^{3-} concentration of the low concentration swine waste at the time of the experiments was 226 mg/L. As with the high waste, the PO_4^{3-} concentration decreased with increasing reaction time (Figure 2A and 2B). The resulting PO_4^{3-} concentrations were 21 and 1.7% higher after 10 min into the reaction than after 40 min into the reaction. Soluble phosphorus concentration in the waste decreased as pH increased from 7.5 to 8.6. At a pH of 7.5, 77.3% of the PO_4^{3-} was removed, and at a pH of 8.6, 95.6% of the PO_4^{3-} was removed. Molar ratio tests were performed at a pH of 8.5 using a 10 min reaction time.

Molar ratio tests

As previously indicated in the Methods section of this paper, both the high and low concentration waste were magnesium deficient and lacked the 1:1:1 molar ratio necessary for struvite precipitation. Additionally, previous research has shown that additional amounts of Mg^{2+} are required to overcome the effects of complexing agents that can bind to

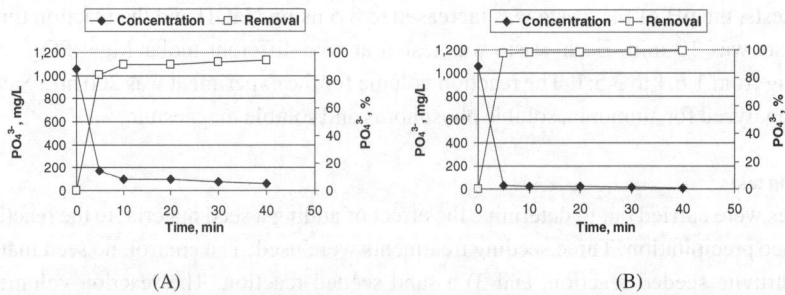

Figure 1 The effect of reaction time and pH on soluble phosphorus removal from the high concentration waste at a $Mg^{2+}:PO_4^{3-}$ molar ratio of 1.6:1. (A) Without pH adjustment and (B) with pH adjusted to 8.5

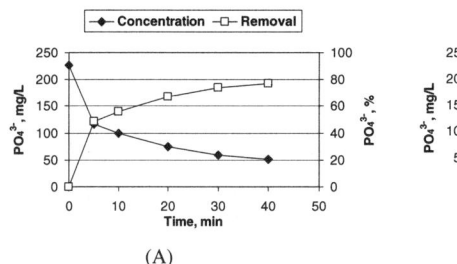
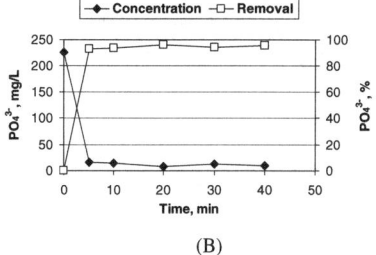

Figure 2 The effect of reaction time and pH on soluble phosphorus removal from low concentration swine waste at a $Mg^{2+}:PO_4^{3-}$ molar ratio of 1.6:1. (A) Without pH adjustment and (B) with pH adjusted to 8.5

magnesium, such as organic material (Schuiling and Andrade, 1999; Nelson *et al.*, 2000; Burns *et al.*, 2001). Both the high and low PO_4^{3-} concentration wastes were reacted using molar ratios of $Mg^{2+}:PO_4^{3-}$ greater than 1:1. For these experiments, the initial PO_4^{3-} concentration of the waste was 590 mg/L for the high concentration waste and 157 mg/L for the low concentration waste. The results indicated that at molar ratios greater than 1:1, high PO_4^{3-} removal rates were obtained (Figure 3A and 3B). In the high concentration waste shown in Figure 3A, a $Mg^{2+}:PO_4^{3-}$ of 1.6:1 was sufficient to achieve 98% PO_4^{3-} removal. Increasing the ratio to 3.2:1, increased removal by 1%. Similarly, the low concentration waste (Figure 3B) was not effected by increasing the ratio from 1.7 to 3.5. The results show that a molar ratio of 1.6:1 ($Mg^{2+}:PO_4^{3-}$) is sufficient to overcome the binding of organic complexing agents to the additional magnesium.

Seeding tests

Seeding studies were performed in an effort to enhance the precipitation and particle size of phosphorus for a precipitate recovery process. In the control, when no seed material was added to the reaction, the particle size did not increase with time. Particle sizes throughout the reaction time ranged from 16 to 30 μm. When struvite was the seed material, the particle sizes did not increase with time. However, there were two ranges of particle sizes present. Throughout the reaction, particles present ranged from 16 to 47 μm and from 63 to 110 μm. The sand seeded reaction did change with time. Initially, for the 1 min and 60 min sample period, there were two particle size ranges present, 16 to 40 μm and 47 to 78 μm. However, by the 120-minute sampling period, only one particle size range was present. These particles ranged from 30 to 47 μm. While there were larger particles in the sand seeded reaction than in the non-seeded control reaction, the larger particles appeared to be independent of the sand particles. Phosphorus removal was enhanced when the sand seed material was used (Figure 4). However, because the removal rates from the other reactions were also high the difference is not notable. Adding seed material to the reaction did appear to

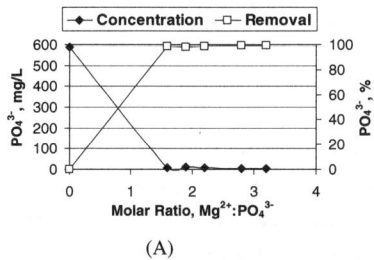
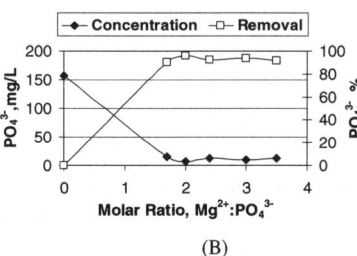

Figure 3 Effect of molar ratio on PO_4^{3-} removal from swine waste. (A) High concentration waste and (B) low concentration waste

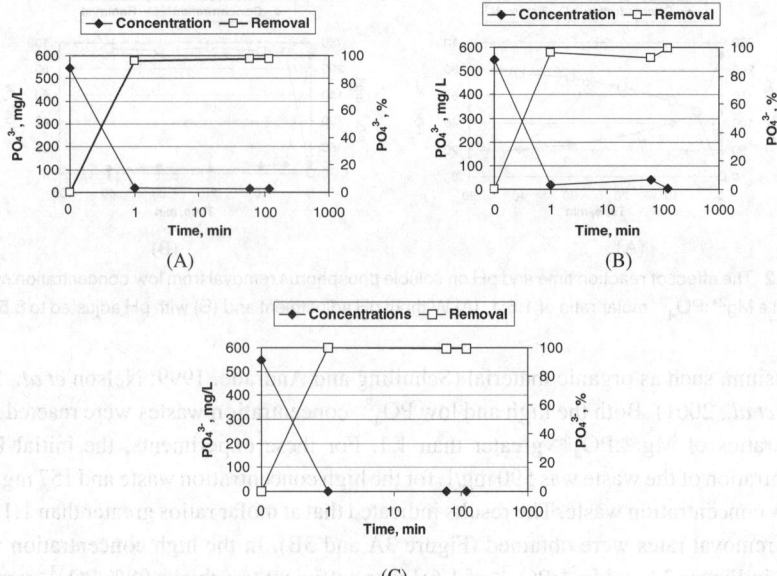

Figure 4 Effect of seed material and reaction time on soluble phosphorus removal. (A) No seed material (B) struvite as seed material (C) sand as seed material

increase the particle size by two to three folds. However, the largest precipitate produced was still only a fraction of the size of that produced in pure solution by Stratful *et al.* (2001), 3 mm.

Precipitate analysis

Precipitate from the high concentration waste has been analyzed using X-ray diffraction. The results show the presence of the minerals struvite and brushite ($CaPO_3(OH) \cdot 2H_2O$). This shows that phosphorus precipitates other than struvite are forming. The precipitate contained 34,250 mg kg^{-1} magnesium, 18,550 mg kg^{-1} of ammonia–nitrogen, 431,480 mg kg^{-1} of phosphate. The molar ratio of $Mg^{2+}:NH_4^+:PO_4^{3-}$ of the precipitate was 1:0.74:3.2. The phosphorus precipitate we have produced is not pure struvite, as the molar ratio of pure struvite is 1:1:1, excluding the hexahydrate. The formed precipitate is enhanced with phosphorus from the formation of brushite and other phosphate containing compounds that may have been formed but not yet identified. Since our overall goal is to recover phosphorus, rather than produce pure struvite, this is a favorable result.

Conclusions

The results of the reaction time portion of this study indicate that a long residence time in a field-scale precipitation unit is unnecessary. While an increase in PO_4^{3-} removal was observed with a 40 min reaction time when compared to the 10 min test, this increase in recovery was not great enough to justify the increased energy and time required for the increased reaction time.

Adjusting the pH in the high and low concentration wastes from 7.4 and 7.5 to 8.5 resulted in an additional 7 and 18% of PO_4^{3-} removal, respectively. These results show the value of increasing swine waste pH to optimize the precipitation of PO_4^{3-} as a recoverable struvite based precipitate. While we used NaOH as a convenient method to increase waste pH in lab studies, we do not believe the use of NaOH will prove economically feasible in a full-scale farm recovery process.

Molar ratios of $Mg^{2+}:PO_4^{3-}$ of 1.6:1 were sufficient to overcome the binding problems presented by the organic material in the high and low concentration waste. Magnesium addition greater than 1.6:1 did not effectively increase PO_4^{3-} removal because PO_4^{3-} removal at 1.6:1 was already high.

The seeding studies discussed here did not provide the results expected. Though the size was increased by two to three folds, the precipitate is still not large enough to warrant enhanced recovery. More work needs to be done to better understand the seeding process to improve the results.

Future work in phosphorus precipitation from swine manures will include work to increase and determine the feasibility of an on-farm recovery process. To make the process more economical, a fast and inexpensive method for increasing the wastewater pH will be studied. Because the precipitate is not pure struvite, the recovered material will be analyzed for fertilizer value and tested in field cropping applications. Also, an economic analysis of the process should be performed to determine the cost per animal and per gallon for operating a precipitation and recovery system.

Acknowledgements

The authors would like to acknowledge the following persons: Mrs. Galena Melnichchenko for laboratory analysis and Dr. William Klingeman for assistance with particle size measurement.

References

Battistoni, P., Boccadoro, R., Pavan, P. and Cecchi, C. (2001). Struvite crystallization in sludge dewatering supernatant using air stripping: the new-full scale plant at Treviso (Italy) sewage works. *Proceedings of the 2nd International Conference on Phosphorus Recovery for Recycling from Sewage and Animal Wastes*. Noordwijkerhout, Holland. March 12–14.

Beal, L.J., Burns, R.T. and Stalder, K.J. (1999). Effect of anaerobic digestion on struvite production for nutrient removal from swine waste prior to land application. Presented at the *1999 ASAE International Meeting* in Toronto, Canada. Paper No. 994042. ASAE St. Joseph, MI.

Burns, R.T., Moody, L.B., Walker, F.R. and Raman, D.R. (2001). Laboratory and in-situ reductions of soluble phosphorus in liquid swine waste slurries. *Environmental Technology*, **22**, 1273–1278.

Burns, R.T., Cross, T.L., Stalder, K.J. and Theurer, R.F. (1998). Cooperative Approach to Land Application of Animal Waste in Tennessee. *Proceedings of the Animal Production Systems and the Environment: An International Conference on Odor, Water Quality, Nutrient Management and Socioeconomic Issues Meeting*. 1:151–156 Des Moines, Iowa.

Gaterell, M.R., Gay, R., Wilson, R. and Lester, J.N. (2001). An economic and environmental evaluation of the opportunities for substituting phosphorus recovered from wastewater treatment in existing UK fertilizer markets. *Proceedings of the 2nd International Conference on Phosphorus Recovery for Recycling from Sewage and Animal Wastes*. Noordwijkerhout, Holland. March 12–14.

Greaves, J., Hobbes, P., Chadwick, D. and Haygarth, P. (1999). Prospects for the recovery of phosphorus from animal manures: a review. *Environmental Technology*, **20**, 697–708.

Kalyuzhnyi, S., Skylar, V., Epov, A, Arkhipchenko, I., Barboulina, I., Orlova, O. and Klapwijk, A. (2001). Phosphate recovery via precipitation from anerobically treated pig manure wastewater. *Proceedings of the 2nd International Conference on Phosphorus Recovery for Recycling from Sewage and Animal Wastes*. Noordwijkerhout, Holland. March 12–14.

Kumashiro, K., Ishiwatari, H. and Nawamura, Y. (2001). A pilot plant study using seawater as a magnesium source for struvite precipitation. *Proceedings of the 2nd International Conference on Phosphorus Recovery for Recycling from Sewage and Animal Wastes*. Noordwijkerhout, Holland. March 12–14.

Liberti, L., Petruzzelli, D. and de Florio, L. (2001). REM NUT ion exchange plus struvite precipitation process. *Environmental Technology*, **22**(11), 1313–1324.

Miles, A. and Ellis, T.G. (2001). Struvite precipitation potential for nutrient recovery from anaerobically treated wastes. *Wat. Sci. Tech.*, **43**(11), 259–266.

Mitani, Y., Sakai, Y., Mishina, F. and Ishiduka, S. (2001). Struvite recovery from wastewater having low phosphate concentration. *Proceedings of the 2nd International Conference on Phosphorus Recovery for Recycling from Sewage and Animal Wastes*. Noordwijkerhout, Holland. March 12–14.

Nelson, N.O., Mikkelsen, R.L. and Hesterberg, D.L. (2000). Struvite formation to remove phosphorus from anaerobic swine lagoon effluent. *Proceedings of the 8th International Symposium on Animal, Agricultural and Food Processing Wastes*. October. Des Moines, Iowa. J.A. Moore (ed), ASAE Publications. St. Joseph, MI.

Parsons, S. (2001). Recent scientific and technical developments: struvite (11): precipitation. *Scope Newsletter*, **41**, 15–22.

Piekema, P. and Giesen, A. (2001). Phosphate recovery by the crystallization process: experience and development. *Environmental Technology*, **21**, 1067–1084.

Schuiling, R.D. and Andrade, A. (1999). Recovery of struvite from calf manure. *J. Environ. Technol.*, **20**, 765–768.

Standard Methods for the Examination of Water and Wastewater (1998). 20th edn, American Public Health Association/American Water Works/Water Environment Federation, Washington, DC, USA.

Stratful, I., Scrimshaw, M.D. and Lester, J.N. (2001). Conditions influencing the precipitation of magnesium ammonium phosphate. *Wat. Res.*, **35**(17), 4191–4199.

Ueno, Y. and Fujii, M. (2001). Three years operating experience selling recovered struvite from full-scale plant. *Environmental Technology*, **22**(11), 1373–1381.

Wrigley, T.J., Webb, K.M. and Venkitachalm, H. (1992). A laboratory study of struvite precipitation after digestion of piggery wastes. *Bioresource Technology*, **41**, 117–121.

Selected requirements on a sustainable nutrient management

C. Lampert

Vienna University of Technology; Inst. for Water Quality and Waste Management, Karlsplatz 13, 1040 Vienna, Austria (E-mail: *clampert@iwag.tuwien.ac.at*)

Abstract Nutrients are a limited resource and call for management. A sustainable nutrient management strategy reintegrates nutrients in the environment without accumulating harmful substances above an acceptable level. In this study a methodology to assess the environmental compatibility was developed. For this assessment both the (i) enrichment of pollutants in the soils and (ii) the area specific nutrient demand of the crops were taken into account. The method considers, that products applied on soils also contain stable substances, and as a consequence the accumulation of pollutants diminishes. Additionally, it is considered, that increasing substance concentrations in the soil will lead to an increase of substance flows out of the soil by percolation, plant-removal (and erosion). In practice long term management strategies are restricted by the time span considered, the accepted accumulation of substances, the plants real needs and legal constraints. The rating of various goods can be made with the ratio of the added nutrients, considering the pollution criteria, the legal constraints and the plants real needs.
Keywords Accumulation; dilution; nutrient management; pollution; soil

Introduction

Nutrients are a limited resource and need to be managed properly. The phosphate-reserves will last according to different estimations between 88 (Global 2000, 1976) and 500 years (Finck, 1992), and those having low cadmium concentrations are even more limited (Semi Island Kola: 1 mg Cd/kg DM, Taiba/Senegal: 68–111 mg Cd/kg DM) (Sauerbeck and Rietz, 1980). Additionally, the production of mineral N- and P-fertilizers demands (fossil) energy input.

Sewage sludge contains considerable amounts of nutrients that can be reintegrated into the nutrient cycle. Together with the nutrients sewage sludges contain potentially hazardous substances having different origins. The heavy metal contents of sewage sludge can be highly influenced by diffuse sources (corrosion of roofings, etc.) (Zessner and Lampert, 2002). These substances might be accumulated in the long run in the environment if their quantity is not properly considered.

Therefore the challenge of a sustainable nutrient management strategy is to reintegrate nutrients in the environment without accumulating harmful substances above an acceptable level. In order to optimize the nutrient management the plants real needs have to be considered.

This paper is focused on heavy metals and their accumulation in agricultural soils.

Substance concentrations of goods (nutrient concentrations, concentrations of potential hazardous substances) are not sufficient to derive nutrient management concepts. The combination of nutrients and heavy metals such as the ratio of Cd and P which is often used to characterize sewage sludges enables us to assess comparable goods (e.g. various sludges). However, this ratio is not an effective tool to compare different goods like compost, sludges and manure. In Table 1, where all goods depicted represent a Cd : P ratio of 100 mg Cd/kg P, this can be seen clearly:

It is obvious that the good "soil" has the lowest potential to accumulate Cd in soil

Table 1 Identical Cd:P ratios of different goods

	100 mg Cd/kg P	
	mg Cd/kg DM	mg P/kg DM
Soil	0.1	1,000
Compost	0.6	6,000
Sewage sludge	1.5	15,000
Mineral P-fertilizer	12.6	126,000

DM = dry matter

without increasing the P-content. On the other hand mineral P-fertilizers are to be preferred to achieve a high fertilizing effect.

In the following section aspects of a sustainable nutrient management are presented, which are not restricted to sewage sludge only.

Method

The methods used are based on the methodology of materials accounting (Baccini and Brunner, 1991). To calculate the accumulation of pollutants in soils, the additional inputs and the counteracting outputs have to be considered. Inputs can be the application of manure, of mineral fertiliser, of compost, of sewage sludge, pesticides and deposition. The output-flows from the soil are plant-removal, percolation to the underground/groundwater and erosion. If the total input of a substance exceeds the total output, a stock is formed, increasing the substance concentration in the soil.

To estimate substance accumulation in the soils, often linear models have been used until now (e.g. Chawla *et al.*, 1976; von Steiger and Baccini, 1990; Wintzer *et al.*, 1996). These models do not consider the "diluting effect" of the stable matrix applied on the soil and the increasing outflows of substances due to their increased concentration in the soils. In the following, the outline of a dynamic model is shown.

Diluting effect

The goods applied on agricultural soils consist of organic and inorganic matter. The content of organic matter in these goods differ in a broad range: mineral fertilizer 0%, compost 30%, sewage sludge 50%, manure 90%. Depending on the type of organic matter (easily degradable –stable) the organic matter will be mineralised partly or completely within a certain time span. The solubility of the inorganic components varies (slightly soluble –

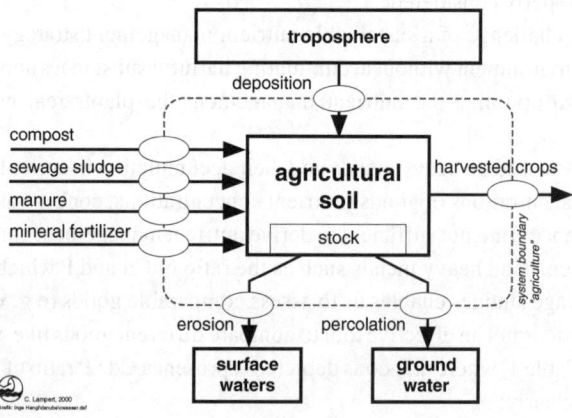

Figure 1 System agricultural soil

non-soluble) too. In the long run, most of the organic matter will be decomposed and the inorganic matter will be partly dissolved. The remaining part ("stable matrix") has a "diluting effect" on the heavy metal load of the goods applied on the soils. Therefore, it is not the total heavy metal load applied which causes accumulation.

Depending on the good a different amount of "stable matrix" is applied. Assuming that the substance concentration within the "stable matrix" (final concentration) was similar to a "geogenic" (or background) substance concentration no accumulation would take place due to the application of this good. The net accumulation is only due to the difference between the concentration in the stable matrix (final concentration) and the geogenic concentration (background concentration), i.e. the resulting substance load (= anthropogenic additional load (Lampert and Brunner, 1999)).

The anthropogenic additional load can also be related to one nutrient unit to give the specific anthropogenic additional load; e.g. mg Zn/1kg P) (Lampert, 2001). The accumulation-effective pollutant load can be calculated using this unit if the amount of nutrients applied on the soil is known.

In Table 2, Zn- and P-concentrations and the anthropogenic loadings of the goods "mineral P-fertilizer", "pig manure", "sewage sludge" and "compost" are compared. As can be seen, the impact of the "diluting effect" for compost is the highest. In the example chosen, mineral fertilizer has the lowest risk of Zn-accumulation.

Output flows depending on the stock

All output-flows are related to the substance concentration in the soil. The relation, especially for the plant removal depends on the substance considered, the soil conditions, the type of crops, etc. In general, an increase in the substance concentration of the soil causes a concomitant increase in that of the output flows.

In order to calculate the resulting substance concentration in the soil, the "layer model" was developed. The "diluting" effect of the stable matrix can be considered if the reference layer of the soil is kept at a constant thickness (e.g. 30 cm plough layer). This is required if the concentration of the substances e.g. within the plough layer is of interest. In fact the "layer model" has some similarities with the "Dynamic Soil Composition Balance" developed by (Moolenaar et al., 1997).

The layer model includes:
- if a "stable matrix"-layer is applied to the "top" of the soil, the soil-compartment is "growing" in height. To keep the compartment constant at 30 cm, a layer of an equal dimension has to be reduced from the "bottom" of the reference layer.
- If erosion takes place, again the reference layer has to be kept constant. Therefore a layer equal to the thickness of the eroded soil layer is to be added at the bottom of the reference layer. Two cases can be distinguished: the thickness of the soil layer eroded is higher than the application of "stable matrix" and vice versa.

Table 2 Impact of the diluting effect on the accumulation effectiveness

Goods	Zn-conc. in mg/kg DM	OM in %	Anth. add. load. MgZn/kgDM	P-conc. mg/kg DM	mgZn/kg P	spec.anth. add. load mgZn/kgP
Min. P-fertilizer	220	0	193	32,700	6,628	5,900
Pig manure	1,300	90	1,294	26,500	50,189	48,850
Sewage sludge	800	50	772	20,000	40,000	38,575
Compost	200	30	160	4,000	50,000	40,025

Assumptions: background Zn-concentration in the soil: 60 mgZn/kgDM; Decomposition of the organic matter (OM): all 100%; easily soluble inorganic matter : mineral fertilizer 55%, all others 5%

A general formula for the calculations used:

- aB applied "stable matrix" in 10^{-1} m
- eB thickness of the soil layer eroded in 10^{-1} m
- E substance output by erosion in kg/10^{-2} km^2
- h heights of the soil compartment considered (e.g. plough layer) in 10^{-1} m
- mx_a annual applied amount of substance x in kg/10^{-2} km^2
- mx_n amount of substance x at the time n in kg/10^{-2} km^2
- mx_{n-1} amount of substance x at the time n minus 1 in kg/10^{-2} km^2
- P substance output by percolation in kg/10^{-2} km^2
- R substance output by plant removal in kg/10^{-2} km^2

$$mx_n = mx_{n-1} * \frac{h - aB}{h} + mx_a - E - P - R + mx_{n-1} * \frac{eB}{h}$$

Results and discussions

Figure 2 underlines the importance of the diluting effect through the example of compost application. The diluting effect for sewage sludge would be lower as the non degradable substance is less and the substance concentration is higher. In each case 4 kg Zn/ha.y (1 ha = 0.01 km^2) are applied on the soil. In Figure 2 neither additional input flows, nor the output flows of zinc from the soil are considered.

Although the same annual zinc load of 4 kg Zn/ha is applied, the accumulation differs up to 30% after a time span of 300 years. Keeping the Zn-load constant (at 4 kg/ha.y), a reduction in the amount of compost by 50% (from 13.33 Mg DM/ha.y to 6.67 Mg DM/ha.y) yields a difference in the accumulation of 19%. A further reduction from 6.67 Mg DM/ha.y to 3.33 Mg DM/ha.y changes the increase of the Zinc-stock by only 9%. The calculated accumulations have significantly different gradients and change of the gradient of the stock-increase.

The final concentration (gradient = 0) in this example applying the compost with 300 mg Zn/kg DM would be 430 mg Zn/kg soil, applying the compost with 1,200 mg Zn/kg DM it would be 1,725 mg Zn/kg soil.

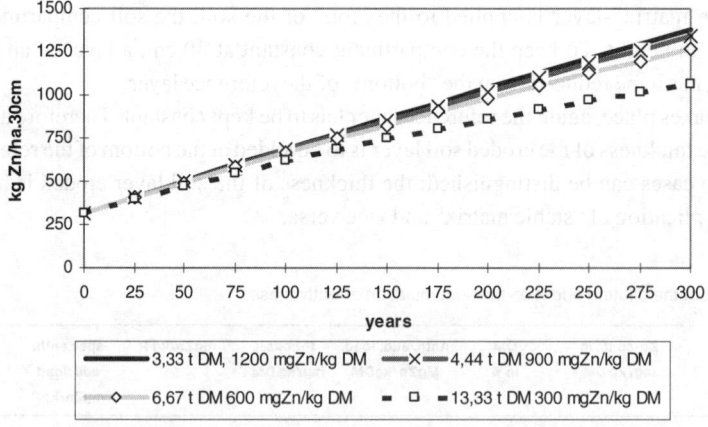

Figure 2 Change of the Zinc-stock in the soil due to the application of compost containing 4 kg Zn/ha.y (different substance concentration and related amount of dry matter (DM))
Further assumptions: 30% organic matter, out of this 10% non degradable; 5% of the inorganic matter is slightly soluble; background concentration in the soil: 70 mg Zn/kg DM (equal to a Zn stock in the soil of 315 kg Zn/ha.y, at a plough layer of 30 cm and a soil density of 1.5 kg/dm^3) (t = Mg)

The following conclusion can be drawn: the "diluting effect" is the higher the (i) more the portion on "stable matrix" is and the (ii) more similar the natural background concentration and the "final" substance concentration (substance concentration of the stable matrix) in the good applied is.

In the following example a "pure" Zinc load of 4 kg Zn/ha.a was applied and the output-flows percolation, plant removal and erosion have been considered.

According to the a conventional linear model, the annual increase of the Zinc-stock in the soil is constantly 0.46 kg Zn/ha (Figure 3). Using the layer model, the output flows increase from originally 0.35 kg Zn/ha (less than in the linear model, as the net-flow due to erosion is negligible) to 1.35 kg Zn/ha after 300 years (about 4 times higher as in the first year respectively about 3 times higher as in the conventional calculation). The Zinc-stock amounts to about 1,400 kg Zn/ha.30 cm after 300 years using the linear model, but 1,200 kg using the layer model (therefore more than 15% less even if the output by erosion is not corrected in the conventional calculation).

As shown in Figure 3 the substance output due to erosion only amounts to the difference of the substance concentration on the top and on the bottom of the reference layer. This means, in many cases the output–load due to erosion with respect to the substance concentration in the reference layer is only of slight importance. Erosion would be of greater concern in other environmental compartments such as receiving waters.

In the example given, the increasing substance outputs diminish the difference between the total input and the total output from initially 3.65 kg Zn/ha.y to 2.6 kg Zn/ha.y, amounting to about 30%. In the conventional calculation, the annual change of stock remains constant with 3.54 kg Zn/ha.y (including the output by erosion) (see Figure 4).

Implementation

Time horizon. In order to implement this model into agricultural practice (e.g. to calculate the "acceptable" amounts applied) a time horizon has to be included. Usually soil standards do not include a time horizon, this means that various constraints like laws, ordinances,

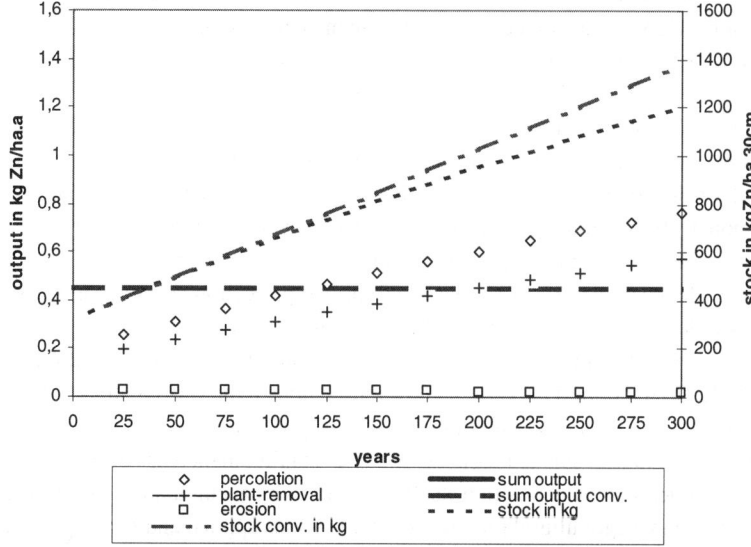

Figure 3 Change of output flows related to the change of the substance concentrations (Lampert, 2001) Assumptions t = 0: 315 kg Zn/ha (equal to 70 mg Zn/kg TS, 30 cm soil depth, soil density 1.5 kg/dm³); erosion: 1.5 t/ha.a; percolation: 0.2 kg Zn/ha.a; plant removal: 0.15 kg Zn/ha.a; percolation, plant removal and erosion increase proportional to the soil concentration

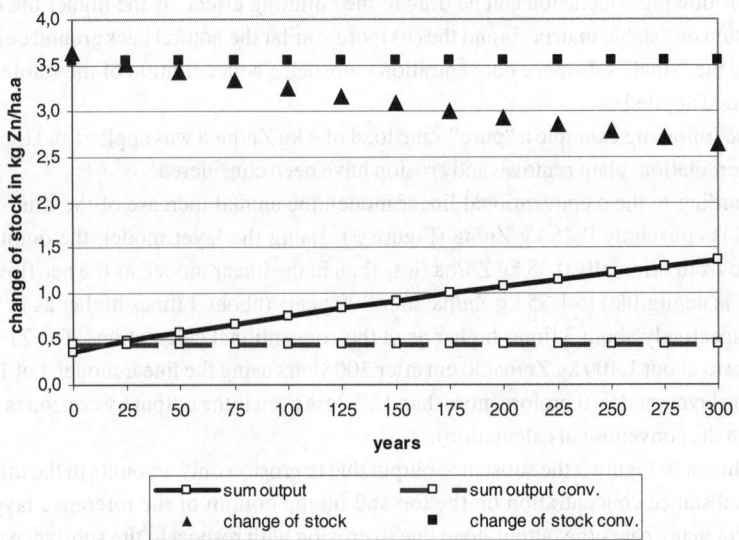

Figure 4 Comparison of the annual change of stock resulting from a conventional linear calculation and a calculation considering the change of output flows depending on the total stock (Lampert, 2001)

guidelines, etc. do not differentiate if the standard will be reached/exceeded in a few, in some or in many generations. For a sustainable soil protection strategy, there is a need to supplement the various constraints by a time horizon – a time span has to be incorporated in all possible strategies. Depending on the time span selected and the acceptable accumulation defined, the annual application rates of goods can be calculated (e.g. in Mg DM/ha.y).

Acceptable accumulation. Substance balances of soils show that for some substances (e.g. Cd, Pb), accumulation in soils takes place even without any agricultural activities. The atmospheric deposition rate exceeds the substance outflows from the soil through plant removal, percolation, and erosion. To postulate "Input flows into the soils have to be equal to the Output flows" is to neglect the actual case in many regions.

Therefore, to enable (traditional) feed and food production, an additional accumulation of potential hazardous substances has to be taken into account. Obviously, discussion on the extent of this accumulation is crucial. We still lack clear evidence on the acceptable levels of substances in the soil. The standards, in terms of maximum concentrations allowed in soil, differ from country to country in a wide range. It seems that these standards are influenced more by the national geological situation than by potential hazards for the biosphere. Therefore, the question on acceptable accumulation is not answered yet.

At least three strategies can be identified for soil protection concerning the accumulation of pollutants which are adapted to the (regional) soil characteristics:

(i) to define a maximum allowable substance concentration in the soil (e.g. 300 mg Zn/kg DM), (ii) to define a maximum allowable change of the various stocks of substances in the soil (e.g. maximum increase of the Cd-stock by 10% in), and (iii) to define an acceptable absolute change of substances in the soil (e.g. maximum increase by 300 kg Zn/ha).

As mentioned above, deposition itself can lead to an accumulation of hazardous substances. To enable agricultural practice for single goods applied and for all goods applied, the acceptable accumulation has to be defined. This acceptable accumulation has to be defined restrictive to be on the "safe side" of soil protection, to remain a buffer space for pollutant accumulation due to unavoidable (?) substance input by deposition.

The time horizon chosen and the definition of the acceptable accumulation characterize the "pollution criteria" and serve as a starting point for the calculations. E.g.: "A 10% increase of the Zinc-concentration within 250 years in the plough layer, through sewage sludge is accepted."

Legal constraints can restrict application rates. In practice, various laws, ordinances and guidelines limit the amounts of goods that can be applied on soils. Regulations on the application of sewage sludge and compost can limit the maximum concentrations, the amount applied on the soils or the accepted pollutant load, like 2.5 Mg DM sewage sludge/year, 8 Mg DM compost/year; 5 g Cd/ha.y. The water act (BGBl. 1990/252) limits the amount of Nitrogen applied on arable land by 175 kg N/ha.y. Therefore, also the legal constraints influence the regional nutrient management.

Orientation of amounts toward the plant's real needs. In addition to the accumulation of pollutants in the soils (pollution criteria) and the legal constraints the amounts of sewage sludge applied should correspond to the plants real needs. This means to consider the nutrient content in the soils, the kind of crops, the yields aimed at, etc. Only this orientation ensures that the limited resource nutrients are not disposed of but used in adequate amounts. As a consequence there will be no general advice on the amounts of goods applied like 2.5 Mg DM/ha.y.

To consider the plant's real needs will lead to the following situation: if the "pollution criteria" of substances in the soils is defined strict (small accumulation, long accumulation period) possibly the "pollution criteria" will be the limiting one. If it is defined permissible (short time period, high accumulation) the crop's real needs will limit the amounts applied.

To rank various competing goods the following procedure can be applied: In the first step for each good the most restrictive constraint is identified (i.e. legal constraints, pollution criteria) and the related nutrient amount is calculated. In the next step the ratio between the calculated nutrient amount and the plants real needs is determined. This ratio can be used as ranking criteria. The ratio varies between 0 and 1. A ratio of 1 would indicate, that the plants real needs can be satisfied completely without an unacceptable pollutant accumulation and considering legal constraints.

Conclusions

To calculate the amount of goods (sewage sludge, compost, manure, etc.) that can be applied annually, (i) an acceptable concentration in the soil and (ii) a time horizon has to

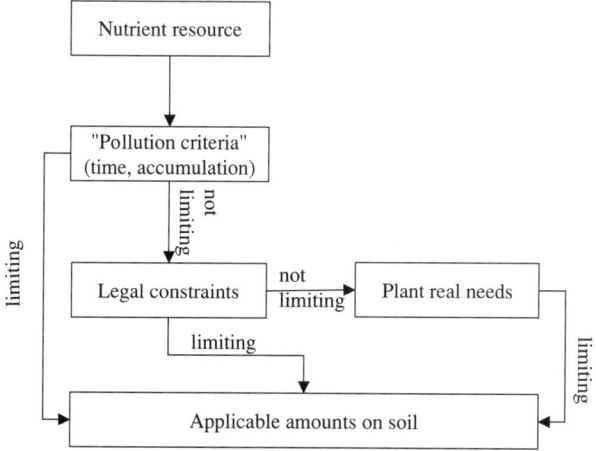

Figure 5 Calculation of applicable amounts

be defined. In addition, the calculations have to consider (iii) the diluting effect of the "stable matrix" applied as well as (iv) that the output-flows will increase if the substance concentration in the soil increases. As a consequence identical loads of pollutants applied may cause different accumulation in the soil and a "steady state" of the pollutant concentration in the soils will be achieved in the long run. (v) Legal restrictions can limit the application of goods. (vi) The amounts applied should be oriented towards the crop's real needs. The amounts of dry matter applied can be restricted by the "pollution criteria", by the plant's real needs or by legal constraints.

The ratio between the nutrient amount applied (considering the pollution criteria and the legal constraints) and the plants real needs allows the ranking of various goods.

References

Baccini, P. and Brunner, P. (1991). *Metabolism of the Anthroposphere*. Berlin, Heidelberg, New York. Springer Verlag.

BGBl. 1990/252. Novelle des Wasserrechtsgesetzes.

Chawla, V.K., Cohen, D.B. and Bryant, D.N. (1976). Environmental Canadas Research and development activities in land application of sludges. in: Land as a waste management alternative, ed. by Raymond C. Loehr. *Proceedings of the 1976 Cornell Agricultural Waste Management Conference*. Ann Arbor Sciences, Publishers Inc.

Finck, A. (1992). *Dünger und Düngung*. 2. Aufl., VCH-Verlagsgesellschaft mbH, Weinheim.

Global 2000 (1976). Der Bericht an den Präsidenten. Verlag Zweitausendeins, Frankfurt am Main.

Lampert, C. (2001). Decision support on the management of biogenic materials regarding aspects of resource conservation and environmental compatibility. PhD thesis, Institute for Water Quality and Waste Management, Vienna University of Technology.

Lampert, C. and Brunner, P.H. (1999). Erhöhung der Aussagekraft von Analyseparametern durch Berücksichtigung der anthropogenen Additionsfracht; in: 111 VDLUFA-Kongress in Halle/Saale; "Richtwerte, Vorsorgewerte und Grenzwerte – Bedeutung für Landwirtschaft, Ernährung und Umwelt"; VDLUFA-Schriftenreihe 52/1999; VDLUFA-Verlag Darmstadt.

Moolenaar, S.W., Lexmond, T.M. and Van der Zee, S.E.A.T.M. (1997). Calculating heavy metal accumulation in soil: a comparison of methods illustrated by a case study on compost application. *Agriculture, Ecosystems and Environment*, 66, 71–82.

Sauerbeck, D. and Rietz, E. (1980). Zur Cadmiumbelastung von Mineraldüngern in Abhängigkeit von Rohstoff und Herstellungsverfahren. *Landw. Forschung, Sdhft*. 37, pp. 685.

von Steiger, B. and Baccini, P. (1990). Regionale Stoffbilanzierung von Landwirtschaftlichen Böden mit messbarem Ein- und Austrag. *Bedeutung der Stoffbilanzierung für den qualitativen Bodenschutz. Bericht 38 des Nationalen Forschungsprogramms "Boden"*. Liebefeld-Bern.

Wintzer, D., Leible, L., Rösch, C., Bräutigam, R., Fürniß, B. and Sardemann, G. (1996). *Wege zur umweltverträglichen Verwertung organischer Abfälle. Abfallwirtschaft in Forschung und Praxis, Band 97*. Erich Schmidt Verlag.

Zessner, M. and Lampert, C. (2002). The use of regional material balances in water quality management. *Urban Water*, 4, 73–83.

Effect of lime stabilisation of enhanced biological phosphorus removal sludges on the phosphorus availability to plants

D. Seyhan* and A. Erdincler**

* Vienna University of Technology; Inst. f. Water Quality and Waste Management, Karlsplatz 13, 1040 Vienna, Austria (E-mail: *d.seyhan@iwa.tuwien.ac.at*)

** Institute of Environmental Sciences, Bogazici University, 80815 Bebek Istanbul, Turkey (E-mail: *erdincle@boun.edu.tr*)

Abstract This study investigates the phosphorus (P) availability in lime stabilised biological phosphorus removal sludges. Lime-stabilised sludge amendments (LS), non-stabilised sludge amendments (S) and amendments with a chemical fertiliser (TSP) were compared through plant uptake of P and Olsen-extractable P for this purpose. In the first part of the study, pot experiments were performed, where a dewatered biological phosphorus removal sludge was applied to pots at increasing rates of P. A P-deficient, alkaline soil was used in the experiments and *Lollium perenne* was the testing plant. In the second part (incubation tests), the waste activated sludge from an Enhanced Biological Phosphorus Removal (EBPR) process was mixed with the same soil at a pre-determined P-based rate. The pot experiments showed that, the efficiency of the fertilising materials, based on the minimum P applied to reach the maximum yield, was in the following order: S~LS>TSP. However, the P concentration in the plant tissue was in the order of TSP>S>LS for all P application rates. In the incubation tests, the EBPR sludge raised the soil P-level from the low range to the medium range. The P-availability in TSP decreased rapidly with time whereas that in S and LS remained almost constant.

Keywords Agricultural use; lime stabilisation; nutrient recycling; phosphorus; sludge

Introduction

Future projections show that sludge utilisation is going to be the prime route for disposal, with 45% by 2005 (IAWQ, 1996). This practice returns some valuable elements into the material cycle and provides a long-term solution for the ultimate disposal problem that also meets the ecological concerns. The sludges are worth recycling especially because of their phosphorus (P) content, as the phosphate reserves are being exhausted. P is the most expensive of plant nutrients applied to agricultural soils and is a non-renewable resource (White, 1981). Thus, the practice can be seen as "recycling" of a waste material rather than a "disposal" method.

Pertaining to the nutrient management, phosphorus is being given priority over nitrogen by both environmental scientists and soil scientists in the last decades. Environmental engineers are trying to reduce P concentrations in treatment plant effluents in an attempt to control eutrophication. This attempt leads to high P concentrations in the sludge. As a good example of this, EBPR is a recently developed technique for the removal of dissolved phosphorus from wastewater. It is based on the biological uptake of P in excess of the normal metabolic requirements and wastage of excess sludge removes the phosphorus from the system. Soil scientists, on the other hand, are trying to establish favourable P levels in arable soils. However, P added to the soil turns into insoluble forms within a short time and becomes slowly available or totally unavailable. If fertilisers are added annually and excessively to get a small amount of available P, the P accumulates in the soil (Brady, 1990), again threatening water quality.

The sludge application rates to the agricultural land should be based on satisfying the nutrient requirements, especially P needs, of the plants (USEPA, 1983; ASCE, 1987). Since most crops require more N than P, there is an inherent safety factor in the approach. Although nitrogen-based management has been practiced for many years, P-based agricultural management is now promoted also by the United States Department of Agriculture (Sharpley et al., 1999).

In comparison with the chemical fertilisers, sludge does not provide readily available nutrients, and is a slowly available source of these materials. Certain sludges might show poor nutrient availability despite their relatively high P content by analysis. Thus, in order to quantify the fertilising value of biosolids and as part of the P-management, it is crucial to accurately identify the availability of P in biosolids applied to land.

When the land application of sewage sludges is concerned, the need for the stabilisation of the sludge must be considered as well. As an economic stabilisation alternative and regarding its benefit of improving the soil structure and decreasing the heavy metal availability to plants, lime stabilisation seems to be the most reasonable choice. Many studies have been carried out to assess the P-availability in untreated or digested secondary and tertiary treatment sludges. The effect of lime stabilisation on P-availability, on the other hand, has not been studied thoroughly.

This study investigates the effect of lime-stabilisation on the availability of P in Enhanced Biological Phosphorus Removal (EBPR) sludges. The objectives are to;
- compare the yield increase of *Lollium perenne*, grown on a P-deficient soil in pots, by supplying P through sludge (S), lime-stabilised sludge (LS) and triple super phosphate (TSP) at increasing rates.
- measure the P-availability in S, LS and TSP, in terms of plant tissue P, and compare their fertilising performance.
- monitor the change in available P with time, applied at the optimum rate on the same soil via non-stabilised and lime-stabilised EBPR sludge and TSP, through incubation tests.

Materials and methods

In the first part of the study, pot experiments were carried out in order to assess the fertilising effect of the sludge in terms of yield, and P-availability to the plant. In the second part of the study, the change of available P with time was observed in EBPR sludge amended soil, through incubation tests. The same soil was used in both pot experiments and incubation tests.

Analyses

The dewatered biological phosphorus removal sludge, used in the first part of the study, was obtained from a rather young treatment plant, Paşaköy Biological Phosphorus Removal Plant, in Istanbul. At the time of the sampling and analyses, the P-removal efficiency of the plant was low. The dewatered sludge applied to the pots had a total solids concentration of 26% and N/P/K ratio of 5/4.5/1 (Table 1). The lime dosage to meet the USEPA requirement of holding the pH of sludge above 12 for 2 hours was found to be 3% (w/w).

Table 1 Properties of the dewatered sludge and lime stabilised sludge used in pot experiments

Parameter	pH 2h*	TS	P_2O_5	TKN	K	T.Org.	Fe	Cd	Zn	Ni	Cu
		%					ppm				
S	7	26	4.5	4.9	1	51	31,650	–	949	134	880
LS	12.3	73	3.9	3.6	0.8	44	21,700	–	774	111	660

* after 2 h in deionised water

The waste activated sludge (WAS), used in the second part of the study, was taken from a laboratory-scale sequencing batch reactor working for enhanced biological phosphorus removal from a synthetic wastewater. WAS samples were collected daily and refrigerated at 4°C until a sufficient amount was obtained. The batch reactor that was cycled through anaerobic and aerobic conditions was fed daily with acetate having a COD of 500 mg/l for 10 days and the release and uptake of P was observed. The supernatant of the settled sludge was filtered before the analysis for ortho-P. When the effluent P concentration in the filtered supernatant was as low as 4 mg/l and the TP of the mixed liquor was 290 mg/l, the sludge was assumed to store most of the dissolved phosphorus. The sludge was thickened then by gravity settling. The solids and phosphorus contents of both sludges are presented in Table 2. The lime (CaO) dosage to meet the USEPA requirements of holding the pH above 12 for 2h. was found to be 0.5% (weight/weight).

The soil used in the pot experiments and the incubation tests was selected to be one with a low background P concentration and organic content in order to observe the net effect of sludge amendments. Supplied from the Çukurova Region, this entisol known as Karaburun Serial has the properties presented in Table 3. The soil was taken from the 0–20 cm depth, A1 level, sieved to 2 mm before being used in the analyses and experiments. The methodologies for all sample analyses are presented in Table 4.

Set-up

Pot experiments. Sludge, lime stabilised sludge and triple super phosphate were applied to pots at 5 different rates supplying 0, 100, 200, 300, 400 ppm P. As P is an immobile nutrient,

Table 2 Properties of the sludge in the batch experiments

Sludge type	Volume	Suspended solids	Total phosphorus
Mixed liquor	1,000 ml	2,500 mg/l	290 mg/l (11.6%)
Thickened WAS	300 ml	7,000 mg/l	600 mg/l (8.6%)

Table 3 Soil properties

pH	TP	Available P	N	K	Fe	Cd	Zn	Cu	Organics*	Texture*
				ppm					(%)	
7.8	350	3	3,075	2,820	10,100	0.1	50	25	2	CL

* as determined in Soil Science Department of Çukurova University

Table 4 Methodology for sample analysis

Parameter	Analytical method	Instrument
Soil and sludge		
Total phosphorus	$H_2SO_4 + H_2O_2$ digestion Colorimetry (Ascorbic Acid)	Spectrophotometer
Available phosphorus	Olsen Method Colorimetry (Ascorbic Acid)	Shaker, Spectrophotometer
TKN	$H_2SO_4 + H_2O_2$ digestion Colorimetry (Nesslerization)	Spectrophotometer
Total organics	Incineration at 550°C	Muffled Furnace
Total solids	Evaporation + Gravimetry	Oven
pH	1:1 deionised water (2 h)	pH-meter
Metals	$H_2SO_4 + H_2O_2$ digestion	Atomic Absorption S.
Plant tissue		
Total phosphorus	Dry ashing + dissolving in 1/3 HCl + Colorimetry	Spectrophotometer

sludges and TSP were band-applied to pots at 0.5 cm below the seed, without any contact. There were three replicate pots with 1 kg soil content and five application rates for each fertiliser material (S, LS, TSP), making up to 45 pots overall. The lime treated sludge having 73 per cent TS, was grounded prior to pot application, whereas the dewatered sludge having 26 per cent TS was applied as it is. While preparing the TSP pots, the required amounts of the chemical fertiliser were dissolved in 50 ml. of distilled water for each pot. S and LS pots were added to the calculated amounts of distilled water to make up their water content to 50 ml. as well (Blank pots were only watered with 50 ml distilled water). Nine replicate pots were prepared as blanks for higher precision. After 21 days of a conditioning period, the pots were seeded. *Lollium perenne*, being a quick growing plant, was selected to be the test plant. Seeding was at a high rate (1.3 g/pot) to impose a large demand for P. The seeding was performed in September, following a 3-week conditioning period of sludge-soil mixtures and the sprouting took 4 days. The harvesting of the plant at a level of 2 cm above the soil surface was done on the 20th day after seeding.

Incubation tests. Thickened WAS (see Table 2), lime stabilised WAS and TSP were mixed with 300 g portions of soil, so that each mixture has 200 ppm P. This dosage was the optimum concentration to quench the P-need of the subject soil as determined during the pot experiments. Total phosphorus analyses in all mixtures (S, LS and TSP) revealed that the preperation of the sludge-soil mixtures, the sampling and the analyses of the samples were carried out successfully. For the incubations, the prepared soils were divided into three replicates of 100 g portions and loaded into plastic containers. Incubation lasted for 60 days at 21°C. Samples taken from the pots were extracted with Olsen reagent (Olsen and Sommers, 1982) every tenth day to estimate the available phosphorus content. Each pot was watered with 15 ml of deionised water 5 days after each extraction in order to maintain the soil moisture content at the same level, again with 10-day intervals and no leachate was formed.

Results and discussions
Pot experiments

The fertilising potential of the sludge, obtained from Paşaköy Biological Treatment Plant was measured in terms of plant yield and plant uptake of P. Additions of P increased the yield in the pots (Figure 1). Yield increase in TSP verified that the soil was inadequate in supplying P for the plant growth, as our soil background testing also suggested (Table 3).

In the literature, there are conflicting findings about the performances of chemical fertilisers (TSP, MCP) and various sludges in terms of yield (Gestring and Jarrel,1982; McCoy *et al.*, 1986; Frossard *et al.*, 1996). In Figure 1, the performances of TSP, S and LS are compared through the arithmetic means of the three parallel runs and no data were omitted. Our results can be evaluated in three parts, namely, at applications of P less than 200 ppm, at 200 ppm and more than 200 ppm. The maximum yield was attained at 100 ppm P for S and LS, whereas in the case of TSP, this optimum dosage was 200 ppm P. The presence of growth stimulating additives in sludge other than phosphorus might be the reason for that, leading to more restricted application of phosphorus to the agricultural land in order to get the same yield. This, in return, can minimize the phosphorus transport to the surface waters through the runoff. At an application of 200 ppm P, all treatments gave the same yield. This implies that surmounting the P-deficiency in such soils through any means can maximise the yield. Therefore, 200 ppm happens to be the dosage quenching the P-deficiency of the soil. At application rates higher than 200 ppm P, the yields of S and LS treatments were suppressed due most probably to the over-application of nutrients and other additives from sludges into the soil (Figure 1).

It is to be expected that the availability was highest in TSP, as this chemical fertiliser

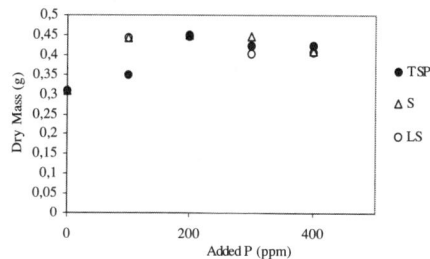

Figure 1 Change in yield with increasing P supply

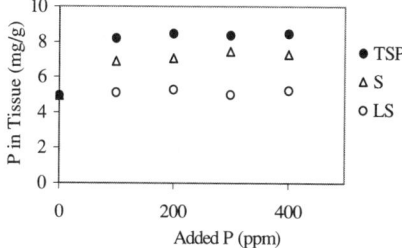

Figure 2 Average Phosphorus concentrations in plant tissue

provides readily soluble and highly available inorganic phosphates (Figure 2). The available P in the sludge amendments, showing the trend of a typical saturation function, was less than that in TSP for the equal amounts of total phosphorus applied. The reason for that is the presence of the different forms of phosphates in the sludge other than water-soluble phosphorus. Their transformation into available forms through mineralisation is too slow.

On the other hand, lime stabilised sludge did not increase the P-uptake when compared to control samples containing soil only. The difference between the results of S and LS may be explained with their different water contents. Water-soluble P makes up the largest portion of available P. Thus, reduction in the water content leads to the formation of more complex P forms. Researchers suggested that P was less available in dried sludges. In our case, total solids contents of S and LS applied to pots were 26% and 73% respectively. During the preparation of LS, strong pellet-like clumps were formed upon the lime stabilisation of the dewatered sludge. These clumps could not be dissolved in the soil and made the P slowly available to the plants. Formation of less soluble and more stable crystalline phosphates on drying must have decreased the P-utilisation by the plant.

Another reason for the reduced availability of P in LS might be the addition of Ca through lime. In alkaline soils $CaCO_3$ governs the P-availability providing extra sorption sites for P. Researchers claim that this bonding was a low energy bonding (Soon and Bates, 1981). Yet, the change of $H_2PO_4^-$ ion to tricalcium phosphate $[Ca_3(PO_4)_2]$ and to even more insoluble compounds (Brady, 1990), with time, decreases the availability to plant drastically. Liming can decrease the availability of P also by increasing the pH. At high pH values, phosphorus makes very stable compounds with heavy metals and its availability to plants is reduced.

Incubation tests

In incubation tests, the change in the plant available P with time was observed in terms of Olsen available P. Olsen Method provides a relative measure of labile soil P (Thomas and Peaslee, 1973). As phosphorus is immediately involved in certain physicochemical reactions in the soil, a reliable estimate of the actual initial concentration could not be obtained. Therefore, data obtained in 40 h were used as the initial conditions. Figure 3 shows the averages of 3 parallel runs of S, LS and TSP. Only a small portion of added 200 ppm total phosphorus gets into available form when applied to the soil. This portion was initially about 50 per cent in the case of TSP but finally dropped to the 23 per cent of the added P. TSP showed a steady decrease throughout the incubation period, whereas both sludges released some of their phosphorus to the soil solution.

According to Larsen (1976), a 1st order reaction model can describe the transport of P from labile to the non-labile pool for readily soluble fertilisers like TSP. In this study, the reaction rate constant and the half-life for the logarithmic reduction of labile phosphorus sourced from TSP were estimated and found to be $0.0136 d^{-1}$ and 51 days respectively.

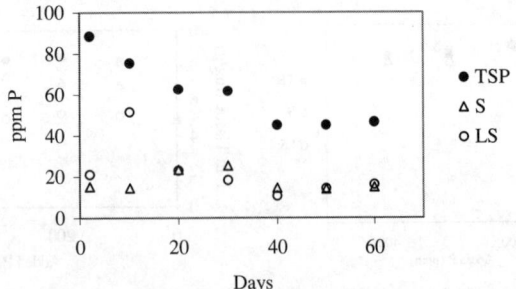

Figure 3 Comparison of the incubated soil material treated with TSP, S and LS in terms of the change in available P over time

Barrow (1976) reported that, in short-term, the most efficient sources of phosphate are the soluble fertilisers. In this case, phosphate is, in effect, added directly to the soil solution and it precipitates or is adsorbed readily when applied to the soil (Brady, 1990). The only way that can be delayed is to delay the addition to the soil solution – for example by using slowly soluble sources. The trade-off involved is increased long-term effectiveness for decreased short-term effectiveness (Barrow, 1976).

Sludge can be seen as a low-analysis, slow-release phosphorus fertiliser (McLaughlin and Champion, 1987; Sui et al., 1999; Lu and O'Connor, 2001). In our study, about 10 per cent of the added P was in available form at the beginning and at the end of the testing period. The reason for the P-release in between might be the organic matter decomposition that may lead to the formation of macromolecules like carboxylic acids. These molecules can complex metals, blocking the P sorption sites and P is released to the soil solution.

Figure 3 also reveals the effect of lime on the availability of phosphorus. LS showed a strong P release until the 10th day of the incubations. Plant-available phosphorus increased up to 55 ppm P, and again turned back to the initial value as in the case for S.

Application of LS increased the calcium content in soil leading to the formation primarily of calcium phosphates. As stated before, fresh precipitates of $Ca_3(PO_4)_2$ are still plant available due to the improper crystallisation and small surface area. The increased availability in LS might indicate the predominance of calcium phosphates to other forms of P. Furthermore, Olsen reagent ($NaHCO_3$) mostly extracts the P bound to Ca and it might give higher available P values for soils treated with lime stabilised sludge.

Mineralization of the organic P in the soil takes a very long time (Holford, 1976). However, considerable amounts of organic material are added through the sludge to the soil. When these organics start to decompose in the soil, with the accelerating effect of lime, inorganic phosphorus might be formed in a shorter term. Yet, the rapid increase in available P might be attributed to the pH effect rather than the mineralization of the organic P in the sludge. Lime added through LS must have caused the desorption of the added P as well as the native P in soil by increasing the pH. Anion exchange and desorption replace phosphates, with OH^-, HCO_3^- and organic anions, releasing the P into the soil solution (Schachtschabel et al., 1998).

After reaching the highest availability, the trend of LS follows the same adsorption pattern as the chemical fertilisers. Starting on the 10th day, phosphorus availability of P is reduced exponentially due to aging and crystallisation.

Conclusions

In this study, all the pot experiments and incubation tests were performed by using an alkaline, P-deficient soil to see the net effect of the P-amendment. Based on the results of the study the following conclusions can be drawn:

Pot experiments

- In the pot experiments of 20 day growth period, sludge amendment (S) increased the amount of plant available P. Plant available P was not increased in the lime-stabilised sludge amended soil (LS). It seems that, lime has a retardation effect on the availability of P in sludge when added to soil. However, the phosphorus in LS may be available in the longer term.
- The efficiency of the fertilising materials, based on the minimum P applied to reach the maximum yield, was found to be in the following order: S~LS>TSP. However, the P concentration in the plant tissue was in the order of TSP>S>LS for all P application rates.
- In LS application, the stimulation of the crop production was independent of the phosphorus enrichment of the soil, other factors like nitrogen, micronutrients, organic content, soil conditioning effect of sludge, effect of lime, must have been responsible for this result.
- Results of the pot experiments showed that, applying sludges at P-based rates could prevent the over-application of any material, providing an appreciable yield comparable to or even higher than that from TSP application. After determining the soil background levels, and especially the available P in soil and sludge, the average yearly application should be based on P-demand. P-based application can prolong the useful life of sites (5 to 20 years with most sludges) as well.

Incubation tests

- The incubation tests showed that EBPR sludge can quench the P-need of a P-deficient soil. The background Olsen-available P concentration of 3 ppm (low range) in the soil was initially increased to 15–20 ppm (medium range) when EBPR sludge was applied to it.
- The available P concentration in TSP amendments decreased exponentially and achieved the half-life in 51 days. Yet, TSP supplied more plant available P than both sludges in this short-term. Biological P removal sludges can be accepted as low analysis P-fertilisers.
- Lime stabilisation decreased the P-availability when dewatered sludge was used (pot experiments). However, on the same soil, P availability was not decreased due to lime stabilisation when the sludge was undewatered (incubation tests). This result suggests that lime stabilisation of the sludge does not necessarily decrease the availability of its P. Therefore, pre-lime stabilisation might be preferred to post-lime stabilisation when EBPR sludges are to be applied on P-deficient soils.

Acknowledgement

This study has been supported by B.U. Research Fund under project number 01HY101.

References

ASCE (1987). *Land Application of Wastewater Sludge: A Report of the Task Comitee on Land Application of Sludge*. Ed. T.M. Younos. Published by American Society of Civil Engineers. ISBN 0-87262-622-9.

Barrow, N.J. (1976). The Adsorption and Desorption of Phosphate. In: *Proceedings of a Symposium Held at the University of New England, Armidale, Australia: Prospects for Improving Efficiency of Phosphorus Utilisation*, Blair, G.J. (Ed.), pp 83–88.

Brady, N.C. (1990). *The Nature and Properties of Soils*, 10th ed., McMillan Publishing Company.

Frossard, E., Sinaj, S., Zhang, L.-M. and Morel, J.L. (1996). The Fate of Sludge Phosphorus in Soil-Plant Systems. *Soil Sci. Soc. Am. J.*, **60**, 1248–1253.

Gestring, W.D. and Jarrel, W.M. (1982). Plant Availability of Phosphorus and Heavy Metals in Soils Amended With Chemically Treated Sewage Sludge. *J. Environ. Qual.*, **11**, 669–675.

Holford, I.C.R. (1976). Factors affecting the accumulation and availability of residual fertiliser phosphate. In: *Proceedings of a Symposium Held at the University of New England, Armidale, Australia: Prospects for Improving Efficiency of Phosphorus Utilisation*, Blair, G.J. (Ed.), pp 41–45.

IAWQ, EWPCA, WEF (1996). *Global Atlas of Wastewater and Biosolids Use and Disposal*. IAWQ Scientific and Technical Report Number 6, P. Matthew (ed), London, UK.

Larsen, S. (1976). Evaluation of native and residual phosphorus in soils as a source of phosphorus for plants. *Proceedings of a Symposium Held at the University of New England, Armidale, Australia: Prospects for Improving Efficiency of Phosphorus Utilisation*, Blair G.J. (Ed.), pp 31–33.

Lu, P. and O'Connor, G.A. (2001). Biosolids Effects on Phosphorus Retention and Release in Some Sandy Florida Soils. *J. Environ. Qual.*, **30**, 1050–1063.

McCoy, J.L., Sikora, L.J. and Weil, R.R. (1986). Plant Availability of Phosphorus in Sewage Sludge Compost. *J. Environ. Qual.*, **15**, 403–409.

McLaughlin, M.J. and Champion, L. (1987). Sewage Sludge As a Phosphorus Amendment for Sesquioxic Soils. *Soil Science*, **143**, 113–119.

Olsen and Sommers (1982). Phosphorus. In: *Methods of Soil Analysis*, Part 2. Page, A.L., Miller, R.H. and Keeney, D.R. (Eds.). American Society of Agronomy, Inc. and Soil Science Society of America, Inc. Publisher, Madison, Wisconsin.

Schachtschabel, P., Blume, H.P., Brümmer, G., Hartge, K.H. and Schwertmann, U. (1998). *Lehrbuch der Bodenkunde*, 14. Auflage, Ferdinand Enke Verlag, Stuttgart.

Sharpley, A.N., Daniel, T., Sims, T., Lemunyon, J., Stevens and Perry, R. (1999). *Agricultural Phosphorus and Eutrophication*. USDA, ARS-149.

Sui, Y., Thompson, M.L. and Shang, C. (1999). Fractionation of Phosphorus in a Mollisol Amended with Biosolids. *Soil Sci. Soc. Am. J.*, **63**, 1174–1180.

Thomas, G.W. and Peaslee, G.E. (1973). Testing Soils for Phosphorus. In: *Soil Testing and Plant Analysis*, L.M. Walsh and J.D. Beaton (eds), Soil Sci Soc Am, Madison Wrs, pp 115–132.

US EPA (1983). *Process Design Manual-Land Application of Municipal Sewage Sludge*. EPA 625/11-83-016. Municipal Environmental Reasearch Laboratory, Cincinnati, Ohio.

White, R.E. (1981). Pathways of Phosphorus in Soil. In: *Phosphorus in Sewage Sludge and Animal Waste Slurries*. T.W.G. Hucker and G. Catroux (eds), Reidel, Dordrecht, Holland, pp 21–46.

Development of a high-efficiency phosphorus recovery method using a fluidized-bed crystallized phosphorus removal system

K. Shimamura*, T. Tanaka*, Y. Miura* and H. Ishikawa**

* Ebara Corporation, 4-2-1 Honfujisawa, Fujisawa-shi, Kanagawa-ken 251-8502, Japan
(E-mail: *shimamura.kazuaki@ebara.com*; *tanaka.toshihiro@ebrara.com*; *miura.yukiko@ebara.com*)
** Ebara Corporation, 1-6-27 Konan, Minato-ku, Tokyo 108-8480, Japan
(E-mail: *ishikawa.hideyuki@ebara.com*)

Abstract The authors have been engaged in the research and development concerning the recovery of MAP (Magnesium Ammonium Phosphate) using a fluidized-bed crystallized phosphorus removal system. In the reactor of the fluidized-bed crystallized phosphorus removal system, seed crystals (of MAP) are fluidized previously and new MAP crystals are produced on the seed crystal surfaces. Conventionally, the reactor consisted of one reaction tank only, but this practice had the problem that as the crystallization progresses, the seed crystal is grown excessively and as a result, the effective reaction surface areas are decreased and the fluidization effect is degraded, causing the recovery ratio to be decreased. Recently, the authors have devised a two-tank type reactor by adding a sub reaction tank to the reactor (now the main reaction tank) so that the MAP particle size in the main reaction tank may be kept constant making the recovery ratio stable. They conducted a demonstration test with a pilot experimental system of the 2-tank type reactor. For raw water T-P 111 to 507 mg/L, the main reaction tank treated water T-P 14.0 to 79.5 mg/L and phosphorus recovery ratios 84 to 92% were obtained. Because the mean MAP particle size in the main reaction tank could be kept constant, the phosphorus recovery ratio could always be above 80%, realizing stable treatment.
Keywords Crystallization; fluidized-bed; magnesium ammonium phosphate

Introduction

Phosphorus is a limited resource which is anticipated to be exhausted in the 21st century. Japan is short of the phosphorus resource and imports it mostly as fertilizers, industrial chemicals, foods and feeds. On the other hand, in the living environmental system, much phosphorus is discharged and red tides and other eutrophication phenomena have become a serious problem in closed waters, such as lakes and ponds, inland bays, etc. (Matsumiya *et al.*, 2000) The authors have considered it important to recover phosphorus in an easily usable form in spite of mere removal from wastewaters and they have been investigating recovering phosphorus by the crystallization method. The phosphorus recovery using the crystallization method has the following advantages.

1. Phosphorus can be recovered in an easily reusable form.
2. The sludge generation resulting from coagulants in the wastewater treatment plant is reduced.

Recently in Japan, MAP (magnesium ammonium phosphate: struvite) recovery facilities making use of fluidized-bed reactors have been in operation and the recovered MAP has been reused as fertilizer (Ishizuka *et al.*, 1998; Tomoda, 1999; Ueno and Fujii, 2001). However, with the conventional fluidized-bed reactors, as the crystallization progresses, the particle sizes of MAP in the reactor become large and it is difficult to keep them constant. Moreover, excessive growth of MAP crystals has such problems that the fluidization effect is degraded, the MAP surface areas necessary for reactions are reduced and as a result, the MAP recovery ratio is decreased (Shimamura *et al.*, 2001). Then, this time, the authors added a sub reaction tank and used this two-tank type reactor in our MAP recovery experiment. This system has such features that the seed crystals of relatively small particle sizes prepared in the sub reaction tank are added from time to time into the main reaction tank so that the particle sizes of MAP in the main reaction tank may be adjusted not to become too large. The purposes of this experiment are to realize stable treatment with high phosphorus recovery ratios and also to obtain the recovered products that are easy to reuse and having stable properties. In this paper, the influence of the phosphorus concentration in the raw water influent portion and the phosphorus surface area loading on the phosphorus recovery ratio is investigated and the treatment performance of the two-tank type reactor is confirmed.

Methods
Experimental apparatus
MAP is formed when phosphate ions, ammonium ions and magnesium ions exist in the solution in excess of the solubility products. The reaction for MAP formation is shown by Eq. (1).

$$HPO_4^{2-} + NH_4^+ + Mg^{2+} + OH^- + 5H_2O^- \rightarrow MgNH_4PO_4 \cdot 6H_2O \tag{1}$$

Figure 1 shows the experimental apparatus used in this experiment. This apparatus consists of a main reaction tank and a sub reaction tank, and each reaction tank has the following roles.

Main reaction tank. Relatively large MAP particles (mean particle size 0.5 to 1.5 mm) are fluidized in this tank, and new MAP particles are precipitated on the surfaces of MAP particles already fluidized in the main reaction tank, and through this process, phosphorus is recovered. MAP in the main reaction tank is withdrawn from time to time through the bottom of the main reaction tank. The mean particle sizes of MAP in the main reaction tank are kept constant by supplying the seed crystal (mean particle size about 0.3 mm) at a frequency of once in 3 to 4 days from the sub reaction tank.

Sub reaction tank. Fine MAP particles (mean particle size 0.05 to 0.25 mm) floating in the upper portion of the main reaction tank are transferred at a frequency of once in 3 to 4 days

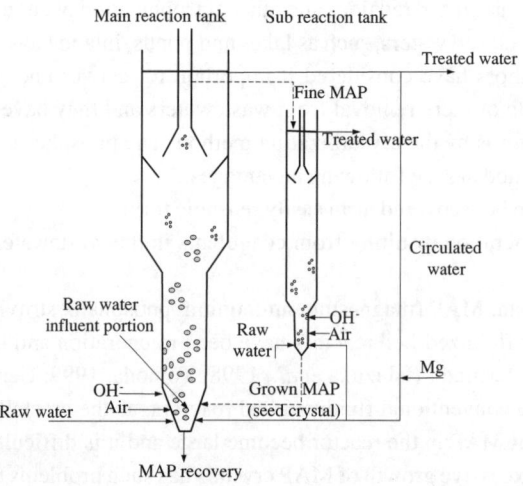

Figure 1 Experimental apparatus

into the sub reaction tank where they are grown up to about 0.3 mm. These grown MAP particles are returned entirely into the main reaction tank where they are used as the seed crystals.

The raw water and the circulated water are applied in up-flows through the raw water influent portion of the reaction tank bottom in both the main reaction tank and the sub reaction tank. The water application ratio of the raw water is about 9:1 between the main reaction tank and the sub reaction tank. Magnesium was supplied in the circulated water. As the source of magnesium, magnesium chloride was used. The ratio of magnesium addition was 1.3 to 1.9 in Mg/P ratio by weight. To each reaction tank, air for MAP agitation and alkali for pH adjustment were supplied. As the raw water for this experiment, the actual wastewater, anaerobically treated, was used. The mean particle size in this report is shown on a weight basis.

Investigation on the fundamental operating conditions

Before conducting a continuous experiment with the two-tank type reactor, the fundamental operating conditions were investigated separately.

Relation between the raw water influent portion phosphorus concentration and the phosphorus recovery ratio. In general, crystallization reactions produce fine crystals when the degree of supersaturation is high. The fine crystals settle slowly and therefore, flow out easily with the treated water and when the degree of crystallization is large, the recovery ratio is degraded. In the case of this reactor, the degree of supersaturation is highest in the raw water influent portion of the reactor bottom where the raw water is mixed with magnesium and alkali. Then, in order to grasp the conditions of phosphorus concentration under which the formation of fine MAP particles is minimum, the phosphorus recovery ratio was compared by changing the PO_4-P concentration in the raw water influent portion. The raw water flow rate was 4.4 m^3/d and the raw water PO_4-P was about 150 mg/L. The PO_4-P concentration in the raw water influent portion was adjusted to 150 to 50 mg/L by returning the circulated water to the raw water influent portion at a circulation ratio of 0 to 2 without changing the raw water flow rate. The mean MAP particle size used was 0.3 to 0.8 mm.

Relation between the phosphorus surface area loading and the phosphorus recovery ratio and the MAP growth rate. The phosphorus surface area loading (L_{sur}) means the amount of phosphorus supplied per unit MAP surface area (NOTE 1). The relation of the phosphorus recovery ratio and the crystallization ratio to the L_{sur} was investigated. Moreover, because the L_{sur} gives a large influence on the MAP growth rate, the relation between the L_{sur} and the MAP growth rate was also investigated. By changing the raw water flow rate to 0.3 to 0.8 m^3/d and the charged MAP mean particle size to 0.2 to 0.4 mm while keeping raw water T-P 100 mg/L and charged MAP amount 120g constant, respectively, L_{sur} 5 to 30 g-P/m^2·d was obtained. Here, the phosphorus recovery ratio means the T-P recovered in proportion to the raw water T-P (NOTE 2), and the crystallization ratio shows the PO_4-P crystallized in proportion to the raw water PO_4-P (NOTE 3).

NOTE 1: $L_{sur} = W_P/S_{MAP}$
NOTE 2: Recovery ratio $= \left\{ \dfrac{(T-P)_{Rw} - (T-P)_{Tw}}{(T-P)_{Rw}} \right\} \times 100$

NOTE 3: Crystallization ratio $= \left\{ \dfrac{(PO_4-P)_{Rw} - (PO_4-P)_{Tw}}{(PO_4-P)_{Rw}} \right\} \times 100$

Where L_{sur} : phosphorus surface area loading, g-P/m²·d
W_P : phosphorus influent flow rate, g-P/d
S_{MAP} : total MAP surface area in the reactor, m²
T-P : total phosphate, mg/l
PO_4-P : orthophosphate, mg/l
Subscripts: $_{Rw}$ raw water
$_{Tw}$ treated water

Fine MAP particle growth rate. In order to determine the retention time for the fine MAP particles in the sub reaction tank transferred from the main reaction tank, the fine MAP particle growth rate was measured. The retention time was set at a value at which the mean fine MAP particle size was above 0.3 mm. The conditions of water application were raw water T-P about 300 mg/L, raw water flow rate 0.13 m³/d and circulated water flow rate 0.87 m³/d.

Continuous water application experiment

The treatment performance of the two-tank type reactor was confirmed. The items investigated include the treated water quality and, in addition, the stability of the mean MAP particle size in the main reaction tank and the changes of the phosphorus recovery ratio when the mean MAP particle size in the main reaction tank was changed. The continuous water application experiment was conducted under four experimental conditions, Test 1, Test 2, Test 3 and Test 4. Table 1 shows these experimental conditions. In any of the four Tests, the PO_4-P concentration in the raw water influent portion was adjusted to about 50 mg/L by circulating the treated water. The phosphorus concentration of the raw water was 100 mg/L in Test 1, 300 mg/L in Test 2 and Test 3, and 500 mg/L in Test 4. The mean particle size of MAP in the main reaction tank was 1 mm in Test 1 and Test 2 and 0.5 mm in Test 3 and Test 4.

Results and discussion

Investigation on the fundamental operating conditions

Relation between the raw water influent portion phosphorus concentration and the phosphorus recovery ratio. Figure 2 shows the relation between the PO_4-P concentration in the raw water influent portion and the phosphorus recovery ratio. The phosphorus recovery ratio was 92% and 66% when the PO_4-P concentration in the raw water influent portion was PO_4-P 50 mg/L and 150 mg/L, respectively; that is, there was a tendency that the phosphorus recovery ratio decreased when the PO_4-P concentration in the raw water influent portion increased. In the raw water influent portion, the condition of supersaturation occurred

Table 1 Experimental conditions

		Test 1	Test 2	Test 3	Test 4
Raw water properties	T-P (mg/L)	111	330	298	507
	NH_4-N (mg/L)	177	619	514	862
Main reaction tank	Raw water flow rate (m³/d)	6.7	1.5	2.2	1.1
	Circulation ratio (–)	1.0	8.2	4.9	10.6
	Mean MAP particle size				
	Max. (mm)	0.85	0.94	0.41	0.55
	Min. (mm)	1.18	1.43	0.67	0.68
Sub reaction tank	Raw water flow rate (m³/d)	0.32	0.11	0.13	0.14
Raw water split stream ratio (main to sub)		95:5	93:7	94:6	89:11
Addition of Mg (Mg/P by weight) (–)		1.5	1.9	1.3	1.4

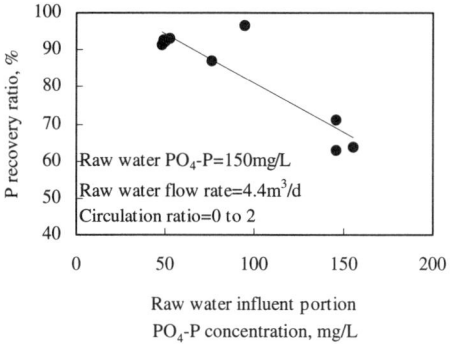

Figure 2 Relation between PO$_4$-P concentration and P recovery ratio

locally because of the high PO$_4$-P concentration, and fine MAP particles were precipitated in large numbers. Based on the results of this experiment, the PO$_4$-P concentration in the raw water influent portion was lowered by diluting with the circulated water in the continuous water-application experiment using a two-tank type reactor.

Relation between the L_{sur} and the MAP growth rate and the phosphorus recovery ratio.
Figure 3 shows the relation between the L_{sur} and the crystallization ratio and the phosphorus recovery ratio. The phosphorus recovery ratio was 80%, 75% and 60% when the L_{sur} was 10 g-P/m²·d, 20 g-P/m²·d and 30 g-P/m²·d, respectively; that is, there was a tendency that the phosphorus recovery ratio decreased when the L_{sur} increased. The crystallization ratio was about 90% with almost no change at any value of the L_{sur}. Based on the fact that the difference between the crystallization ratio and the recovery ratio increased with increasing L_{sur}, it is considered that as the loading increased, fine MAP particles were precipitated in larger numbers and flowed away with the treated water. It was found that, in order to raise the phosphorus recovery ratio, it was necessary to lower the L_{sur}. In the continuous water application experiment using the two-tank type reactor, the recovery ratio was compared when the L_{sur} was changed by changing the mean particle size of MAP to be charged. Figure 4 shows the relation between the L_{sur} and the MAP growth rate. The MAP growth rate was 0.05 mm/d, 0.11 mm/d and 0.19 mm/d when the L_{sur} was 10 g-P/m²·d, 20 g-P/m²·d and 30 g-P/m²·d, respectively; that is, there was a tendency that the MAP growth rate increased when the L_{sur} increased. As the L_{sur} was increased, the phosphorus recovery ratio was lowered because the MAP growth rate was increased and fine MAP particles were precipitated in larger numbers.

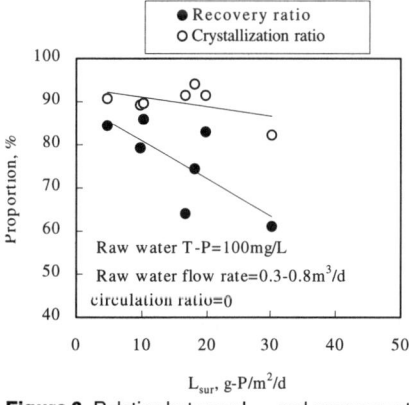

Figure 3 Relation between L_{sur} and recovery ratio and crystallization ratio

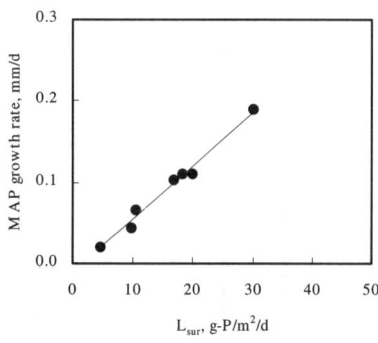

Figure 4 Relation between L_{sur} and MAP growth rate

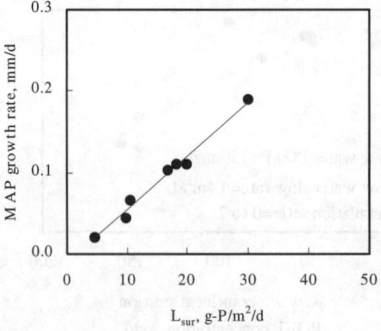

Figure 5 Growth of fine MAP particles

Fine MAP particle growth rate. Figure 5 shows the increase of the mean fine MAP particle size. The mean particle size of fine MAP particles transferred from the main reaction tank was 0.17 to 0.26 mm. The growth rate of fine MAP particles was 0.025 to 0.057 mm/d and the mean growth rate was 0.046 mm/d. When the retention time was longer than 3 days, the mean particle size of fine MAP particles was mostly larger than 0.3 mm. In the continuous water application experiment, the retention time of fine MAP particles in the sub reaction tank was 3 to 4 days.

Continuous water application experiment

Figure 6 shows the changes with the passage of time (d) of the treated water quality in the continuous treatment experiment and the changes with the passage of time (d) of the mean MAP particle size in the main reaction tank.

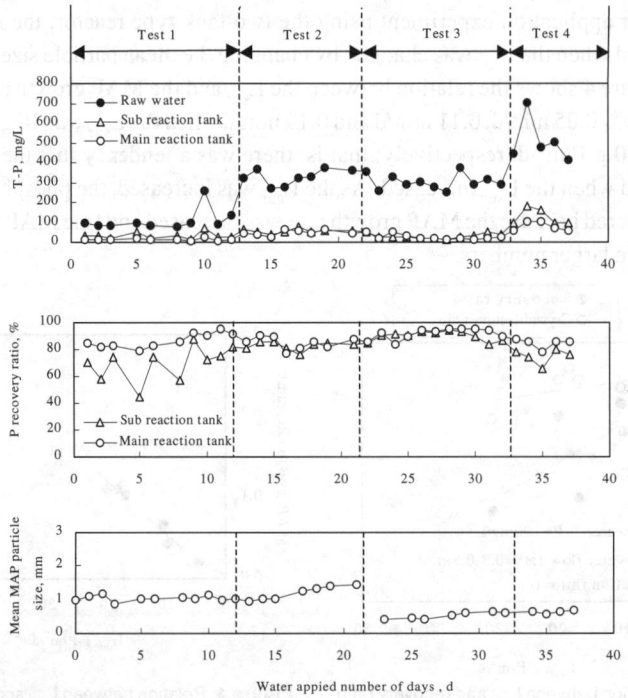

Figure 6 Changes with the passage of time (d)

Particle size of MAP in the main reaction tank. When the water was applied under the same conditions as in Test 1 in the reactor of the conventional one-tank type, the mean particle size was 1.5 mm, 1.8 mm and 2.1 mm after 3 days, 7 days and 14 days, respectively, when the mean seed crystal particle size (at the beginning of the experiment) was 0.79 mm; that is, the mean particle size increased by 1.3 mm for 2 weeks. When the two-tank type reactor was operated this time, the MAP mean particle size was 0.85 to 1.43 mm in Test 1 and Test 2, and 0.41 to 0.68 in Test 3 and Test 4, respectively, as seen in the changes with the passage of time (d) of the mean MAP particle size in Figure 6; that is, the mean MAP particle size in the reactor was stable. It was confirmed that the MAP particles in the main reaction tank could be kept approximately constant without excessive growth by supplying the seed crystal from time to time from the sub reaction tank.

Treated water quality. Table 2 shows the average water quality of the raw water, the main reaction tank treated water and the sub reaction tank treated water.

When the raw water had T-P 111 to 507 mg/L, the main reaction tank treated water had T-P 14.0 to 79.5 mg/L and PO_4-P 7.1 to 18.5 mg/L. The treated water T-P increased with increasing raw water T-P, but the treated water PO_4-P hardly changed irrespective of the raw water T-P. The phosphorus recovery ratio was 84 to 92%. The sub reaction tank treated water had T-P 27.8 to 128 mg/L and PO_4-P 4.0 to 12.1 mg/L. Similarly to the case of the main reaction tank, the treated water T-P increased with increasing raw water T-P, but the treated water PO_4-P hardly changed irrespective of the raw water T-P. The phosphorus recovery ratio was 71 to 91%, and it was lower than in the main reaction tank in any Test. In the case of the main reaction tank, the mean MAP particle size in this tank was stable, and therefore, the phosphorus recovery ratio was above 80% at all times; that is, high recovery ratios can be maintained.

Mean MAP particle size and phosphorus recovery ratio. In Figure 3, it was confirmed that, as the L_{sur} was decreased, the phosphorus recovery ratio increased. Then, in Test 2 and Test 3, the L_{sur} was changed by changing the mean MAP particle size in the main reaction tank

Table 2 Experimental results

Reaction pH (–)		Test 1	Test 2	Test 3	Test 4
		8.4	8.4	8.3	8.4
SS	Raw water (mg/L)	66.0	70.4	209	224
	Sub reaction tank (mg/L)	112.0	203	310	630
	Main reaction tank (mg/L)	53.0	222	285	509
T-P	Raw water (mg/L)	111	330	298	507
	Sub reaction tank (mg/L)	32.5	56.3	27.8	128
	Main reaction tank (mg/L)	14.0	49.2	23.2	79.5
PO_4-P	Raw water (mg/L)	90.1	274	247	427
	Sub reaction tank (mg/L)	8.5	4.0	5.6	12.1
	Main reaction tank (mg/L)	7.5	7.1	9.6	18.5
NH_4-N	Raw water (mg/L)	177	619	514	852
	Sub reaction tank (mg/L)	153	467	395	632
	Main reaction tank (mg/L)	154	492	412	673
Mg	Raw water (mg/L)	5.3	6.6	9.7	11.3
	Raw water* (mg/L)	140	527	331	609
	Sub reaction tank (mg/L)	32.1	190	126	260
	Main reaction tank (mg/L)	50.3	209	140	312
Recovery ratio					
	Sub reaction tank (%)	71	83	91	75
	Main reaction tank (%)	87	85	92	84

* After $MgCl_2$ addition (calculated)

while keeping the amount of MAP charged constant. In Test 2, the mean particle size was 0.94 to 1.43 mm and the L_{sur} was 4.8 g-P/m^2·d and in Test 3, the mean particle size was 0.41 to 0.67 mm and the L_{sur} was 3.3 g-P/m^2_d. As seen in Table 2, the treated water T-P of the main reaction tank was 49.2 mg/L and 23.2 mg/L in Test 2 and Test 3, respectively, when the raw water T-P was about 300 mg/L; that is, the concentration in Test 3 was about half that in Test 2. The phosphorus recovery ratio was 85% and 92% in Test 2 and Test 3, respectively; that is, the value in Test 3 was 7 points higher. It was confirmed that the L_{sur} was lowered and the phosphorus recovery ratio was raised by reducing the mean particle size without changing the amount of MAP charged.

Mass balance for P, N and Mg

The mass balance for P, N and Mg in Test 1 was investigated by comparing the decrease in PO_4-P, NH_4-N and Mg concentrations in the raw water with the composition of recovered MAP. The composition of recovered MAP was measured by XRF. While the raw water quality is PO_4-P 90.1 mg/L, NH_4-N 177 mg/L and Mg 140 mg/L (after $MgCl_2$ addition), the treated water quality is PO_4-P 7.5 mg/L, NH_4-N 154 mg/L and Mg 50.3 mg/L. The decrease ratios of component concentrations, P:N:Mg, were 1.0:0.28:1.1. On the other hand, XRF indicates that the composition of recovered MAP is P-16 wt%, N 4.8 wt% and Mg 16 wt% and therefore, the composition ratios, P:N:Mg, were 1.0:0.30:1.0. Since the decrease ratios of raw water concentrations and the composition ratios of recovered MAP approximately agree with each other, it could be confirmed that, by the growth of MAP, the raw water concentrations were decreased.

Conclusions

With the conventional fluidized-bed reactors, as the crystallization progressed, the particle sizes of MAP in the reactor became large and it was difficult to keep them constant. Moreover, excessive growth of MAP crystals had such problems that the fluidization effect was degraded, the MAP surface areas necessary for reactions were reduced and as a result, the MAP recovery ratio was decreased. This time, the authors constructed a two-tank type reactor consisting of a main reaction tank and a sub reaction tank and operated it in such a manner that the mean particle size of MAP in the main reaction tank might not be changed. It was confirmed that stable treatment with high recovery ratios was possible. Moreover, the MAP could be recovered in a form easy to reuse because of the uniform mean particle sizes. It is considered that, in addition to phosphorus recovery by MAP formation, this system can be applied to phosphorus recovery by calcium phosphate formation, water softening by calcium carbonate formation, fluorine recovery by calcium fluoride formation, etc.

References

Durrant, A.E., Scrimshaw, M.D., Stratful, I. and Lester, J.N. (1999). Review of the Feasibility of Recovering Phosphate from Wastewater for Use as a Raw Material by the Phosphate Industry. *Environmental Technology*, **20**, pp 749–758.

Hirasawa, I. and Toya, Y. (1998). Development of ion removal process from wastewater by crystallization. *International Symposium on Industrial Crystallization*, pp 724–739.

Ishizuka, S., Sato, S. and Shibata, M. (1998). Recovery of Fluoride and Phosphate from Wastewater. *International Symposium on industrial Crystallization*, pp 716–723.

Matsumiya, Y, Yamashita, T. and Nawamura, Y. (2000). Phosphorus removal from sidestreams by crystallization of magnesium-ammonium-phosphate using seawater. *J. Chem. Inst. Water Environ.*, **14**, 291–296.

Shimamura, K., Homma, Y., Watanabe, A. and Tanaka, T. (2001). Research on MAP recovery conditions using a fluidized-bed crystallized phosphorus removal system. *Asian Waterqual 2001*, pp 843–848.

Tomoda, M. (1999). Utilization of MAP as fertilizer in Fukuoka-city, *Journal of Japan Sewage Works Association*, **36**, 437, pp 42–46.

Ueno, Y. and Fujii, M. (2001). Three years' experience of operating and selling recovered struvite from full-scale plant. *Environmental Technology*, **22**, 1373–1381.

Removal and recovery of phosphate and ammonium as struvite from supernatant in anaerobic digestion

M. Yoshino*, M. Yao*, H. Tsuno** and I. Somiya**

* Maezawa Industries, INC., 5-11, Naka-cho, Kawaguchi-shi, Saitama 332-8556, Japan
(E-mail: *masaaki_yoshino@maezawa.co.jp*; *makoto_yao@maezawa.co.jp*)
** Graduate School of Engineering, Kyoto University, Yoshida-honmachi, Sakyo-ku, Kyoto 606-8501, Japan
(E-mail: *tsuno@water.env.kyoto-u.ac.jp*; *isoumiya@rins.ryukoku.ac.jp*)

Abstract Removal of phosphorus and nitrogen is required to prevent eutrophication problems in lakes and enclosed coastal seas. And recovery of phosphorus from wastewater has been attracting attention because of lack in phosphorus resources in the near future. In this study, reaction kinetics and design parameters of struvite production are experimentally investigated by using basic reaction type and a draft-tube type reactors. Struvite production rate, which is a very important parameter in reactor design and efficiency estimation, is formulated in an equation consisting of a rate constant (k_2), and magnesium, phosphate and ammonium concentrations. The value of k_2 is shown to be increased with struvite concentration and mixing intensity in the reactor. The developed equation is applied to the results obtained from the draft-tube type reactor experiments and verified for its applicability. High struvite concentration of 10–25% is maintained in the draft-tube reactor experiments. 92% removal and recovery efficiency with effluent phosphorus concentration of 17 mg/L is achieved under the conditions of 4 minutes reaction time, pH of 8.5 and Mg/P molar ratio of 1.1.

Keywords Draft-tube type reactor; nitrogen removal; phosphorus removal; struvite; supernatant in anaerobic digestion

Introduction

Removal of phosphorus and nitrogen is required to prevent eutrophication problems in lakes and enclosed coastal seas. As the reserves of phosphorus rock are limited (Steen, 1998), a recovery of phosphorus from wastewater has been also attracting attention. The removal of phosphate for recycling from wastewater has been practiced widely and a number of effective and reliable techniques have been employed. Although these are already put in practical use, improvement for cost reduction is also required. In this study, reaction kinetics and design parameters of struvite production are experimentally investigated to seek an optimum method of struvite production. The experiments were conducted by using basic reaction type and draft-tube type reactors. The problems of struvite grown on the walls of the pipes in anaerobic digestion systems have been reported (Borgerding, 1972). Therefore, the wastewater subject to phosphorus recovery in this study is the supernatant in anaerobic digestion of sludge from a sewage treatment plant.

Struvite production rate, which is a very important parameter in reactor design and efficiency estimation, is formulated in an equation consisting of a rate constant (k_2), and magnesium, phosphate and ammonium concentrations. The developed equation is applied to the results obtained from the draft-tube type reactor experiments and verified for its applicability. Design parameters, such as reaction time (hydraulic retention time), magnesium addition ratio and pH are discussed and the relationships among these parameters required to meet the effluent phosphorus concentration are drawn based on this equation.

The constitution of the obtained crystal are checked with Electron Spectroscopy for Chemical Analysis (ESCA).

Reaction kinetics

Struvite is generated by phosphate, ammonia nitrogen, magnesium ion and hydroxide ion. Its reaction formula is shown as follows:

$$Mg^{2+} + NH_4^+ + HPO_4^{2-} + OH^- + 5H_2O \rightarrow MgNH_4PO_4 \cdot 6H_2O \tag{1}$$

The value of K, which is solubility product, is calculated by the following equation, when reaction (1) is at equilibrium:

$$K = [Mg^{2+}][NH_4^+][HPO_4^{2-}][OH^-] \tag{2}$$

The struvite generation rate can be obtained by using k ($L^3/mol^3 \cdot min$) as a rate constant under the condition that temperature and agitation are constant.

$$dX/dt = k[Mg^{2+}][NH_4^+][HPO_4^{2-}][OH^-] \tag{3}$$

where X represents the generation weight of struvite in mol/L, and t represents the time in minutes. When the concentration of NH_4-N is much higher than the concentration of PO_4-P and changes little before and after the reaction, as it is in the case of supernatant in anaerobic digestion, Eq. (3) can be simplified as follows, if pH is maintained at a fixed level:

$$dX/dt = k_2[Mg^{2+}][HPO_4^{2-}] \tag{4}$$

where k_2 (L/mol·min) is the reaction rate constant. Factors influencing the value of k_2 are agitation intensity and the concentration of the struvite in the reaction section.

If we suppose that, in the batch experiment, there are a mol/L of Mg^{2+} and b mol/L of HPO_4^{2-} when the time is zero and that both Mg^{2+} and HPO_4^{2-} react by an amount of x mol/L at a certain time t with struvite generation in an amount of X mol/L, Eq. (4) can be expressed as follows:

$$dX/dt = k_2 (a-x-a_0)(b-x-b_0) \tag{5}$$

where a_0 is the equilibrium concentration (mol/L) of Mg^{2+} and b_0 is the equilibrium concentration (mol/L) of HPO_4^{2-}.

Under the initial conditions that $x = 0$ at $t = 0$, the solution of this differential equation becomes:

$$[1/(a-a_0) - (b-b_0)] \ln [(b-b_0)(a-x-a_0)]/[(a-a_0)(b-x-b_0)] = k_2 t \tag{6}$$

Methods

Two types of experiments were conducted in this study: a batch experiment using artificial wastewater, which was synthesized as the supernatant in anaerobic digestion, and a continuous struvite generation experiment using the actual supernatant in anaerobic digestion. A batch experiment was conducted to understand the mechanisms of struvite generation and factors that influence them. Figure 1 shows the experimental apparatus used in the batch experiment. The apparatus shown in Figure 1(a) is a paddle agitation reactor, in which a jar tester was used for agitation and a 1 litre beaker was used as a reaction tank. In the apparatus shown in Figure 1(b), aeration was used for agitation of the mixture. A 0.6 litre glass tank (height of 145 mm; inner diameter of 96 mm; inner diameter of 26 mm of the bottom opening for aeration) was used as the reactor. KH_2PO_4 solution, NH_4Cl solution, and $MgSO_4.7H_2O$ solution were added to ion-exchanged water in a 1 litre measuring flask to get a 1 L mixture of a given set of concentrations of PO_4-P, NH_4-N, and Mg^{2+}, which was promptly poured into the beaker used as a reactor in which a pH meter was installed. The mixture was then agitated by the jar tester at a fixed rotation speed. The pH level was adjusted by adding 1 mol l^{-1} NaOH solution. The PO_4-P concentration and the NH_4-N concentration were set to be 100 mg/L and 500 mg/L, respectively, which are the same level as those in the supernatant in anaerobic digestion. The experimental conditions set were as follows: concentration of Mg^{2+} to be added to the mixture, 100 mg/L; pH level, 9.0; rotation speed, 200 r.p.m. Agitation intensity, pH and struvite concentration in the reaction tank were considered to be operational parameters which influence the reaction rate constant. Aeration agitation as well as paddle agitation were tested. The experimental conditions are shown in Table 1.

Based on the results of the batch experiment, a continuous struvite generation experiment using supernatant in anaerobic digestion was conducted. Figure 2 shows the schematic diagram of the experimental apparatus used in the study. It consists of a center section, which is designed as a draft-tube type reactor, and a peripheral section, where generated products are separated. The supernatant as the influent and Mg^{2+} solution were continuously added into the part of the draft tube. The pH level in the reaction section was adjusted automatically by a pH controller with 1 mol l^{-1} NaOH solution. A paddle agitator was installed in the inner tube of the draft tube to create a descending current in the inner tube and an ascending current in the outer tube of the draft tube. Thus, the influent, Mg^{2+} solution and struvite were mixed well in the reactor. Generated struvite was separated from effluent in the peripheral section outside the draft tube by gravity separation.

Table 2 shows the quality of the supernatant in anaerobic digestion used as the influent after sedimentation for about 60 minutes. Table 3 shows the experimental conditions. The struvite concentration in the reaction section was expressed as apparent volume ratio of struvite in the mixed liquor of the reactor, which was measured by sampling of the mixed liquor and sedimentation for 5 minutes in a graduated measuring cylinder. The effective capacity of the reaction section was calculated by deducting the volume of sedimented struvite at the bottom of the reaction section and the volume of struvite floating inside the reaction section from the total volume of the reaction section.

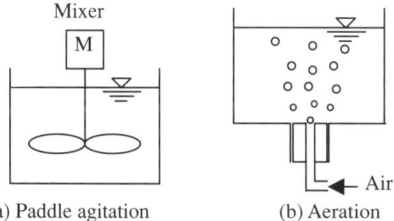

(a) Paddle agitation   (b) Aeration

Figure 1 The apparatus used in the batch experiment

Table 1 Experimental parameters of the batch experiment

Method for mixing	G-value	Aeration volume	pH	Concentration of the struvite in the reaction section
	(1/sec.)	(m³-air/m³-reactor/min)	(–)	(mg/L)
Paddle agitation	4~133	–	8.8~9.0	55~1,380
Aeration	–	1.67	9.0	400~3,200

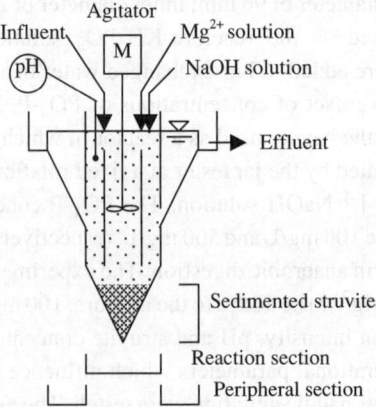

Figure 2 The apparatus used in the continuous experiment

Table 2 The quality of the influent of the continuous experiment

	Run 1	Run 2
pH (–)	7.5~8.0	7.5~7.7
Water temperature (c)	19.1~25.0	8.1~17.1
SS (mg/L)	18~52	38~97
PO_4-P (mg/L)	198~242	266~290
NH_4-N (mg/L)	441~602	511~591
Mg^{2+} (mg/L)	5.0~19	7.0~11
Ca^{2+} (mg/L)	8.0~16	–

Table 3 The experiment conditions of the continuous experiment

Item			Run 1	Run 2
Center section	Volume	(L)	2.1~2.3	2.5
	pH	(–)	8.4~8.5	8.2~8.3
	Mg/P	(–)	0.8~1.2	0.7~1.1
	Concentration of the struvite	(%)	5~25	7~12
		(mg/L)	37,000~140,000	36,000~64,000
	Retention time	(min)	3.9~12.1	13.2~15.8
	R.p.m.. of the paddle agitator	(r.p.m.)	250	200
	G-Value	(1/sec.)	178~219	132~142
Peripheral section	Overflow rate	(cm/min)	0.4~1.2	0.3~0.4

Results and discussion

Batch experiment

Figure 3 shows an example of change in soluble concentration of PO_4-P and NH_4-N with reaction time in the batch experiment. This was the result obtained at a low rotation speed of 20 r.p.m. which corresponds to the G value of 4 (1/sec). The reaction was considerably slower as compared with the reaction seen at standard rotation speed (200 r.p.m.). However, the reaction was fast in the initial stage, when the concentration of PO_4-P was high. Meanwhile, NH_4-N concentration changed little. Based on the data obtained after a lapse of 2 hours when chemical equilibrium had almost been achieved, the product of concentration values $[Mg^{2+}][NH_4^+][HPO_4^{2-}]$ was calculated. Figure 4 shows the relationship between $[OH^-]$ and the obtained product of concentration values $[Mg^{2+}][NH_4^+][HPO_4^{2-}]$. A straight line, with an inclination of –1 corresponding to the experimental data, was drawn in Figure 4, to obtain k (the product of solubility values) in Eq. (3) at the equilibrium state

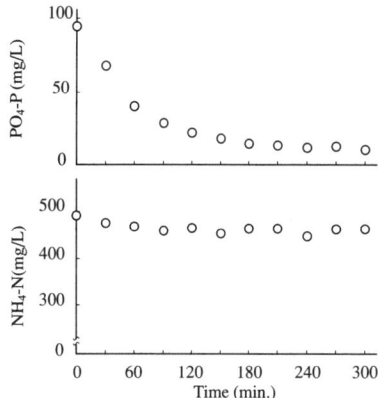

Figure 3 The change of PO_4-P and NH_4-N concentration

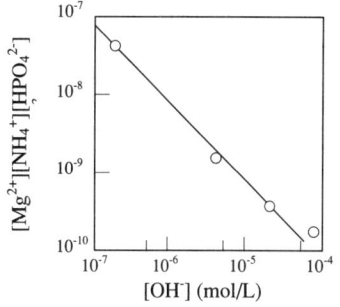

Figure 4 The relationship between [OH⁻] and $[Mg^{2+}][NH_4^+][HPO_4^{2-}]$

($dX/dt = 0$). The obtained value for k was 7.8×10^{-15} $(mol/L)^{-4}$. Figure 5 also shows how to calculate the reaction rate constant k_2 based on Eq. (6).

The results of experiments done under the conditions indicated in Table 1 are also shown in Figures 6 and 7. Figure 6 shows the result of the experiment using paddle agitation. When the agitation intensity reached 40 (1/sec), its effect seemed to have reached its upper limit. Figure 7 shows the relationship between the concentration of struvite in the reaction section and the reaction rate constant k_2. Data obtained when the G value was 4 (1/sec) was removed from Figure 7 because of rate limitation by agitation. As Figure 7 shows, it was found that the reaction rate constant k_2 is proportional to the concentration of struvite in the reaction section.

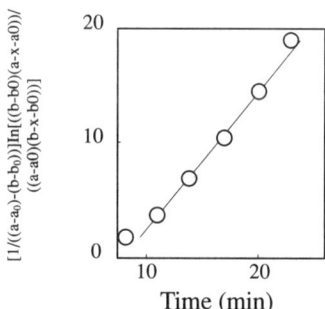

Figure 5 The calculation of the rate constant k_2

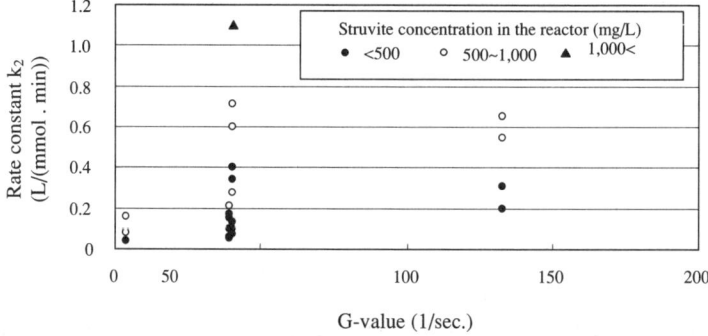

Figure 6 The relationship between mixing speed and rate constant k_2

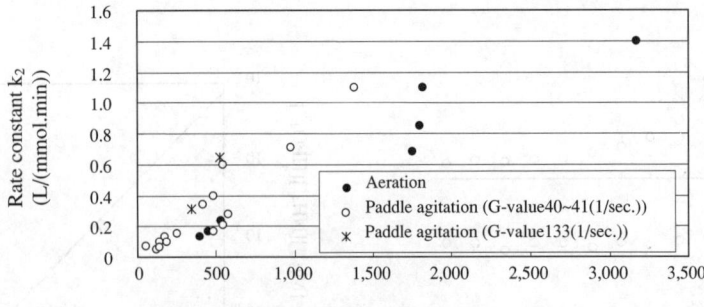

Figure 7 The relationship between the concentration of struvite in the reactor and rate constant k_2

Continuous struvite generation experiment

It was found, from the result of the batch experiment, that the generation rate of struvite is considerably affected by the concentration of struvite in the reaction section. Thus, a continuous struvite generation system using paddle agitation with variable rotation speed was designed for this experiment. Figure 8 shows the results of the experiment conducted under the conditions shown in Table 3 (RUN1). As shown in Figure 8, the product of concentration values became smaller, which means larger values of k, as the hydraulic retention time became longer. The concentration of struvite in the reaction section was high, reaching 5 to 25 percent (37,000 to 140,000 mg/L), when the experiment was conducted. The reaction was faster when the concentration of struvite was 10 to 25 percent than it was at a concentration of 5 to 10 percent. In RUN 1, the molar ratio of Mg/P was changed from 0.8 to 1.2. The concentration of PO_4-P in the effluent was within the range of 9.9 to 52 mg/L. As shown in Figure 8, the variation range of the product of concentration values was very small. This indicates that the product of concentration values was an appropriate parameter, proving the generation efficiency of struvite. And it was possible to obtain a concentration of 17 mg/L and a removal rate of 92 percent for PO_4-P in the effluent, even when the hydraulic retention time was as short as 3.9 minutes, if the pH level was set at 8.5 and the molar ratio of Mg/P at 1.1.

A continuous experiment was also conducted for a longer period of time under the conditions shown in Table 3 (RUN 2). Struvite crystals were allowed to grow in the reaction section. Figure 9 shows the change in the struvite crystal grain size distribution.

Struvite crystals had been growing from the beginning of the experiment up until 18 hours. Figure 9 also indicates that struvite crystals with a grain diameter of more than 0.6 mm sedimented to the bottom of the reaction section, because of the higher sedimentation

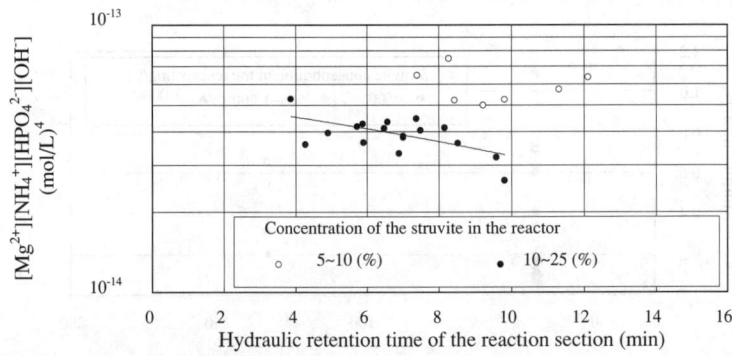

Figure 8 The relationship between the hydraulic retention time of the reaction section and the product of concentration values $[Mg^{2+}][NH_4^+][HPO_4^{2-}][OH^-]$

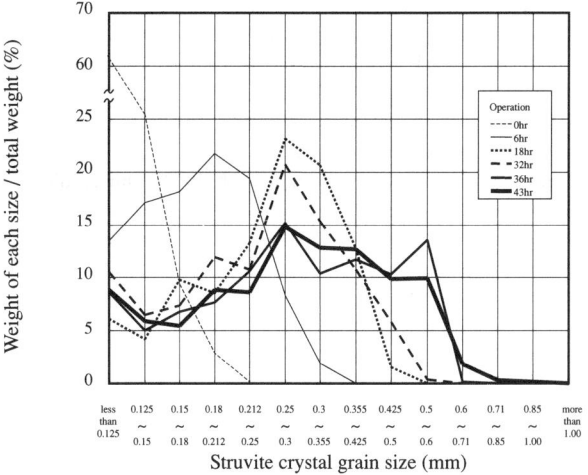

Figure 9 The change in the struvite crystal grain size distribution

rate than the speed of the ascending current in the draft tube. The constitution of the obtained crystal was checked with Electron Spectroscopy for Chemical Analysis (ESCA) and proved to be struvite.

Conclusion

In this study, reaction kinetics and design parameters of struvite production are experimentally investigated by using basic reaction type and draft-tube type reactors, the main results obtained were as follows.

1. Based on the batch-experiment data obtained after a lapse of 2 hours when chemical equilibrium had almost been achieved, the product of solubility values K was calculated to be 7.8×10^{-15} (mol/L)$^{-4}$.
2. The struvite production rate can be obtained by using k_2 (L/mmol.min) as a rate constant, if the concentration of NH_4-N is much higher than the concentration of PO_4-P, as it is in the case of supernatant in anaerobic digestion.

$$dX/dt = k_2[Mg^{2+}][HPO_4^{2-}]$$

3. As a result of the experiment using paddle agitation, when the agitation intensity reached 40 (1/sec), its effect seemed to have reached its upper limit. And it was found that the reaction rate constant k_2 is proportional to the concentration of struvite in the reaction section.
4. High struvite concentration of 10–25% is maintained in the draft-tube reactor experiments. 92% removal and recovery efficiency with effluent phosphorus concentration of 17 mg/L is achieved under the conditions of 4 minutes reaction time, pH of 8.5 and Mg/P ratio of 1.1.
5. Using the draft-tube reactor, struvite crystals had been growing from the beginning of the experiment up until 18 hours. And struvite crystals with a grain diameter of more than 0.6 mm sedimented to the bottom of the reaction section, with a higher sedimentation speed than that of the ascending current in the draft tube.
6. The constitution of obtained crystal was checked with Electron Spectroscopy for Chemical Analysis (ESCA) and proved to be struvite.

References

Borgerding, J. (1972). Phosphate deposits in digestion systems. *Journal WPCF*, **44**(5), 813–819

Driver, J., Lijmbach, D. and Steen, I. (1999). Why recover phosphorus for recycling, and how? *Environmental Technology*, **20**(7), pp. 651–662.

Liberti, L. (1986). The $10^3 h^{-1}$ Rim-Nut demonstration plant at westbari for removing and recovering N and P from wastewater. *Water Research*, **20**(6), pp. 735–740.

Salutsky, M.L., Dunseth, M.G., Ries, K.M. and Shapiro, J.J. (1972). Ultimate disposal of phosphate from waste water by recovery as fertilizer. *Effluent and Water Treatment Journal, Oct.*, pp. 509–519.

Steen, I. (1998). Phosphorus availability in the 21st century: Management of a non-renewable resource. *Phosphorus & Potassium September–October 1998*, pp.25–31.

A new phosphate-selective sorbent for the Rem Nut® process. Laboratory investigation and field experience at a medium size wastewater treatment plant

D. Petruzzelli, L. De Florio, A. Dell'Erba, L. Liberti, M. Notarnicola and A.K. Sengupta*

Department of Civil and Environmental Engineering, II Faculty of Engineering, The Polytechnic University of Bari, 8, Viale del Turismo, 74100 Taranto, Italy (E-mail: *l.liberti@poliba.it*)

* Department of Civil and Environmental Engineering, Lehigh University, 3, Packer Avenue, Bethlehem, PA, USA (E-mail: *aks0@lehigh.edu*)

Abstract P-control technologies for municipal wastewater are essentially based on "destructive" methods, that lead to formation of concentrated solid-phases (sludge), usually disposed-off in controlled landfills. Ion exchange, as a "non-destructive" technology, allows for selective removal and simultaneous recovery of pollutants, which can be recycled to the same and/or related productive lines. In this context, the REM NUT® process removes nutrient species ($HPO_4^=$, NH_4^+, K^+) present in biologically oxidised municipal effluents and recovers them in the form of struvites ($MgNH_4PO_4$; $MgKPO_4$), premium quality slow release fertilisers. The main limitation to the extensive application of this ion exchange based process is the non-availability of selective exchangers for specific removal of nutrient species. This paper illustrates laboratory investigation and pilot scale development of a so-called "P-driven" modified REM NUT scheme based on a new phosphate-selective sorbent developed at Lehigh University, PA, USA.

Keywords Eutrophication; ion exchange; nutrients; P-driven Rem Nut® process; struvite

Introduction

The presence of trace amounts of phosphates in treated wastewater is often responsible for eutrophication phenomena in lakes, reservoirs and coastal seawaters (Vollenweider, 1968). Chemical precipitation and biological removal are among the most common technologies for phosphate control in wastewater (Metcalf & Eddy, 1991). Both methods, however, transfer the problem from a diluted liquid-phase (wastewater) to a more concentrated solid-phase (sludge) which must be disposed-off properly or reused (Gaastra, 2001) in the form of P-containing raw material. Incidentally, these processes are highly sensitive to diurnal and seasonal changes in the feed composition as well as to operative variables such as plant hydraulic load and temperature.

Extensive R&D has been carried-out worldwide to explore the effectiveness of fixed-bed systems and the use of selective sorbents for phosphate removal because of their operational simplicity and their adaptability to changing wastewater composition, flow-rate and temperature (Gregory, 1976; Liberti *et al.*, 1979). However, critical shortcomings limiting the application of sorbents in the present context can be summarised as follows (Liberti, 2001): a) poor selectivity toward phosphates over competing anions such as SO_4^{2-}, Cl^-, HCO_3^-, and, consequently, low operative retention capacity; b) inefficient regeneration; c) gradual loss of the sorbents' loading capacity due to fouling phenomena associated with irreversible adsorption of bio-persistent organics and/or physical deposition of residual suspended matter still present in the treated wastewater.

In this context, the REM NUT® process allows for simultaneous removal of phosphate, ammonium and potassium ions from municipal wastewater, followed by their recovery in the form of a premium quality slow-release fertiliser, i.e. struvites: $MgNH_4PO_4$ and $MgKPO_4$ (Liberti *et al.*, 1984, 1989). In spite of the promising results achieved at demonstration level, the process never reached full-scale application, primarily due to the poor selectivity of phosphate ions in the anion exchanger and secondly due to related technical-economic inefficiencies (Liberti *et al.*, 2001).

A new class of phosphate-selective polymeric sorbent was recently developed at Lehigh University, PA, USA. The sorbent material is based on a commercial weak base anion exchanger (Dowex M-4195, from Dow Chemical Co., USA) converted to Cu-form, where after chelation of the metal species on the amino (pyridyl) functional groups, the resin developed a special selectivity toward phosphate ions (Zhao and Sengupta, 1998). The promising results obtained at laboratory scale were presented at the previous edition of this conference (Sengupta, 2001). The basis for international cooperation was set therein, with specific reference to testing modified "P-driven" schemes of the REM NUT® process and the use of the aforesaid sorbent. Extensive application of the reference exchanger to phosphate removal from municipal wastewater was carried out in the laboratory and later on a pilot plant (0.1 m^3/d) that was designed, assembled and run at Grottaglie-Monteiasi Wastewater Treatment Plant, S.E Italy (3,000 m^3/d; 20,000 I.E.). In the present contribution preliminary results from 3 months of continuous operation at the pilot unit are presented along with data from laboratory investigation.

The "P-driven" REM NUT® scheme

The REM NUT® process is based on two commercial exchangers, i.e. a strong base anion resin and a natural zeolite, removing phosphate, ammonium and potassium ions respectively from biologically oxidised municipal effluents. Nutrient species are subsequently recovered from the spent regeneration elutes in the form of ammonium and/or potassium struvites (Liberti *et al.*, 1979). As mentioned, the reference process never reached full scale application due to the following limitations: a) unavailability of true selective exchangers for retaining phosphate ions from secondary effluents, b) imbalance of the molar N/P ratio (ave. 10/1 in municipal effluents) for struvite precipitation. Imbalance between NH_4^+/HPO_4^{2-} for struvite precipitation may be truly overcome by "driving" the process towards recovering *all* phosphate and *just* equimolar amounts of ammonium ions from sewage, thus leaving the excess NH_4^+ removal for other conventional treatment processes like bio (de) nitrification, breakpoint chlorination, ammonia stripping and so forth. Through the adoption of the "*P-driven*" scheme (Figure 1) approximately 1:1 stoichiometric amounts of P-PO$_4$ and N-NH$_4$ are retained by selective ion exchange and later precipitated as struvite, with addition of Mg to ion exchanger regeneration eluate. Besides, the final effluent is acceptable in terms of eutrophication control and improved overall characteristics such as COD and SS.

Figure 1 outlines the typical flow sheet of the "*P-driven*" process, where for continuous operation, two ion exchange units for cation and anion exchanges respectively are operated in parallel.

In the present context, the true problem lies in the lack of a truly selective sorbent material that can ensure successful technical-economic performance of the new process scheme.

Laboratory investigation and pilot plant design

At the department of Civil and Environmental Engineering of Lehigh University, PA, USA, a new class of sorbents was developed that showed good selectivity towards phosphate ions at the ionic composition of biologically treated municipal effluents (Zhao and Sengupta, 1998). The specific sorbent was based on a commercial Cu-form weak base anion exchanger (Dowex M4195, from Dow Chemical Co., USA) that was especially selective towards phosphate ions through ligand exchange through chelation of the metal

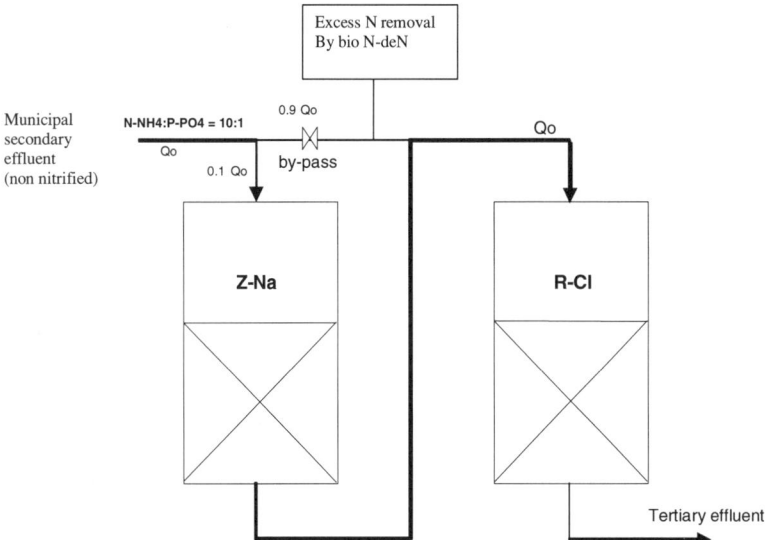

Figure 1 Schematic layout of P-driven REM NUT process (Z-Na = sodium-form zeolite; R-Cl = chloride-form anion resin)

species on the resin amino (pyridyl) functional groups (Outola *et al.*, 2001). The resin was submitted to preliminary laboratory experiments before pilot plant investigation. Figure 2a shows resin exhaustion performance towards phosphate ions after percolation of synthetic solutions that represent average composition of secondary effluents. The solution was prepared in tap water spiked with phosphate ion concentration exceeding 4.4 mgP/L.

In the experiment, 4 g of Cu-form resins samples (10 cm^3) were loaded into glass columns and percolated at different flow-rates (3 to 20 BV/h) (BV = resin Bed Volumes) to optimise hydrodynamic performance of the column. More than 400 BV were percolated through the column with a leakage below 0.1 mgP/L, in compliance with the more stringent limits imposed by the current EU legislation. 20 BV/h also represented the maximum flow-rate that still ensured sufficient column bed contact time for the phosphate retention

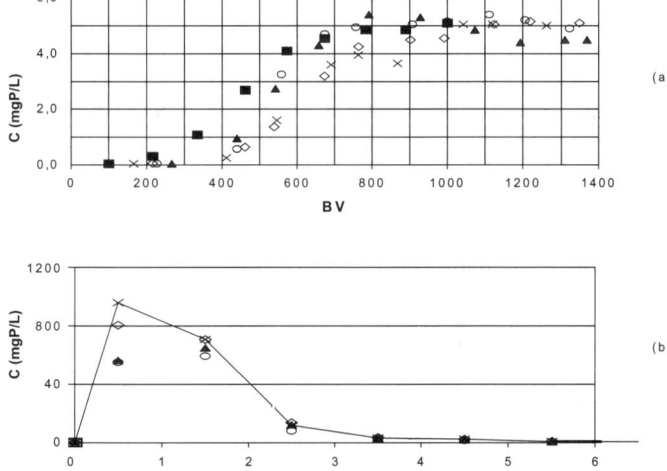

Figure 2 Performance of resin Dowex M-4195 toward phosphate ions. a) Exhaustion (○ 3B/h; ▲ 5 BV/h; × 7 BV/h; ◇ 10 BV/h; ■ 20 BV/h); b) Regeneration (○,× 2,5 BV/h; ◇,▲ 5 BV/h; NaCl 1 mol l^{-1}, pH 4.3)

reaction to be completed. Figure 2b shows the corresponding phosphate regeneration curve for regeneration with acidic (pH 4.3) 1 mol l^{-1} NaCl solution. Only 3 BV of the indicated solution were sufficient for quantitative recovery (>95%) of the phosphate ions retained, independent of the column contact time, which varied from 2.5 to 5 BV/h. The resin loading capacity was 15 gP/L and the phosphate ion concentration in the effluent regenerant solution exceeded the influent solution concentration by a P-concentration factor of over 100. Laboratory data encouraged pilot investigation, which was carried-out with a 0.1 m^3/d dephosphation unit installed at Grottaglie-Monteiasi Wastewater Treatment Station, S.E. Italy. Figure 3 shows the plant layout, Table 1 summarises its main characteristics.

Pilot plant performance

Figure 4 shows phosphate performance of the resin Dowex M-4195 after 5 exhaustion-regeneration cycles carried out during the pilot investigation at Grottaglie-Monteiasi Wastewater Treatment Station.

Although the resin is still able to meet the legislative requirements, presence of competitor ions such as sulphate, nitrate and bicarbonate ions reduces the performance efficiency leading to resin throughput volumes (leakage < 0.1 mgP/L) lowered to 150 BV and operative phosphate retention capacity averaging 1.2 gP/L (≈10% of the figure obtained in laboratory experiments). Despite irreversible sorption of bio-persistent organics and/or mechanical deposition of micro-dispersed suspended matter, no fouling of the polymer matrix was observed. Neither was significant copper bleeding during resin exhaustion seen for constant and reproducible performance of the resin. Investigations are still on course to

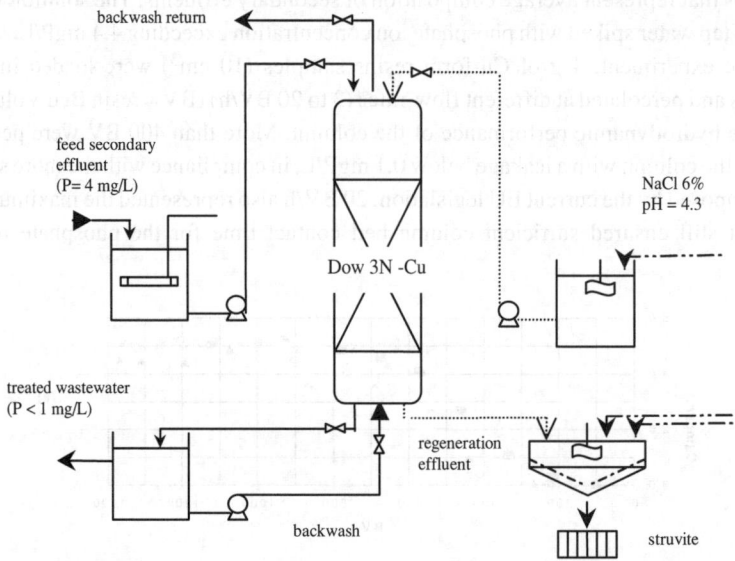

Figure 3 Layout of the dephosphation pilot plant

Table 1 Design details of the pilot plant

Plant potentiality	0.1 m^3/d
Operative retention capacity of the resin	15 gP/Lr
Column flow rates	10 BV/h (exh.); 2.5 BV/h (reg.)
Column height	1.2 m
Resin volume	0.5 L
Throughput volume (leakage <0.1 mgP/L)	400 BV
P-regenerant concentration factor	130

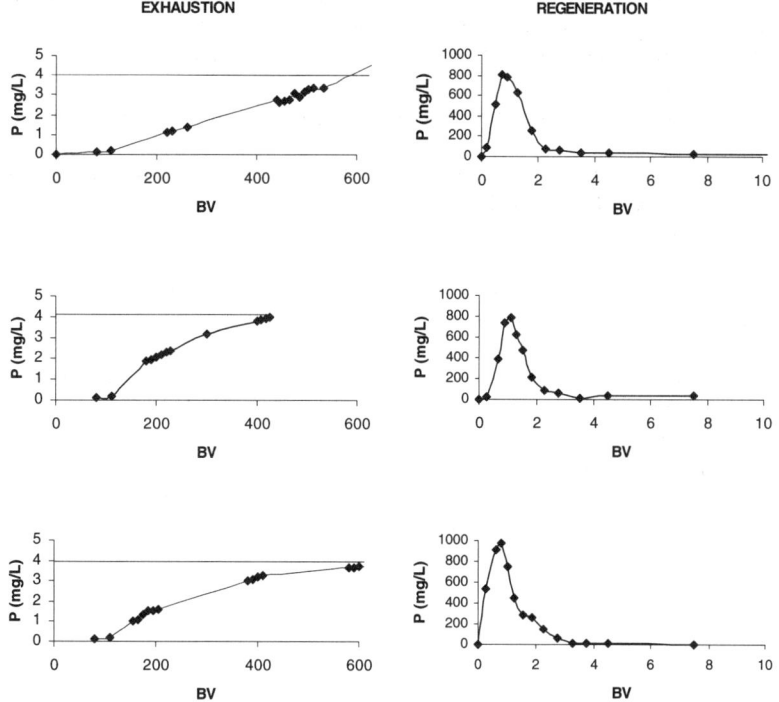

Figure 4 Exhaustion-regeneration performance of resin Dowex M-4195 treating municipal effluent from Grottaglie-Monteiasi wastewater treatment station (F_{exh} = 10 BV/h; Influent P: 4 mgP/L; F_{rig} = 2.5 BV/h; 1 mol l^{-1} NaCl, pH 4.3)

evaluate resin performance in different influent conditions such as variable influent composition and flow rates, by operating the pilot plant at different wastewater treatment installations.

Conclusion

Availability of selective sorbents in fixed-bed unit operations is of paramount importance while investigating the technical-economic feasibility of a given process. Moreover, a proper "tailoring" of the process schematics towards specific objectives (e.g. removal and simultaneous recovery) adds additional appeal to otherwise unattractive technical solutions. With the aim to "add value" to a technical solution proposed more than twenty years ago, the present investigation aimed at finding out a better sorbent for phosphate removal, through properly sized and better organised modifications in the original process scheme. Although preliminary results do not appear very encouraging in terms of performance improvement of the new sorbent, the investigation is continuing to get a better insight into the general behaviour of the process under different operative conditions.

References

Gaastra, S. (2001). Full scale calcium phosphate recovery plant at Geesterambacht in Holland. *2nd Int. Conf. Recovery of Phosphorus from Sewage and Animal Wastes*, Noordwijkerhout, The Netherlands, March 12–13.

Gregory, J. (1976). Anion exchange equilibria and kinetics in the ternary system chloride/sulphate/phosphate. *Proc. Int. Conf. Theory and Practice of Ion Exchange*. Society Chemical Industry, Cambridge, UK, Jan. 10–15.

Liberti, L. (2001). Feasibility study on the application of the REM NUT process to phosphate recovery from wastewater. Report for Comité Europeen D'etudes Des Polyphosphates, March 2001.

Liberti, L., Boari, G. and Passino, R. (1979). Phosphate and ammonium recovery from secondary effluents by selective ion exchange with production of a slow release fertiliser. *Wat. Res.*, **13**, 65–73.

Liberti, L., Boari, G. and Passino, R. (1984). Method for removing and recovering nutrients from wastewater. U.S. Pat. no. 4,477,355.

Liberti, L., Boari, G. and Passino, R. (1989). Method for removing nutrients from wastewater, Eur. Pat. no. 114,038.

Liberti, L., Petruzzelli, D. and De Florio, L. (2001). REM NUT ion exchange plus struvite precipitation process. *Environ.Technol.*, **22**, 1313–1324.

Metcalf & Eddy Inc. (1991). *Wastewater Engineering. Treatment Disposal Reuse.* 3rd Ed. McGraw-Hill Pub.Co.NY, NY, USA.

Outola, P., Leinonen, H., Ridell, M. and Lehto, J. (2001). Acid/base and metal uptake properties of chelating and wek base resins. *Solv. Extr. & Ion Exch.*, **19**(4), 743–756.

Sengupta, A.K. (2001). Removal of phosphate from wastewater using a new class of polymeric sorbents. *2nd Int. Conf. Recovery of Phosphorus from Sewage and Animal Wastes*, Noordwijkerhout, The Netherlands, March 12–13.

Vollenweider, R.A. (1968). Water management research. Scientific fundamentals of the eutrophication of lakes and flowing waters with particular reference to nitrogen and phosphorus as factors of eutrophication. O.C.D.E. Tech. Rep. no.194, Directorate for Scientific Affairs, Paris.

Zhao, D. and Sengupta, A.K. (1998). Ultimate removal of phosphate from wastewater using a new class of polymeric ion exchangers. *Wat. Res.*, **32**, 51613–1625.

Phosphate recovery from sewage sludge in combination with supercritical water oxidation

K. Stendahl and S. Jäfverström

Feralco AB, Industrigatan 126, SE-252 32 Helsingborg, Sweden (E-mail: *info@feralco.com*)

Abstract Supercritical Water Oxidation (SCWO) is an innovative and effective destruction method for organics in sewage sludge. The SCWO process leaves a slurry of inorganic ash in a pure water phase free from organic contaminants, which opens possibilities for a simple process to recover components like phosphates from the sewage sludge. In a continuous pilot plant for the SCWO process digested sludge has been treated. The ash has been extracted in lab scale with both caustic and acids in order to recover phosphates. By leaching the ash with caustic, 90% of the phosphorus could be separated as a sodium phosphate solution. By treating the sodium phosphate solution with lime, calcium phosphate was precipitated and caustic recovered and circulated back to the leaching process.
Keywords Phosphorus recovery; sludge; supercritical water oxidation

Introduction

In Sweden, until now, most of the sewage sludge has been recycled to the agriculture and the remainder is generally discharged to landfill.

However, the Farmer's Union and the food industry have placed a ban on recycling sludge to agriculture. This, in combination with a new EU regulation prohibiting disposal of organics to landfill, means that new methods for disposal of sewage sludge must be found.

As phosphate is not an endless resource, the Swedish Environmental Protection Agency is studying the possibility to put a demand for phosphate recovery before final disposal of sewage sludge.

Stockholm Water (the local water and sewage company in Stockholm) has therefore shown interest in the Aqua Reci process where supercritical water oxidation is used to decompose organic contaminants. The oxidation is followed by a chemical process in order to recover components in the inorganic rest ash, like phosphates and coagulants.

Supercritical water oxidation (SCWO)

At a temperature above 375°C and a pressure above 220 bars, water is entering a fourth phase – *the supercritical phase* – with properties between those of a gas and a liquid, Figure 1.

The change in magnitude of different properties is very rapid at entrance into the supercritical region. The most important change is the change in solubility. The organic compounds increase to almost 100% solubility while inorganic compounds become more or less insoluble, Figure 2.

Gaseous oxidants such as oxygen are completely miscible in supercritical water. As water density is lower, diffusivity and ion mobility therefore are higher. Combined with a high temperature, these properties result in rapid oxidation of organic compounds without interfacial mass transfer limitations or sparing availability of the oxidant. The reaction time for complete destruction is below 60 seconds.

SCWO leads to complete destruction: organic carbon is converted into carbon dioxide,

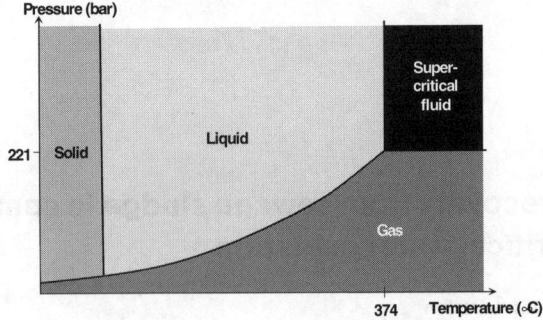

Figure 1 Phase diagram of water

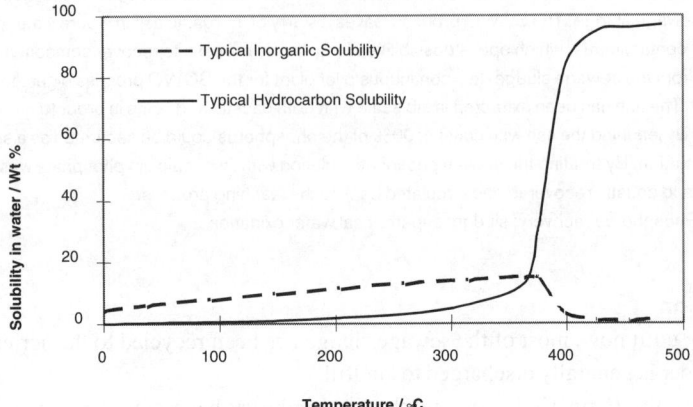

Figure 2 Solubility of organic and inorganics compounds in water

organic and inorganic nitrogen into nitrogen gas (N_2), halogenated organic and inorganic into corresponding acids, sulphonated organic and inorganic into sulphuric acid. Metals are oxidised to their highest valency and phosphorus into P_2O_5. The oxidation reaction is also exothermic, consequently once the feed organic is concentrated enough, the reaction can produce sufficient energy to maintain reactor temperature.

Inorganic ash is very reactive, which means that metal oxides and phosphorus easily will dissolve in acid. Phosphorus can be extracted in caustic and separated from metal oxides due to their insolubility in alkaline conditions.

In a pilot plant located in Karlskoga, Sweden several tests have been conducted on the

Figure 3 Sewage sludge before and after supercritical water oxidation

digested sludge from both Bromma sewage treatment plant in Stockholm and from Karlskoga sewage treatment plant. In both cases the dewatered sludge was diluted to about 17% DS and homogenised in a mass generator before it was continuously pumped into the SCWO process.

The SCWO process, Figure 4, consists of a long pipe where the first part is an economiser in the form of a pipe heat exchanger utilising the heat coming from the reactor. The last part after the reactor is another pipe heat exchanger, which can be used either to produce steam or hot water by utilising the heat coming from the exothermic reaction. At the end of the process, carbon dioxide and nitrogen are separated in an open separator tank. The total retention time is about 5 minutes, where 60 seconds represent the reactor part.

Results

Sludge used for the tests was diluted from 25–30% to about 17% DS and homogenised using a macerator. The reason for the dilution was to obtain a pumpable sludge into the SCWO process. The COD content was on average 125,000 mg/l and the ammonia content 1,400 mg/kg in the 17% sludge.

After the SCWO process, the COD level was, in all tests, reduced to more than 99.9% to a typical level of 40 mg/l COD in the liquid effluent phase. If the temperature in the reactor exceeded 550°C, the ammonia nitrogen could be reduced to more than 99.9% to a value less than 10 mg/l. If the temperature was below 550°C the ammonia nitrogen content reached 150 mg/l.

Outgoing slurry from the process containing the inorganic components of the sludge had a DS content of 7–7.5% and was very easy to thicken to about 30% DS content or filtered to a DS content above 50% (Figure 3).

Table 1 below shows the distribution of the inorganic components in the water phase and respectively the solid phases leaving the reactor.

P-recovery

In lab scale, the filter cake containing about 50% DS was leached with both acid and caustic with a temperature between 80–90°C.

The use of hydrochloric acid after 60 minutes retention time dissolved not only all phosphorus, but also all metal oxides, Figure 5.

By using caustic in excess of about 50% as a 5% solution it was possible to dissolve up to 90% of the phosphorus at a temperature of 90°C, Figure 6.

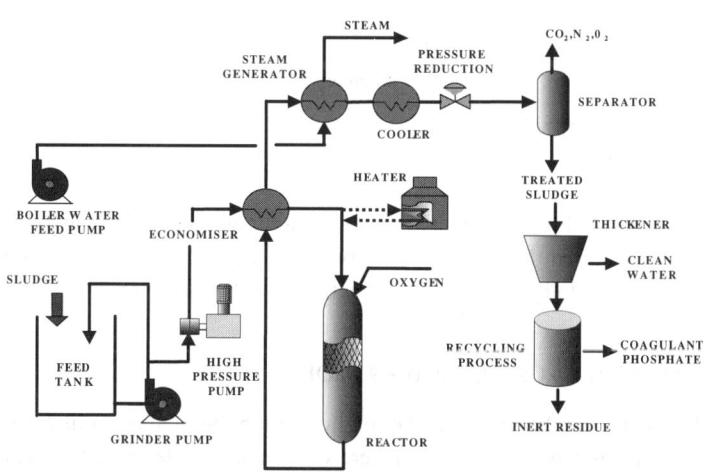

Figure 4 The Aqua Reci process

Table 1 Distribution of the inorganic components and solid phases in the water respectively

Water phase 93% of the total mass	mg/l	Solid phase 7% of the total mass	%	Distribution in %	Solid phase	Liquid phase
Ca	360	SiO_2	35.6	K	66	34
S	320	Fe_2O_3	34.8	Na	70	30
K	110	P_2O_5	15.2	S	74	26
Na	64	CaO	5.4	Ca	87	13
P	24	Al_2O_3	4.8	Mg	97	3
Mg	15	SO_3	3.5	Cu	98	2
Al	2.0	MgO	1.3	P	99	1
Cu	0.8	K_2O	0.4	Fe	100	0
Mn	0.5	Na_2O	0.3	Heavy metals	100	0
Fe	0.4	MeO	0.3			
Heavy metals	ca 0.1					

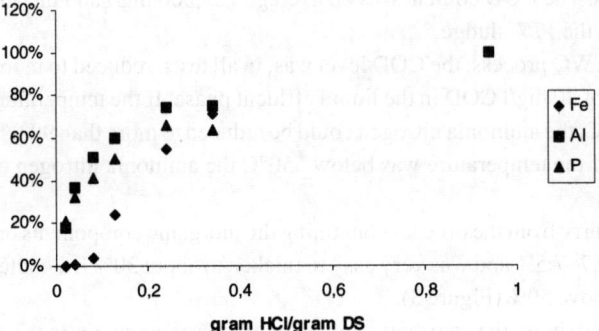

Figure 5 Leaching the ash with HCl

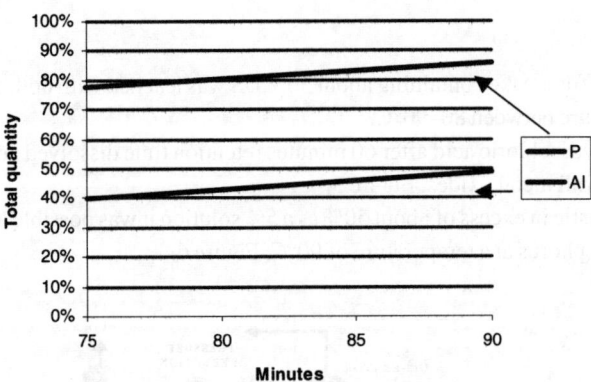

Figure 6 Leaching of ash with caustic

As can be seen from Figure 6 about 50% of the aluminium was following the phosphorus while all other metals in the solution were below 1 mg/kg. The sodium phosphate was precipitated with lime to hydroxy apatite, practically to 100%, according to the following formula:

$$3\,Na_3PO_4 + 5\,Ca(OH)_2 \rightarrow Ca_5(PO_4)_3OH + 9\,NaOH$$

From the formula it can be seen that the total consumption of sodium hydroxide for the phosphate leaching will be recovered and therefore can be recycled back for a new leaching cycle. The only loss will be the remaining sodium hydroxide in the residual sludge after the leaching process, which can be estimated to max. 10% without washing. The only chemical consumption will therefore be the lime required for the production of hydroxy apatite.

The aluminium content dissolved together with the sodium phosphate was found as $KAlSiO_4$ in the precipitate with lime. As aluminium silicate is practically insoluble at a pH above 13, it must be possible to add sodium silicate in combination with the leaching process in order to precipitate all aluminium as silicate. The produced calcium phosphate in the form of hydroxy apatite will therefore be extremely pure.

The use of the Aqua Reci process for Stockholm sewage sludge

Based on the findings from the tests, the following material- and energy balance can be expected by using the Aqua Reci process for the total production of digested sludge from the Stockholm sewage treatment plants, Figure 7. A process based on the alkaline leaching of phosphorus can be designed according to Figure 8.

Cost calculation

A cost estimate has been done for a full-scale operation to cover the total demand of Stockholm sewage works. The calculation is based on an SCWO process consisting of two parallel lines with a capacity of 10,000 ton of DS each per year followed by a process for the recovery of phosphates.

The total investment for the SCWO process has been estimated to 12 MEUR and for the phosphate recovery process 660,000 EUR. The building, with a surface area of 1,000 m², is calculated at a cost of 1.5 MEUR.

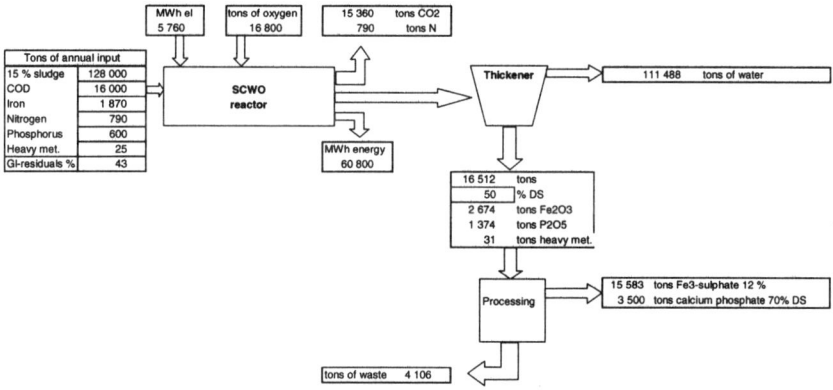

Figure 7 Material- and energy balance for the Aqua Reci process

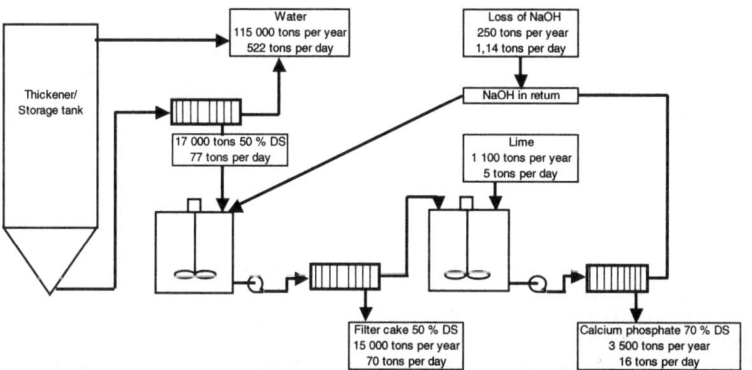

Figure 8 Material flow for phosphate recovery per year resp. day at 220 working days. The calculation is based on 20,000 tons of DS, i.e. 133,000 tons as 15% incoming to the SCWO process

Table 2 Cost calculation for an Aqua Reci process for Stockholm sewage sludge

	Cost in EUR
Energy 6,000 MWh at 44 EUR	264,000
Oxygen 16,800 tons at 78 EUR	1,310,000
NaOH 250 tons at 420 EUR	105,000
Lime 1,100 tons at 110 EUR	121,000
Staff, 2 people	90,000
Miscellaneous	100,000
Total annual operative costs	1,990,000
Operative costs EUR/ton DS	100
Total annual capital costs	1,290,000
Total annual costs	3,280,000
Total costs EUR/ton DS	164
Total costs EUR/ton 25% DS	41

With a depreciation of 15 years and 5% real interest rate for the machinery and 20 years for the building, the total annual capital cost will be 1,290,000 EUR. The total annual cost is illustrated in Table 2.

Future development

In a second step, the filtercake leaving the process can be leached with acid in order to also recover the coagulant. This process, however, has not been further investigated in this study. New pilot tests will be conducted at the end of 2002. The intention is to run the process continuously over a 5-day period. During these 5 days the target will be to form a true picture of the availability of the SCWO process as regards incrust formation and cleaning system. For the second part of the process, the recovery of phosphorus and iron will be further studied and the aim is to find a final use of the inorganic residue.

Conclusions

In pilot scale it has been shown that the supercritical water oxidation process is feasible, not only to decompose organics to more than 99.9%, but also to oxidise ammonium into nitrogen. The total process is a closed process where the main reaction takes place in a pipe reactor without any emission but carbon dioxide and nitrogen. The heat from the oxidation process can be utilised in the form of hot water or for steam production.

The residue from the oxidation process is an inorganic ash in pure water slurry. The ash consists mainly of SiO_2, metal oxide and P_2O_5. Due to the relatively low temperature in the oxidation process, the inorganic ash is very reactive.

Utilisation of caustic allows separation of the phosphate from the inorganic ash as a very pure sodium phosphate that can be precipitated as very pure calcium phosphate. The caustic will be recovered and reused.

The total cost calculated for Stockholm sewage works is 164 EUR/ton of DS or 41 EUR/ton 25% sludge.

References

Anders Gidner, Mats Almemark, Lars Stenmark and Östen Ekengren. Treatment of sewage sludge by Supercritical Water Oxidation. *6th Annual Conference on Sewage Sludge*, 16–17 February, 2000, United Kingdom.

Feralco (2002). Pilotförsök att slutligt omhänderta slam från Stockholm Vattens vatten- respektive avloppsverk genom Aqua Reci-processen. *Stockholm Vatten Rnr* 03, februari 2002.

Patterson, D.A., Stenmark, L. and Hogan, F. (2001). Pilot-scale supercritical water oxidation of sewage sludge. *6th European Biosolids and Organic Residues Conference* 12–14 November, 2001. United Kingdom.

Stark, K. (2002). Phosphorous recovery from sewage sludge by thermal treatment and use of acids and bases. *Conference Proceedings at 10th Gothenburg Symposium on Chemical Water and Wastewater Treatment* 17–19 June, 2002.

High-strength nitrogen removal of opto-electronic industrial wastewater in membrane bioreactor – a pilot study

T.K. Chen*, C.H. Ni**, J.N. Chen* and J. Lin**

* Institute of Environmental Engineering, National Chiao Tung University, No. 75, Poai Street, Hsinchu 300, Chinese Taiwan (E-mail: *tkchen.ev88g@nctu.edu.tw*; *jnchen@green.ev.nctu.edu.tw*)
** Green Environmental Technology CO., Ltd., 4F, 100, Sec. 2, Jung-Shan Rd., Pan Chiao City, Taipei Hsien, Chinese Taiwan (E-mail: *chni01@ms14.hinet.net*; *tsailin@ms17.hinet.net*)

Abstract The membrane bioreactor (MBR) system has become more and more attractive in the field of wastewater treatment. It is particularly attractive in situations where long solids retention times are required, such as nitrifying bacteria, and physical retention critical to achieving more efficiency for biological degradation of pollutant. Although it is a new technology, the MBR process has been applied for industrial wastewater treatment for only the past decade. The opto-electronic industry, developed very fast over the past decade in the world, is high technology manufacturing. The treatment of the opto-electronic industrial wastewater containing a significant quantity of organic nitrogen compounds with a ratio over 95% in organic nitrogen (Org-N) to total nitrogen (T-N) is very difficult to meet the discharge limits. This research is mainly to discuss the treatment capacity of high-strength organic nitrogen wastewater, and to investigate the capabilities of the MBR process. A 5 m^3/day capacity of MBR pilot plant consisted of anoxic, aerobic and membrane bioreactor was installed for evaluation. The operation was continued for 150 days. Over the whole experimental period, a satisfactory organic removal performance was achieved. The COD could be removed with an average of over 94.5%. For TOC and BOD5 items, the average removal efficiencies were 96.3 and 97.6%, respectively. The nitrification and denitrification was also successfully achieved. Furthermore, the effluent did not contain any suspended solids. Only a small concentration of ammonia nitrogen was found in the effluent. The stable effluent quality and satisfactory removal performance mentioned above were ensured by the efficient interception performance of the membrane device incorporated within the biological reactor. The MBR system shows promise as a means of treating very high organic nitrogen wastewater without dilution. The effluent of TKN, NOx-N and COD can fall below 20 mg/L, 30 mg/L and 50 mg/L.
Keywords Activated sludge; industrial wastewater; membrane bioreactor

Introduction

Development and application of biological nutrient removal processes have accelerated significantly over the past decade due to more stringent nutrients (nitrogen and phosphorus) discharge limits being imposed on wastewater treatment plants. Thus, to meet stringent nutrient removal requirements as well as water reuse standards, the conventional activated sludge process has to be enlarged and combined with a membrane unit. The integration of membranes with bioreactors for treatment of wastewater has presented attractive features that conventional treatment processes cannot achieve (Stephenson *et al.*, 2000). These include reliability, compactness, and excellent treated water quality. It is particularly attractive in situations where long solids retention time are required, such as nitrifying bacteria, and physical retention critical to achieving biological degradation of pollutant, such as opto-electronic industrial wastewater.

The opto-electronic industry, developed very fast over the past decade in the world, is high technology manufacturing. Many countries concentrate on its development because the demand for these products increases quickly. A number of opto-electronic industries consist of the following processes: (1) array, (2) color filter, and (3) liquid crystal. All of these steps need large quantities of organic solvent as developers, strippers and rinses, resulting in the discharge of large amounts of wastewater. In addition to organic carbon, such as dimethyl sulphoxide (DMSO, $(CH_3)_2SO$) and isopropyl alcohol (IPA, CH_3CHOCH_3), the wastewater often contains a significant quantity of organic nitrogen compounds, such as ethanolamine, (MEA, $C_2H_5ONH_2$) and tetra-methyl ammonium hydroxide (TMAH, $(CH_3)_4NOH$). In particular, the organic nitrogen has a ratio to total nitrogen of over 95%, which is different from the ammonia ratio of 80% in animal production plants wastewater and municipal wastewater. If it is not treated, or treated after dilution, it will lead to eutrophication or raise the treatment cost.

To completely remove nitrogen from opto-electronic industrial wastewater, a 2-stage anoxic/aerobic bioreactor combined with a submerged membrane system was applied in this study. Membrane bioreactors (MBR) with an anoxic stage have been used for nitrogen removal in municipal wastewater (Suwa *et al.*, 1992). However, the application of the MBR system in treatment of industrial wastewater is still in its infancy. While there are many similarities in the design parameters for municipal plants, industrial plants show considerable variation in design, control and operational performance. This is attributable to the significantly greater differences between industrial wastewater feed compositions and the associated variation in sludge characteristics. In order to demonstrate the technical and economic advantage of applying the MBR system and to develop full-scale process design information, this research is mainly to discuss the treatment capacity of high-strength organic nitrogen wastewater, and to investigate the nitrogen removal capabilities of the MBR process.

Methods

Experimental setup

A pilot-scale MBR plant shown in Figure 1 was used in this study. The system consisted of a 2-stage anoxic/aerobic bioreactor, which had a 3 m³ anoxic tank and a 10 m³ aerobic tank for carbon and nitrogen removal. A membrane pilot unit with hollow fiber membranes, manufactured by Zenon (Model ZW-500, surface area 46 m², pore size 0.4 μm), was applied for separation of biomass before discharge. Hollow fiber membrane was immersed in the membrane tank, where the air was supplied at the bottom of membrane units in order to supply oxygen for biological treatment and to create cross flow for preventing membrane

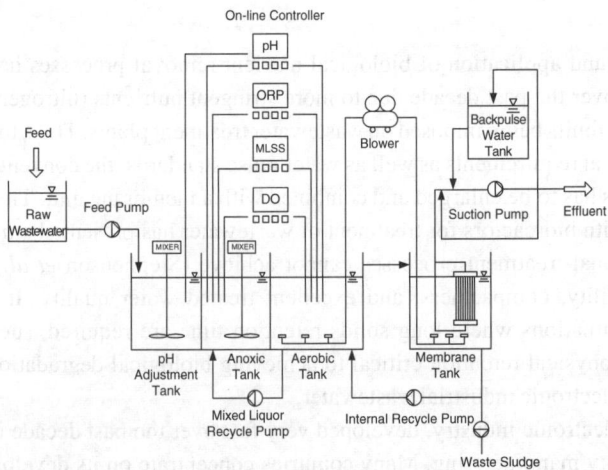

Figure 1 Schematic diagram of the pilot-scale MBR system

fouling. The membranes were backwashed periodically with permeate: 10 minutes of suction were followed by a backwash of about 20 sec with permeate at transmembrane pressure of 1 to 8 psi. The raw wastewater was pumped into the pH adjustment tank for pH neutralization and then went through the anoxic, aerobic and membrane tanks. Mixed liquor in the aerobic tank was re-circulated back to the anoxic tank for denitrification. An internal pump was applied for wasting the sludge, moreover for agitating and balancing the MLSS between the aerobic and membrane tanks. Permeate, being controlled at a constant flow rate, was discharged.

Operating parameters and conditions

After 20 days of seeding and start-up, the MBR pilot plant was in continuous operation for 130 days. The operating parameters are shown in Table 1. The hydraulic retention time (HRT) of the anoxic tank, aerobic tank and membrane tank, determined based on the influent flow rate of the system, was 2.89 hours, 9.66 hours and 0.68 hours respectively. Mixed liquor suspended solids (MLSS) was controlled roughly between 3,000 mg/L and 8,000 mg/L. The average sludge retention time (SRT) was quite long, 30–60 days, as a result of using membranes as the solids-separation devices. The permeate flux of the effluent was 0.18–0.35 $m^3 m^{-2} d^{-1}$. The re-circulation rate of the mixed liquor was set at 300% of the average influent flow rate (1.0~2.0 $m^3 h^{-1}$). The temperature varied within the range of 20 and 35°C.

Control, analysis and monitoring

Dissolved oxygen (DO), oxidation-reduction potential (ORP), pH, MLSS, temperature and flow rates were recorded daily using on-line controllers. Oxygen concentration in the mixed liquor was monitored using a DO meter (Eutech DO1000+DOTPII) and maintained higher than 2.0 mg/L in the aerobic and membrane tanks. Submerged ORP electrodes (Eutech HTPTTSO05B, hydrogen type) monitored the oxidation-reduction potential within the anoxic and aerobic tanks. MLSS concentrations of each tank were monitored using on-line MLSS meter (Zullig Cosmos-25). The influent and effluent were sampled two to three times per week. The analysis of various constituents including total nitrogen (T-N), total Kjeldahl nitrogen (TKN), nitrate (NO_3-N), nitrite (NO_2-N), chemical oxygen demand (COD), total organic carbon (TOC), biochemical oxygen demand (BOD_5), total suspended solids (TSS) and orthophosphate (PO_4^{3-}-P) were performed in accordance with the standard methods (APHA, 1993) and the corresponding instrument instruction manuals.

Results and disscussion

Wastewater characteristics

Real TFT-LCD (thin film transistor liquid crystal display) manufacturing plant wastewater was employed as the influent to the MBR system. The TFT-LCD manufacturing processes including array, color filter and liquid crystal need large quantities of organic solvent as developers, strippers and rinses, result in the discharge of large amounts of wastewater. In addition to organic carbon, such as dimethyl sulphoxide (DMSO, $(CH_3)_2SO$) and isopropyl alcohol (IPA, CH_3CHOCH_3), the wastewater often contains a significant quantity of organic nitrogen compounds, such as ethanolamine, (MEA, $C_2H_5ONH_2$) and tetra-methyl ammonium hydroxide (TMAH, $(CH_3)_4NOH$). The characteristics of the wastewater are indicated in Table 2. Besides, the amount of wastewater varies a lot with the change of manufacturing process and quantity of products. Figure 2 and 3 show the distribution of the COD and T-N concentration by a long-time observation of the influent. Figure 2 indicates that COD varies from 400~1,600 mg/L and the average concentration is about 764 mg/L. The T-N concentration distributes from 100~300 mg/L with an average of about 151 mg/L.

Table 1 MBR pilot plant operating conditions

Design flow	345 l hr^{-1}	Mixed liquor recycle ratio	3
Design peak flow	678 l hr^{-1}	MLSS range	3~8 g/L
Anoxic tank volume	3,000 l	pH range	7~8
Aerobic tank volume	10,000 l	Temperature	20~35°C
Membrane tank volume	700 l	F:M ratio	0.11~0.22
Anoxic tank HRT	2.89 h @ design flow	Volumetric loading rate	0.46~0.91 kgCODm^{-3}d^{-1}
Aerobic tank HRT	9.66 h @ design flow	Membrane surface area	46 m^2
Membrane tank HRT	0.68 h @ design flow	Flux	0.18–0.35 m^{3}m^{-2}d^{-1}
SRT	30~60 days		

Table 2 Components and characteristics of the raw wastewater

Items Components	Stripper (CH$_3$)$_2$SO (DMSO) C$_2$H$_5$ONH$_2$ (MEA)	Developer (CH$_3$)$_4$NOH (TMAH)	Rinse CH$_3$CHOOHCH$_3$ (IPA)	Average –
pH	9–11	10–13	4–10	10–11
SS (mg/L)	<10	<10	<10	<10
COD (mg/L)	800–1,500	100–600	500–3,700	500–2,000
TKN (mg/L)	70–200	60–90	90–240	100–200
NH$_3$-N (mg/L)	0–15	2–15	0.1–10	2
Nox-N (mg/L)	0.1–0.4	0.0–0.3	0.1–1.3	0.2

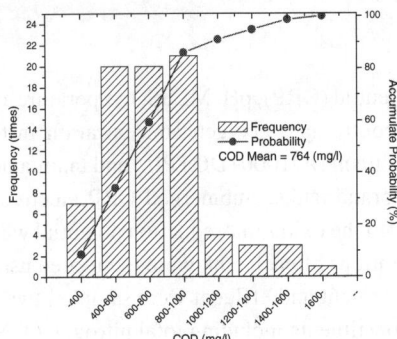

Figure 2 Distribution of the influent COD concentration

Figure 3 Distribution of the influent T-N concentration

The COD/T-N ratio is about 5.1 and organic-nitrogen occupies more than 95% of the total nitrogen which shows the wastewater is highly organic-nitrogen containing.

Monitoring of DO, pH, ORP and temperature

Figure 4 shows the ORP and temperature values recorded during the experimental period. It can be seen that the ORP in the anoxic tank and aerobic tank were maintained at –350~ –450 mV and 50~150 mV respectively. Sillen (1965) suggested that complete reduction of nitrate to nitrite should result in an ORP of –200 mV and that complete reduction of nitrite to nitrogen gas should result in an ORP of –325 mV. At –400 mV, the nitrate is converted first to nitrite then the nitrite is converted immediately to N$_2$ without accumulating. Ideally, complete denitrification should maintain ORP at –375 to –450 mV. With ORP values above 100 mV in the aerobic tank, complete nitrification had occurred. Temperatures were varied between 20°C–35°C. Figure 5 shows DO and pH values recorded during the experimental period. DO and pH were kept at the desired values (DO higher than 2 mg/L in the aerobic tank and lower than 1 mg/L in anoxic tank, pH was kept constant between 7~8).

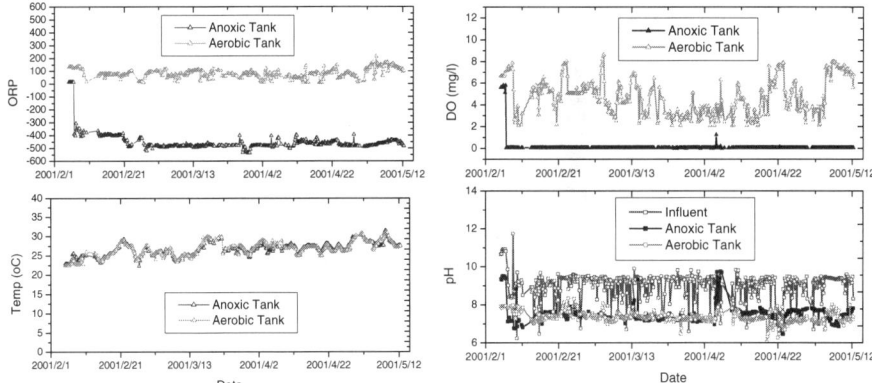

Figure 4 Variation of ORP and temperature during the operation time

Figure 5 Variation of the DO and pH during the operation time

Carbon removal

Figure 6 illustrates the influent and effluent quality of the MBR in terms of COD. More than 94% of soluble COD was removed and the COD concentration of the effluent was between 10~50 mg/L, mean value was about 42 mg/L. The advantage of the MBR system is having a higher biomass concentration, which results in a higher removal of COD. Although the influent COD varied greatly, the effluent COD was really low and stabilized at less than 63 mg/L (Table 3). This result may be attributed to the mass loading being generally low between the range of 0.11~0.22 kgCODkgMLSS^{-1}d^{-1}; because the MBR system can be operated with high MLSS concentration (about 6,000 mg/L) and high volumetric loadings (0.46~0.91 kgCOD/m^{3}d^{-1}) were achieved. Figure 7 shows the COD, TOC and BOD$_5$ concentration in the bioreactors. BOD$_5$ concentrations in the effluent were even at, or close to the laboratory detection limits (2 mg/L). Denitrification was carried out in the anoxic tank, but it needs enough organic substances as the carbon source. The organic substances contained in the wastewater were utilized as carbon source for denitrification. Besides, it was found that over 20% of COD was rejected by comparing the COD concentration between the membrane tank and effluent.

Nitrogen elimination

Figure 8 illustrates the nitrogen concentration in bioreactors. The removal efficiency of T-N and TKN reached 75% and 90% respectively. This result shows that more efficient and stable removal of nitrogen was observed in the MBR system than the conventional

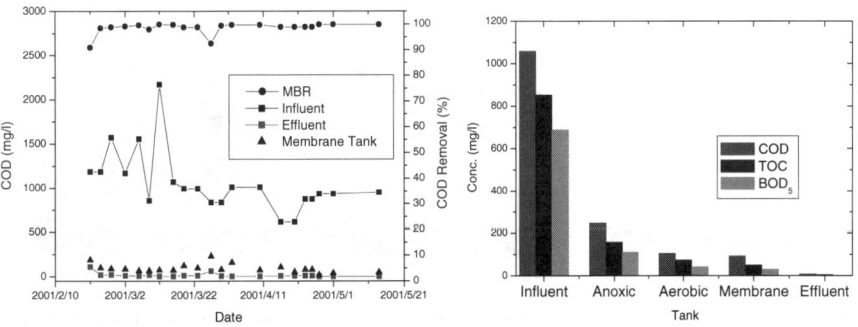

Figure 6 Variation of the COD concentration during the operation time

Figure 7 Variation of the COD, TOC and BOD$_5$ concentration in each bioreactor

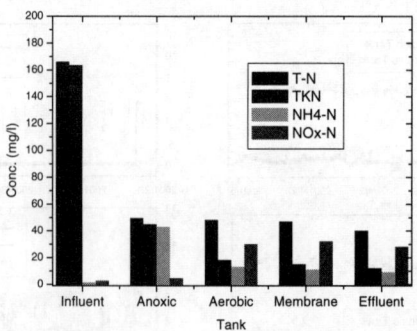

Figure 8 Variation of the T-N, TKN, NH$_4$-N and NOx-N concentration in each bioreactor

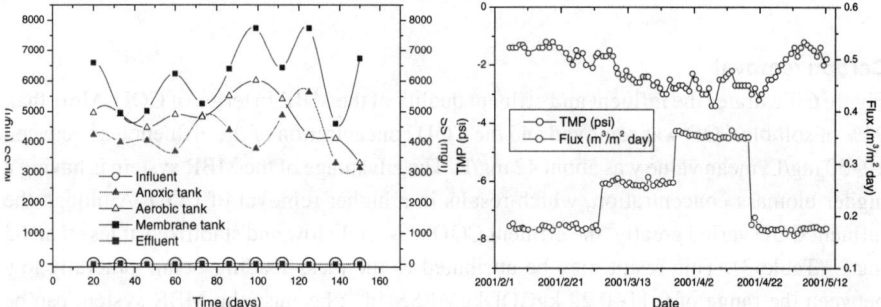

Figure 9 Variation of the MLSS and SS concentration over the operation time

Figure 10 Variation of TMP and flux

Table 3 Performance of MBR pilot plant

	Influent			MBR treated ewter			Removal
	Average	Max.	Min.	Average	Max.	Min.	(%)
Temperature (°C)	27	25	30	28	26	30	–
pH	10.8	9.5	11.2	7.6	7	8	–
SS (mg/L)	4	8	b.d.(<0)	b.d.(<0)	2	b.d.(<0)	–
COD (mg/L)	764	2,300	371	42	63	8	94.5
TOC (mg/L)	550	2,000	320	20	43	5	96.3
BOD (mg/L)	430	1,282	204	10	13	b.d (<2)	97.6
TKN (mg/L)	151	270	90	12	20	6	92.1
NH$_4$-N (mg/L)	2	5	b.d.(<0.2)	9	18	2	–
NOx-N (mg/L)	0.2	1.1	b.d.(<0.1)	28	35	12	–
SDI	–	–	–	3	5	2	–

b.d., below detection limit

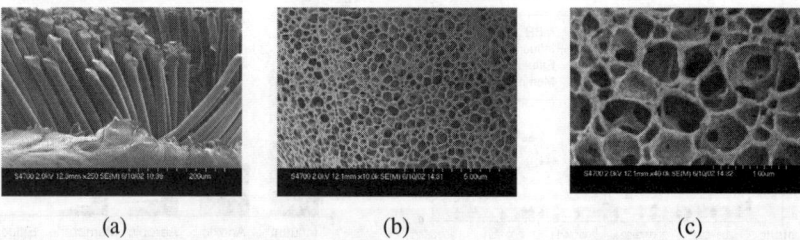

(a)   (b)   (c)

Figure 11 (a) Cross section of hollow fiber membrane. (b) External surface of skin (× 10,000). (c) External surface of skin (× 40,000)

biological treatment processes. The removal efficiency of T-N in the conventional biological treatment process was only about 45–70% (Burica et al., 1996). The complete retention of micro-organisms by the membrane can increase ammonia removal efficiency in the MBR system (Cote et al., 1997). Furthermore, most of the nitrogen content in the treated water was NOx-N, which indicated that denitrification was the rate-determining step in this system rather than nitrification. This may be attributed to the following two points: 1) nitrifying bacteria had sufficiently grown under the long SRT conditions and 2) denitrifying bacteria, depending on the electron donor in organic matter in wastewater, would be deactivated when the COD/T-N ratio was about 5.1.

Variation of MLSS and flux

Figure 9 shows that the plant can be operated with a high biomass concentration by using membranes for sludge separation. MLSS concentration was maintained in the range of 3,000~8,000 mg/L and excess sludge was wasted in accordance with the sludge growth rate (average SRT: 30~60 days). Under the steady state conditions, the sludge production stabilized at a value of 0.047 kgSS/kgCODr, which is 4 to 6 times less than the conventional activated sludge process (0.2~0.3 kgSS/kgCODr). Table 3 shows the overall result of the MBR system. Although the MLSS concentration was higher than 6,000 mg/L, the treated water was free of suspended solids due to a complete rejection of membrane. Besides, the effluent SDI value is under 3, which can be used directly as the feed water to the RO system for recycling without any pretreatment.

Figure 10 shows the variation of TMP by changing the different flux. The TMP varied from –1 to –4 psi with the flux change from 0.18–0.35 $m^3m^{-2}d^{-1}$. No chemical cleaning operation was needed during the period of the test since there was only slight change of TMP. It is proven that less power consumption is needed for the immersed membrane system than the pressure driven type membrane.

SEM analysis of membrane

Figure 11 shows the results of a scanning electron micrograph (SEM) of the membrane. The hollow fiber membrane consists of a very thin skin layer supported by a supporting sub layer (Figure 11(a)). The skin has the main functions of the membrane, since the overall flux and selectivity depend only on the structure of the membrane skin ((Figure 11(b) and (c)).

Conclusion

The opto-electronic industrial wastewater contains a significant quantity of organic carbon and organic nitrogen compounds with a ratio of over 95% in organic nitrogen (Org-N) to total nitrogen (T-N). It was demonstrated that the MBR system is capable to produce an effluent with an excellent removal of 94% and 92% for COD and TKN. The effluent quality of COD and TKN can be dropped lower than 42 mg/L and 12 mg/L respectively. The quality of SS and BOD_5 are also below the detection limit. The MBR system shows promise as a means of treating very high organic nitrogen wastewater and meets the current stringent quality standards around the world. Moreover, the effluent SDI value is under 3 that can be used directly as the feed water to the RO system for recycling without any pretreatment.

References

Burica, O., Strazar, M. and Mahne, I. (1996). Nitrogen removal from wastewater in a pilot plant operated in the recirculation anoxic-oxic activated sludge mode. *Wat. Sci. Tech.*, **33**(12), 255–258.
Cote, P., Buisson, H. and Praderie, M. (1997). Immersed membrane activated sludge for the reuse of municipal wastewater. *Desalination*, **113**, 189–196.

Park, S.J., Yoon, T.I., Seo, H.J. and Park, H.J. (2001). Biological treatment of wastewater containing dimethyl sulphoxide fom the semi-conductor industry. *Process Biochemistry*, **36**, 579–589.

Sillen, L.G. (1965). Oxidation states of earth's ocean and atmosphere: a model calculation on earlier state: the myth of prebiotic soup. *Archiv Kemi Acta*, **24**, 431–456.

Standard Methods for the Examination of Water and Wastewater (1993). 18th edn, American Public Health Association/American Water Works Association/Water Environment Federation, Washington DC, USA.

Stephenson, T., Judd, S., Jefferson, B. and Brindle, K. (2000). *Membrane Bioreactors for Wastewater Treatment*. IWA Publishing, London.

Suwa, Y., Suzuki, T., Toyohara, H., Yamagishi, T. and Urushigawa, Y. (1992). Single-stage, single-sludge nitrogen removal by an activated-sludge process with cross-flow filtration. *Wat. Res.*, **26**, 1149–1157.

Conversion of pollutants to fertilisers: ion exchange synthesis of potassium sulphate from acidic mine waters

D. Muraviev

Department of Chemistry, Autonomous University of Barcelona, E-08193 Bellaterra (Barcelona), Spain
(E-mail: *dimitri@icmab.es*)

Abstract The paper reports the results obtained by the development of ion exchange synthesis of K_2SO_4 from the natural acidic mine waters (AMW) of Rio Tinto area (Huelva, Spain). The process flowsheet includes several sequential stages permitting production of potassium sulphate and desalinated water along with the recovery of four metals.
Keywords Acidic mine waters; ion exchange; metals recovery; potassium fertilisers; synthesis

Introduction

Production of chlorine-free potassium fertilisers such as, e.g. K_2SO_4 is of particular interest as it deals with the problem of effective cultivation of chlorophobic plants such as vegetables, citruses, herbs and some others, which are adversely affected by high chloride concentration (Slack, 1965; Scharf, 1990). Almost 90% of K_2SO_4 demand is for fertiliser but it is also used for manufacture of glass and in medicine. Potassium sulphate is produced in substantial quantities in Europe by the Mannheim process from K_2CO_3 and H_2SO_4 or by reaction of H_2SO_4 with KCl (Scharf, 1990; Elvers *et al.*, 1993). Both versions of the Mannheim process are complicated by problems of utilising gaseous wastes (CO_2 and HCl). In the US and some other countries K_2SO_4 is manufactured by exchange reactions between potassium, sodium and magnesium salts by their dissolution and fractional crystallisation (Scharf, 1990; Elvers *et al.*, 1993). The last process requires utilisation of large volumes of liquid wastes. In other words both processes are characterised by the same drawback: manufacture of fertilisers causes pollution. Hence, any alternative technology, which allows converting pollutants to fertilisers looks most attractive (Muraviev *et al.*, 1998a; Muraviev, 2000). Ion exchange synthesis of K_2SO_4 from metal bearing sulphate waste waters and effluents such as, e.g. acidic mine waters (AMW), is one of the routes for designing technologies of this type.

AMW is a common problem in many mining situations but it is particularly important in pyritic areas. Interaction of pyrite and other sulphide minerals with water and atmospheric oxygen produces acid solutions which can cause serious problems in the environment. Acid mine drainage is perhaps the most serious environmental impact of mining activity. Although efficient and modern engineering design is now incorporated into the mines, ecological problems related to AMW persist, especially after mining activity has stopped. The origin of AMW from pyritic areas is the oxidation of sulphide minerals. A rapid oxidation of sulphide minerals occurs where such minerals in contact with water are exposed to the atmosphere. The sulphide components in pyrite are oxidised to sulphate by acidity generation and release of iron. Both Fe(II) and Fe(III) play a key role in this mineral alteration. These oxidation processes are dynamically accelerated by bacterial activity (Ritcey, 1989; Smith, 1974). Acidic mine water pollution in areas associated with mining can be seen at Rio Tinto (Spain), Neves-Corvo (Portugal), Chuquicamata (Chile), Katanga (Zaire), etc.

Rio Tinto area (Huelva, Spain) is one of the most important pyritic area of the world (see

Figure 1). The mines are rich in deposits of iron pyrite, copper, zinc, manganese and some precious metals and therefore, it is one of the best examples of AMW production. The name Rio Tinto is due to the spectacular reddish colour of the river due to the high concentration of Fe(III) hydroxide produced by pyrite oxidation. The AMW production is independent of the mining activity. Although the pyritic ore deposits are scarcely exploited actually, these continue producing approximately several tens of millions of cubic metres of acidic mine waters every year. These waters flow to Rio Tinto and Odiel rivers that have their origin in the pyritic area and flow into Huelva estuary. In densely populated areas like, e.g. Huelva and other southern provinces of Spain and Portugal, high priority is needed to prevent toxic substances escaping into the environment, and contamination, as a result, of fresh water sources, the number of which is limited. Thus, the ecological impact of acidic effluents produced in the Rio Tinto mines area is evident and its treatment is necessary. The composition and pH of Rio Tinto water samples withdrawn at different periods (during January–May 1993) are shown in Table 1.

AMW are characterised by low pH (around 1.8, see Table 1) and relatively high content of metal ions such as, Fe, Zn, Cu, Mn, Al, Ca and Mg. One paramount feature of AMW of this type is the high content of SO_4^{2-} ions (around 25 kg/m^3) and very low concentration of chlorides. Although during the last decades a lot of attention has been paid to the recovery of metals from AMW (Muraviev et al. 1997a, 1997b, 1997c, 1997d, 1999a) the problem of utilisation of sulphates still remains unsolved. In this context AMW can be considered as a convenient raw material for production of chlorine-free potassium sulphate.

This paper reports the results obtained by the development of ion exchange synthesis of K_2SO_4 from the natural AMW of Rio Tinto area (Huelva, Spain). The process flowsheet includes several stages permitting production of potassium sulphate along with the recovery of four metals from AMWs of this type. The economic estimation of the complex treatment of AMW is presented and discussed.

Methods

The work was performed with samples of the native acidic mine waters of the Rio Tinto area shown schematically in Figure 1. Sulphonate cation exchanger (SR) Lewatit S-100, a macroporous polyacrylic resin (PR) bearing carboxylic groups Lewatit R 250-K, and iminodiacetic (IDA) resin Lewatit TP-207 were kindly supplied by Bayer Hispania Industrial, S.A. All chemicals A.G. quality were purchased from Probus (Spain) and used as received. The concentration of metal ions was determined by atomic emission spectroscopy (ICP-AES technique) using ARL Model 3410 spectrometer (Fisons, USA) provided with minitorch. The uncertainty of metal ions determination was less than 1.5%. Determination of H^+ and OH^- ions was carried out by potentiometric titration using a Crison pH-meter 507 (Spain) provided with a combined glass electrode.

The necessary conditioning of Rio Tinto water samples (RTW) prior to their ion exchange treatment included removal of iron (Muraviev et al., 1997a,b,c). This conditioning involved the adjustment of pH to 3.4–3.5 with concentrated KOH solution followed by

Table 1 Composition of AMW samples from Rio Tinto area (Huelva, Spain)

Date	Concentration (mg/l)						
	SO_4^{2-}	Fe(II)	Fe(III)	Zn	Al	Cu	pH
05.01.1993	26,125	6,479	2,121	1600	490	110	1.80
09.02.1993	25,132	5,965	1,365	1500	550	106	1.60
10.03.1993	23,720	5,764	1,436	1500	480	90	1.70
12.04.1993	20,499	6,433	67	1200	490	70	1.75
19.05.1993	20,662	4,155	945	1100	550	260	2.00

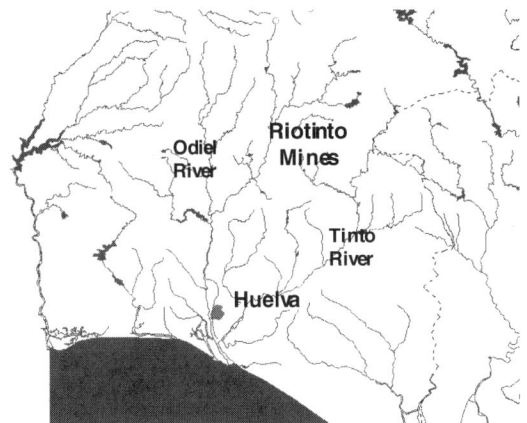

Figure 1 Map of Rio Tinto area. Andalucía, Spain

bubbling of air through RTW sample for several days. Under those conditions the oxidation of Fe(II) to Fe(III) and selective precipitation of Fe(OH)$_3$. took place. These processes are known to be significantly accelerated by *Thiobacillius ferrooxidans* so that the complete oxidation of Fe(II) (and precipitation of Fe(III) hydroxide as a result) can be reduced to several hours (Ritcey, 1989). The final removal of precipitate was carried out by filtration using a sintered glass filter and a water-pump. Composition of RTW samples before and after this treatment is shown in Table 2.

The ion exchange synthesis of chlorine-free K$_2$SO$_4$ from iron-free RTW was carried out under dynamic conditions in columns pre-loaded with SR in the K-form. RTW was passed through the column at a flow rate of 10 cm^3/min and collected in portions followed by the analysis of all metal ions in the eluate. After achieving the equilibrium the column was regenerated with either KCl (2.5–3.0 mol l^{-1}) or K$_2$SO$_4$ (0.67 mol l^{-1}) solution. During the regeneration with KCl the solution leaving the first column was passed through the second column loaded with IDA resin in the K-form, then through the third column with PR and finally through the forth one containing IDA resin. The final eluate (collected from the forth column) was treated with KOH by adjusting pH to 10 or 12 to cause the precipitation of polyvalent metal ions followed by filtration and the analysis of supernatant. RTW metal ions were stripped from PR and IDA resins with 1 mol l^{-1} H$_2$SO$_4$ solution. The stripping solution was collected in portions and analysed.

Results and discussion

The typical concentration-volume histories obtained by the treatment of RTW with SR in the K-form and by regeneration of the resin with 3.0 mol l^{-1} KCl solution are shown in Figures 2a and 2b, respectively. As seen in Figure 2, the exchange of K$^+$ with polyvalent RTW metal ions results in the formation of pure K$_2$SO$_4$ zone. The concentration of the product in this zone corresponds to the total equivalent concentration of all cationic species in the initial RTW. Two additional conclusions follow from the results presented in Figure 2.

Table 2 Composition of natural RTW samples before (a) and after conditioning (b)

AMW sample	Concentration (mg/l)										
	SO$_4^{2-}$	Fe	Cu	Zn	Al	Mn	Mg	Ca	Na	K	pH
(a)	26,745	5,950	225	1,524	633	90	864	378	0	0	1.9
(b)	24,750	3	216	1,585	564	90	896	383	88	11,621	3.5

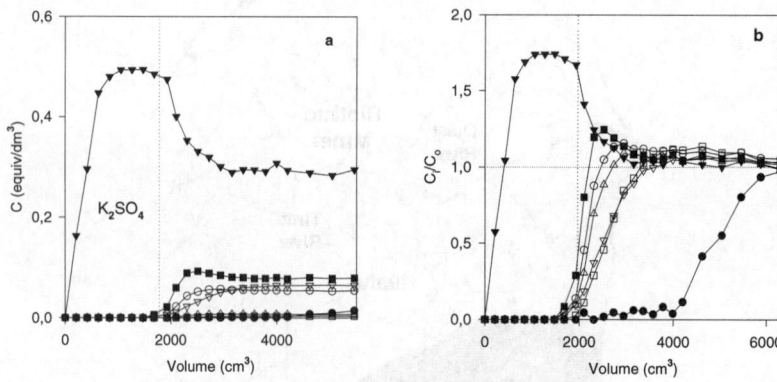

Figure 2 Concentration-volume histories of ion exchange synthesis of K_2SO_4 from RTW plotted in absolute (a) and relative (b) concentration scales: Zn (○); Mn (□); Cu (△); Al (▽); Ca (●); Mg (■); K (▼)

1. The first admixture, which appears in the pure K_2SO_4 zone is Mg^{2+} (see Figure 2b). Hence, the breakthrough of polyvalent metal ions as well as the purity of the product can be controlled by following the concentration of this ionic species by using, e.g. Mg-selective electrode;
2. The product collected from the pure K_2SO_4 zone can be easily concentrated by using electrodialysis techniques. The absence of Cl^- and polyvalent metal ions in product solution significantly simplifies application of this particular concentration method (Strathmann, 1992) and permits yielding of desalinated water as a by-product. This is particularly important for the area that is shown in Figure 1, as the number of fresh water sources in this region is very limited due to the heavy contamination with AMW.

The typical concentration-volume histories obtained by regeneration of SR with KCl and K_2SO_4 solutions are shown in Figures 3a and 3b, respectively. As seen from Figure 3, although the regeneration in both cases leads to substantial concentration of all RTW metal ions in the stripping solution, nevertheless, the use of less concentrated K_2SO_4 solution permits us to significantly improve the efficiency of regeneration process. Indeed, if in the first case (see Figure 3a) more than 3 litres of the stripping solution is required to completely regenerate the resin phase, in the second (see Figure 3b) this volume is more than twice less. As is clearly seen in Figure 2a, Ca^{2+} ions appear to be the most difficult to remove ionic species. Hence, the higher efficiency of regeneration by using K_2SO_4 solution can be

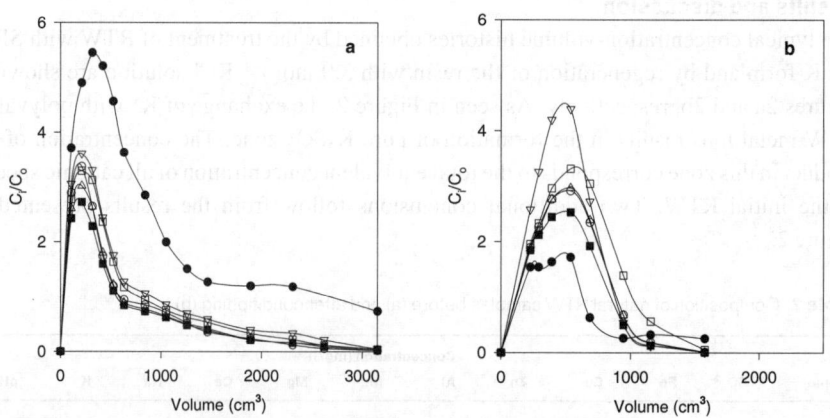

Figure 3 Concentration-volume histories of regeneration of SR loaded with RTW metal ions with 3.0 mol l⁻¹ KCl (a) and 0.67 mol l⁻¹ K_2SO_4 (b) solutions. Symbols are the same as in Figure 2

attributed by the coupling of the Ca^{2+}-K^+ exchange reaction with formation of low soluble calcium sulphate resulting in selective removal of Ca^{2+} ions from the resin phase. The ion exchange equilibrium in systems where ion exchange reaction is accompanied by formation of low solubility substances has been shown to be strongly displaced to the right (Muraviev et al., 1992, 1998b). In certain instances the low solubility substance forms inside the resin bed a stable supersaturated solution, which spontaneously crystallises following the removal from the column. This effect discovered for the first time in 1979 (Muraviev, 1979) is known as Ion Exchange Isothermal Supersaturation (IXISS).

Regeneration of SR with potassium sulphate solution is also accompanied by IXISS of calcium sulphate so that its supersaturated solution does not crystallise inside the column.

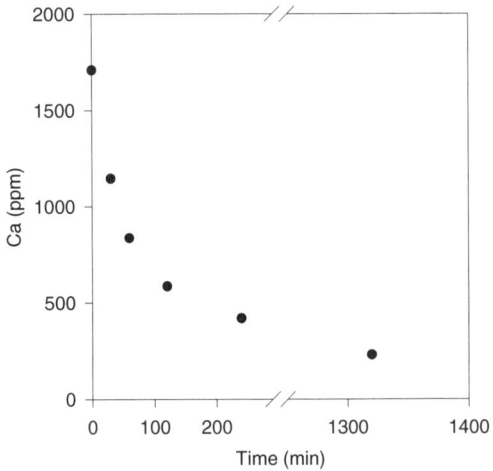

Figure 4 Kinetics of precipitation of calcium sulphate from K_2SO_4 regeneration solution

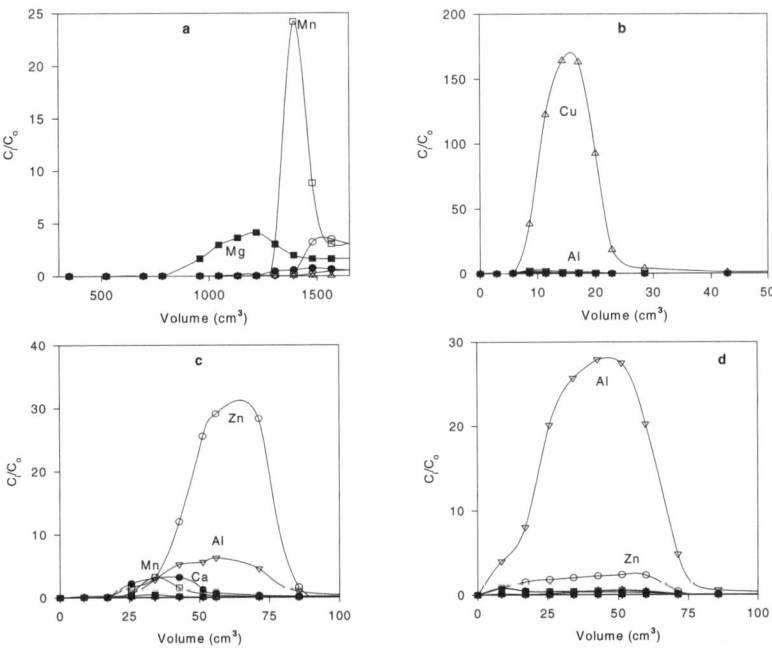

Figure 5 Concentration-volume histories of frontal separation of Mg from KCl regenerating solution (a) and desorption of Cu (b), Zn (c) and Al (d) with H_2SO_4 from second, forth and third column, respectively (see text). Symbols are the same as in Figure 2

Nevertheless, CaSO$_4$ precipitation is observed in the regeneration solution collected. As follows from the results obtained by studying the kinetics of CaSO$_4$ precipitation from regenerating solution presented in Figure 4, the main part of calcium sulphate precipitates during 4–5 hours. The residual concentration of CaSO$_4$ essentially corresponds to its solubility. Optimisation of the regeneration process can be based on the results obtained by regeneration of PR (carboxylic) resin in the Ca-form with a mixture of KCl and K$_2$SO (Muraviev et al., 1999b). It has been shown that the maximal efficiency of regeneration is achieved by using a mixture with KCl to K$_2$SO$_4$ ratio of 1:3.

The results on sequential separation and recovery of RTW metal ions are shown in Figure 5, where the concentration-volume histories of frontal separation (Gorshkov et al., 1999) of Mg (see Figure 5a) from the rest of the RTW metal ions and desorption of Cu from the second (Figure 5b), Zn from the forth (Figure 5c) and Al from the third (Figure 5d) columns are shown. As seen in Figure 5a, Mg^{2+} as the ion of the lowest sorbability is concentrated in the head part of the eluate collected from the last (forth) column forming a zone of the pure component, from which it can be selectively collected and recovered. The second by the order of sorbability component, Mn, is also concentrated in the head portions of the eluate (behind Mg), nevertheless its zone appears to be contaminated by other RTW metals. The other metals (Cu, Zn and Al) are obtained in sufficiently pure state (see Table 3) after the desorption from their respective columns with sulphuric acid solution. The results collected in Table 3 indicate a high efficiency of the separation mode, which permits recovery of four metals from the nano-component mixture at a relatively high purity. This separation technique is known as Tandem Ion Exchange Fractionation (Muraviev et al., 1997d; Gorshkov et al., 1999).

The removal of the residual RTW metal ions remaining in KCl solution after recovery of Cu, Al, Zn and Mg was carried out by adjusting the pH to 10–12 with concentrated KOH solution followed by precipitation of metal hydroxides and their separation by filtration. The final composition of KCl solution is shown in Table 4, from which it follows that this solution contains only Ca^{2+} admixture. The residual calcium can be completely eliminated by addition of the stoichiometric amount of K$_2$CO$_3$ (Muraviev et al., 1999b). The amount of potassium carbonate required for removal of calcium from the regenerating solution is very low (~1.4 kg/m^3) and practically does not affect the economics of the process as a whole. After removal of CaCO$_3$ precipitate by filtration, the regenerating solution appears to be calcium-free and can be reused.

After purification and adjustment of pH (e.g. with H$_2$SO$_4$) from 10–12 to 5–6 KCl

Table 3 Initial, average and maximum purity of products obtained in separation of Cu-Al-Zn-Mg from RTW

Metal	Concentration in RTW, mg/dm³	Initial	Purity, % Average	Maximum
Cu	216	1.5	87	93
Al	564	11	71	79
Zn	1585	4	82	94
Mg	896	6	99	100

Table 4 Composition of KCl regenerating solution after precipitation of polyvalent RTW metal ions with KOH

KCl sample	Fe	Cu	Zn	Al	Mn	Mg	Ca	pH
1	0	0	0	0	0	0	362	10
2	0	0	0	8	0	0	670	12

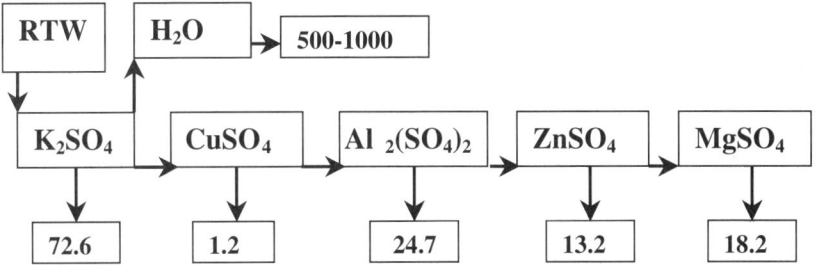

Figure 6 Block-scheme of ion exchange processing of RTW including synthesis of K_2SO_4 and recovery of four metals and desalinated water. Figures show annual productivity of unit (thousand tons) assuming treatment capacity of 1,000 m³ per hour and 50% efficiency of components recovery. Figures for metals refer to hydrated metal sulphates

solution can also be returned back into the process. The precipitates of RTW metal hydroxides removed from KCl solution can be returned back into the reservoir with initial RTW (see Table 2) for dissolution.

The results obtained can be also considered from the practical viewpoint to propose the block-scheme of the ion exchange processing of RTW including the synthesis of potassium sulphate and the recovery of four metals and desalinated water, which is shown in Figure 6 along with estimation of the annual process productivity. As seen in Figure 6, potassium sulphate and desalinated water can be considered as the main products obtained in the course of ion exchange processing of RTW. Recuperation of additional by-products (four metals) gives an additional credit and substantially improves the economy of the whole process.

Conclusions

The results obtained in this study permit us to draw the following main conclusions:
1. Acidic mine waters of RTW type (with high content of sulphate and essential absence of chloride ions) can be successfully used for the ion exchange synthesis of chlorine free potassium fertilisers (K_2SO_4).
2. The proposed treatment technology is based on the "conversion of pollutants to fertilisers" (CPF) approach, whose attractiveness is determined by its relevance to the main Green Chemistry concept known to be "the use of chemistry for pollution prevention". From this viewpoint the technology of this type can be qualified as "green ion exchange technology".
3. One of the main advantages of the proposed CPF technology consists in utilisation of both cationic (metals) and anionic (sulphates) constituents of AMW. This feature makes the CPF approach most attractive when applied to the treatment of metal bearing sulphate effluents.
4. The complex processing of acidic mine waters of RTW type including CPF stage permits to design a flexible technology, which can be oriented to the production of mineral (e.g. potassium) fertilisers, desalinated water and a number of metals or metal salts.

Acknowledgements

The author would like to thank all his co-workers mentioned throughout the text and cited in the various references for their assistance and efforts in making this publication possible. Bayer Hispania Industrial, S.A. is gratefully acknowledged for supplying us with samples of Lewatit resins. Rio Tinto Minera S.A. is also gratefully acknowledged for supplying us with samples of acidic mine waters. This work has been carried out with the financial support of the European Commission (grant EV5V-CT94-0556). Universidad Autónoma de Barcelona and Instituto de Ciencias de Materiales de Barcelona are acknowledged with thanks for their financial support of the author.

References

Elvers, B., Hawkins, S., Russey, W. and Schulz, G., Eds. (1993). *Ullmann's Encyclopedia of Industrial Chemistry*; VCH: Weinheim; 5th Edn.; Vol. A22, p.84.

Gorshkov, V., Muraviev, D. and Warshawsky, A. (1999). Ion-Exchange Methods for Preparative Ultra-Purification of Inorganic, Organic and Biological Substances. In: *Ion Exchange: Theory and Practice. Highlights of Russian Science*, D. Muraviev, V. Gorshkov and A. Warshawsky (eds.), vol. 1, Marcel Dekker, Inc., New York, pp.1–74.

Muraviev, D. (1979). Ion Exchange Isothermal Supersaturation of Aminoacid Solutions. *Zhur. Fiz. Khim. (Russ. J. Phys. Chem.)*, **53**, 438–444.

Muraviev, D. (2000). Ecologically-safe ion-exchange technologies. In: *Encyclopedia of Separation Science*, I. Wilson, C.F. Poole, T.R. Adlard and M. Cooke (eds.), Academic Press, London, pp.2644–2654.

Muraviev, D., Sverchkova, O.Yu., Voskresensky, N.M. and Gorshkov, V.I. (1992). Kinetics of Ion Exchange in Polyphase Systems Including Crystallizing Substance. *React. Funct. Polym.*, **17**, 75–88.

Muraviev, D., Noguerol, J. and Valiente, M. (1997a). Reagentless Concentration of Copper from Acidic Mine Waters by Dual-Temperature Ion-Exchange Technique. In: *Progress in Ion Exchange. Advances and Applications*, A. Dyer, M.J. Hudson and P.A. Williams (eds.), SCI, Cambridge, pp.349–345.

Muraviev, D., Noguerol, J. and Valiente, M. (1997b). Concentration and Recovery of Copper from Acidic Mine Waters by Dual-Temperature Ion-Exchange Technique. *Hydrometallurgy*, **44**, 331–345.

Muraviev, D., Noguerol, J. and Valiente, M. (1997c). Sea Water as Auxilliary Reagent in Processing of Acidic Mine Waters. *Environ. Sci. Technol.*, **31**, 379–384.

Muraviev, D., Noguerol, J. and Valiente, M. (1997d). Tandem Ion-Exchange Fractionation: New Preparative Mode for Separation of Multicomponent Ionic Mixtures. *Anal. Chem.*, **69**, 4234 4240.

Muraviev, D., Khamizov, R.Kh., Tikhonov, N.A., Krachak, A.N., Zhiguleva, T.I. and Fokina, O.V. (1998a). Clean ion-exchange technologies. I. Synthesis of chlorine-free potassium fertilisers by ion-exchange isothermal supersaturation technique. *Ind. Eng. Chem. Res.*, **37**(5), 1950–1955.

Muraviev, D., Khamizov, R.Kh. and Tikhonov, N.A. (1998b). Ion Exchange Isothermal Supersaturation. *Solv. Extr. Ion Exch.*, **16**(1), 151–221.

Muraviev, D., Noguerol, J. and Valiente, M. (1999a). Dual-Temperature Ion-Exchange Fractionation. *Solv. Extr. Ion Exch.*, **17**(4), 767–849.

Muraviev, D., Noguerol, J., Gaona, J. and Valiente, M. (1999b). Clean Ion-Exchange Technologies. III. Temperature Enhanced Conversion of Potassium Chloride and Lime Milk into Potassium Hydroxide on Carboxylic Ion Exchanger. *Ind. Eng. Chem. Res.*, **38**(11), 4409–4416.

Ritcey, G.M. (1989). *Tailings Management, Problems and Solutions in the Mining Industry*, Elsevier, Amsterdam, p. 270.

Scharf, H.J. (1990). *Environmental Aspects of K-Fertilisers in Production, Handling and Application. Development of K-Fertiliser Recommendations*, International Potash Institute, Worblaufen-Bern, p. 395.

Slack, A.V. (1965). *Fertiliser Development and Trends*, Noyes Development Corp., Park Ridge, NJ.

Smith, M.J. (1974). Acid production in mine drainage systems. In: *Extraction of Minerals and Energy: Today's Dilemmas*, R.A. Dejn (ed.), Ann Arbor Science, Ann Arbor, Michigan, pp. 57–75.

Strathmann, H. (1992). Electrodialysis. In: *Membrane Handbook*, W.S. Winston Ho and Kamalesh K. Sirkar (eds.), Van Nostrand Reinolds, New York, p.217.

Carbon and nitrogen removal from tannery wastewater with a membrane bioreactor

A. Goltara, J. Martinez and R. Mendez*

Chemical Engineering Department, Institute of Technology, Campus Sur, University of Santiago de Compostela, Santiago de Compostela E-15782. Spain (*E-mail: eqrmndzp@usc.es)

Abstract A 3.5 L Membrane Sequencing Batch Reactor (MSBR) was used for the treatment of a wastewater coming from the beamhouse section of a tannery. The wastewater, produced after the oxidation of sulphide compounds, contained average COD and ammonium concentrations of 550 and 90 mg/L respectively. The system was operated for a period of 150 days, with no sludge removal during the whole period of operation. The biomass concentration inside the reactor varied considerably, with maximum values close to 10 g/L at the end of operation. Low biomass yield values were achieved probably due to the low feed/microorganisms (F/M) ratio. An important accumulation of organic matter in the reactor was noticed, although the COD effluent was not affected due to the permeation through the membrane. The nature of this organic matter is finally discussed. Removal efficiencies close to 100% in ammonium and 90% in COD were achieved and the TN removal efficiency ranged from 60 to 90%.
Keywords Beamhouse wastewater; membrane bioreactor; sludge production; tannery

Introduction

Industrial tanneries produce great amounts of effluents of a different nature, most of them highly pollutant. The wastewater comes from two main processes: beamhouse and tanning. In the beamhouse process highly loaded effluents are generated, containing organic matter, suspended solids (as a result of unhairing and skinning), sulphides (used in the unhairing) and chlorides (from the salt present in the hides). At this stage bactericides (to prevent degradation of the hide) and surfactants (to improve soaking) are also used.

In the tanning process the main pollutant is chromium, while other compounds may be used in the retanning process, such as tannins. Many authors have done a thorough characterization of the different streams present in industrial tanneries. (Tünay *et al.*, 1995; Orhon *et al.*, 1999; Rivela *et al.*, 2002).

Due to the high variability in flow and in composition of the tannery effluents and the high organic loads, the use of a Sequencing Batch Reactor (SBR) presents the following advantages:

- Operation during most of the time with higher concentrations than those in the effluent, making it possible to achieve higher removal rates than in continuous systems.
- To treat in a single reactor organic matter, nutrients and even phosphate, by alternating aerobic and anoxic stages. Carucci *et al.* (1994) studied the dynamics of organic matter and phosphorus in an SBR, achieving high levels of phosphate removal.
- Flexibility of operation, made it possible to easily adjust the reaction time and the volume of reactor used, depending on the influent conditions. Besides this, the settle and draw steps are made also in the same tank.

Previous results obtained by Carucci *et al.* (1999) in an SBR treating tannery effluents for a long period of time showed that the SBR is an efficient system with respect to COD removal and nitrification.

Coupling a submerged membrane to an SBR would have the following advantages:
- Total retention of solids in the reactor, making it possible to increase the biomass

concentration to higher values than in conventional systems. With high biomass concentration, higher volumetric rates can be reached, as well as lower feed to microorganisms ratio, decreasing sludge production.
- High quality of the effluent, free of suspended solids, bacteria, and turbidity.
- The possibility of treating recalcitrant compounds due to a higher retention time inside the reactor independently of the hydraulic retention time (HRT).

Lübbecke et al. (1995) and Nah et al. (2000) operated membrane bioreactors with very high solids retention times, reaching concentrations up to 40 and 12 g/L respectively.

Materials and methods
Wastewater characteristics
The wastewater fed to the reactor was taken from a local tannery factory. The beamhouse effluent passes through a screen and a tank where the oxidation of sulphide compounds takes place, using MnO as a catalyst. The outlet of this tank is the influent used for our study.

The beamhouse is a discontinuous process, namely in this factory it is done twice a day. The characteristics of the effluent are strongly dependent on the moment of plant operation. For practical reasons the effluent was collected in the factory when it reached its maximum concentration and later it was diluted to the desired operational concentration. The average characteristics of the bulk beamhouse effluent are shown in Table 1.

Experimental set-up and operation
An MSBR of 3.5 L was built including all the accessories presented in the experimental set-up (Figure 1). It was equipped with a submerged hollow fibre membrane (Zenon Environmental Inc.), with 0.10 square metres of surface and the average and maximum pore sizes were 0.04 and 0.1 micrometres respectively. The effluent was permeated through the hollow fibre membrane by means of a peristaltic pump. Pumping direction was periodically reversed, so that permeate was backpulsed through the membrane into the reactor, in order to reduce membrane fouling. Permeation lasted 150 seconds and backpulsing 45 seconds.

To reduce the formation of a cake around the membrane, the module was equipped with an air diffuser, which provides bubbles to maintain the membrane fibres in movement during the permeation. Due to the progressive organic fouling, the membrane also needed a periodical chemical cleaning by immersion in a diluted solution of NaClO during 12 hours.

The cycle pattern lasted 8 hours, with 20 minutes for feeding, 4h 45 minutes of aeration, 1h 15 minute anoxic stage, 30 minutes of reaeration and 1h 10 minutes of permeation. The cycle pattern was controlled with a commercial PLC. Samples of the feed, the mixed liquor just after feeding and the effluent permeating through the membrane were analysed every two days. The daily volume fed to the reactor was 3.5 L, the HRT being 24 hours and the volumetric exchange ratio (VER) on each cycle 1/3.

Table 1 Bulk beamhouse wastewater characteristics

Compound	Concentration	Compound	Concentration
TSS (mg/L)	400–1,720	TKN (mg/L)	52–165
VSS (mg/L)	330–1,150	Sulphide (mg/L)	0.0–0.5
Total COD (mg/L)	732–1,576	P-orthophosphate (mg/L)	0.0–1.7
Soluble COD (mg/L)	320–1,280	Sulphate (mg/L)	510–1,220
Soluble BOD/COD	0.33–0.96	Chloride (mg/L)	900–2,500

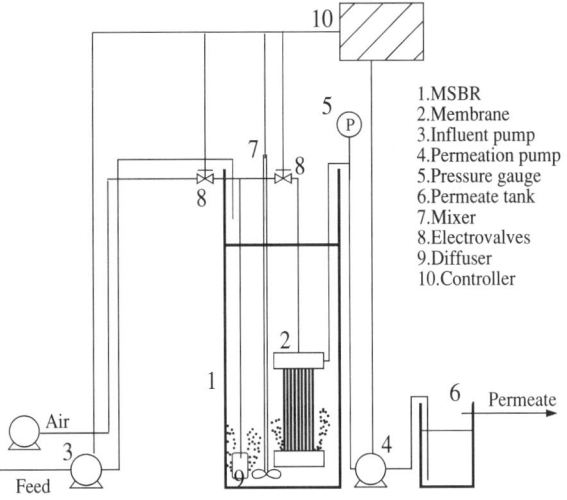

Figure 1 Experimental set-up

Analytical methods

Ammonium-nitrogen ($N\text{-}NH_4^+$) and nitrite-nitrogen ($N\text{-}NO_2^-$) were measured by spectrophotometry according to Standard Methods (APHA, 1995). Nitrate-nitrogen ($N\text{-}NO_3^-$), chloride (Cl^-) and sulphate (SO_4^{-2}) were measured by ionic chromatography. Organic matter (COD) was measured as chemical oxygen demand (COD) by closed reflux digestion and a titrimetric method (APHA, 1995). Biological oxygen demand (BOD) was measured with 5-Day BOD test, with a manometric measuring device Oxitop IS12. Total-, Inorganic- and total organic-carbon (TOC) were measured with a Shimadzu TOC-5000 analyser. Total, inorganic and total Kjeldhal nitrogen (TKN) were measured with a Dohrman-Rosemat analyser, DN1900. Total suspended solids (TSS) and volatile suspended solids (VSS) were determined using 0.45 micrometre filters, according to *Standard Methods* (APHA, 1995).

Biomass nitrification activity was measured with an Ammonium NITrifying Analyser (ANITA, Austep).

Results and discussion

Influent conditions: nitrogen and organic loads

The membrane bioreactor was operated for a period of 150 days. The evolution of nitrogen compounds and organic matter are represented in Figures 2 and 3. The organic loading rate (OLR) applied was around 0.5 gCOD/L·d between days 40 and 120. The ammonium loading rate (ALR) was maintained near 0.09 gTKN/L·d between days 40 and 70 and close to values of 0.06 gTKN/L·d till day 120. Due to the high removal efficiencies reached the OLR was increased till values between 0.7–0.8 gCOD/L·d maintaining the ALR close to 0.11 gTKN/L·d from day 120.

Nutrient and organic matter removal

The evolution of removal efficiencies of COD and nitrogen are presented in Figures 4 and 5. The first 30 days of operation correspond to the acclimation period, where no high removal efficiencies were achieved. After this period, the ammonium removal was almost complete during the whole research (always near 100%), achieving maximum nitrification rates of 6 mgN-NH_4^+/gVSS·h. It seems that it was possible to achieve an active population of nitrifying bacteria. Due to the feeding in anoxic conditions, a part of the organic matter was consumed in denitrification. Thus the C/N ratio in the aerobic period was more favourable to allow the growth of nitrifying bacteria.

The TN removal varied from 65 to 85% between days 45 to 65, and from 80 to 90% per cent from days 65 to 125. After this period, there was a decrease to 50 per cent of removal, due to the increase in ammonium load. The TN removal was afterwards increased because nitrification was restored completely, but an important accumulation of nitrite was noticed. This accumulation could be a consequence of the inhibition of nitrite oxidisers. The origin of this inhibition was not clear, but the low dissolved oxygen concentration during the aerobic stage in this period seemed to be the main reason. The lower affinity of oxygen for nitrite oxidisers compared to ammonium oxidisers can restrict the growth of nitrite oxidisers (Van Loosdrecht and Jetten, 1998). These microorganisms can also be inhibited by free ammonia, but pH values were always under 8.5. The TN finally decreased to 40 per cent because nitrite accumulation increased from 25 to 50 mg/L. However, denitrifying organisms are able to use nitrite instead of nitrate (Wiesmann, 1994), so it is not clear why denitrification decreased. In general, nitrite is inhibiting for microorganisms (Henze *et al.*, 1995), and this can be the explanation of this drop in TN removal.

The organic matter removal varied from 60 to 100% during the acclimation period. After this period, the COD removal was always between 85 and 95%.

Suspended solids evolution

The biomass concentration varied considerably during the whole experiment. One of the aims of this study was to minimise the sludge production, so the membrane bioreactor was operated with no sludge removal, as other authors have previously done (Muller *et al.*, 1995; Nah *et al.*, 2000). For this reason, the theoretical sludge age was infinite, because there were no sludge losses, except from the samples taken to measure the VSS and TSS, and the possible losses of biomass produced during the membrane manual cleaning.

Lübbecke *et al.* (1995) showed that the high biomass concentration reached in the reactor could hinder the mass transfer, both of oxygen and substrates, this agrees with the profiles of dissolved oxygen concentration observed sometimes during some periods of operation. The biomass population did not reach steady state conditions during the operation due to the high sludge retention time (SRT) in the reactor. In general, it is considered

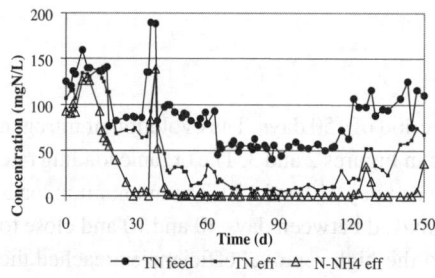

Figure 2 Evolution of nitrogen compounds

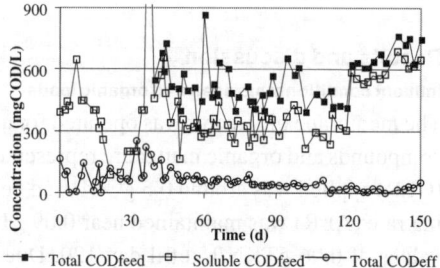

Figure 3 Evolution of organic matter

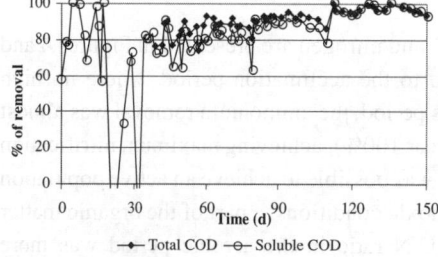

Figure 4 Removal of organic matter

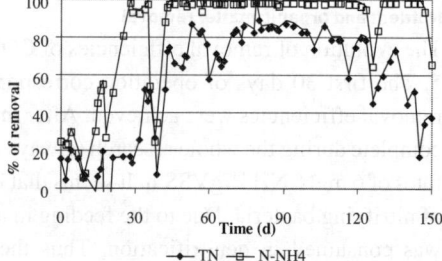

Figure 5 Removal of nitrogen compounds

that steady state conditions are achieved after 2 or 3 SRTs (Cicek *et al.*, 1998; Oh and Silverstein, 1999; Cicek *et al.*, 2001). Although the operation at steady-state would be more instructive to see if the process could actually be applied to beamhouse wastewater, it was preferred to check if it was possible to perform the process without production of sludge waste to avoid sludge disposal.

At the beginning, the VSS concentration increased very fast, decreasing only sporadically, as can be seen in Figure 6. A possible explanation of these punctual decreases of VSS could be that some organisms in the reactor were adapted to degrade non active biomass. Van Loosdrecht and Henze (1999) showed that predation by protozoa could contribute to the significant drop in solids concentration. Although protozoa were observed in our system, their proliferation is not necessarily related to the decreases of biomass. This decrease in VSS concentration in systems operated without purge of sludge was also reported by other authors (Choo and Stensel, 2000; Nah *et al.*, 2000).

From day 80, the TSS increased from values of 4.5 g/L to values close to 10 g/L, the VSS/TSS ratio always remaining near to 0.9 as can be observed in Figure 6.

Biomass yield

The biomass yield (Y) reported for conventional activated sludge systems is around 0.4 gCOD/gCOD (Henze *et al.*, 1995). The lower the yield, the lower the production of sludge. One way to reduce this sludge production is to operate with low F/M ratio. The MSBR makes it possible to achieve high concentrations of biomass, so the F/M ratio can be lower than in conventional systems, achieving lower biomass yields. The observed yield values (Y_{obs}) for our system were quite low, always lower than 0.12 gVSS/gCOD, reaching minimum values of 0.08 gVSS/gCOD (0.17 and 0.11 gCOD/gCOD assuming biomass COD as 1.42 gCOD/gVSS, Henze *et al.*, 1995).

Orhon *et al.* (1998) calculated an average yield of 0.64 gCOD/gCOD for biomass fed with tannery wastewater, operating in a conventional activated sludge system with a sludge retention time (SRT) of 15 days. This yield is much higher than the one obtained in our system. Cicek *et al.* (2001) evaluated the evolution of yield as a function of SRT in a MSBR. The higher the SRT, the lower the yield, achieving yield values of 0.27 gVSS/gCOD when the SRT was 30 days. The theoretical SRT in the MSBR is infinite, but estimating the losses of biomass due to sampling and membrane cleaning an SRT of 400 days is calculated. It seems reasonable then to obtain such low values in the system. Choo and Stensel (2000) obtained yield values even lower than 0.08 gVSS/gCOD, operating with an estimated SRT of 1,400 days.

Organic matter accumulation

An accumulation of organic matter inside the bioreactor was noticed during an important part of the research. Due to the permeation through the membrane, there was a fraction of organic matter that could not pass through the membrane pore. The COD concentration was

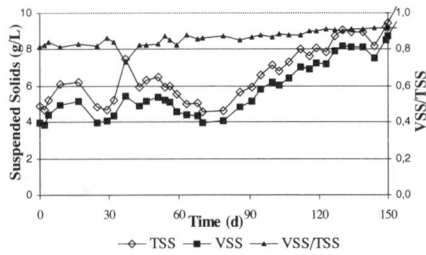

Figure 6 Suspended solids evolution

different if the effluent was permeated through the membrane (0.04 micrometres of average pore size) or if it was taken inside the reactor and filtrated through a 0.45 micrometre filter. An increase in the difference between these CODs was observed, reaching maximum values of 400 mg/L. This means that there was an important amount of organic compounds of size between 0.45 and 0.04 micrometres. The latter value could be even lower, because the biomass attached to the membrane can also act as a biomembrane, retaining compounds of lower molecular size, even viruses (Ueda and Horan, 2000).

This organic matter was later progressively degraded with time, as is shown in Figure 7, probably because of the acclimation of the biomass to degrade these high molecular weight compounds.

This COD accumulation was detected while estimating the COD in the reactor just after the feed stage (COD_{estim}):

$$COD_{estim} = 1/3\ COD_{feed} + 2/3\ COD_{perm}$$

where COD_{feed} is the soluble COD in the feed and COD_{perm} is the soluble COD in the permeate.

The actual soluble COD after the feed stage would be:

$$COD_0 = 1/3\ COD_{feed} + 2/3\ (COD_{perm} + COD_{0.45-0.04\mu m}) - COD_{denit} - COD_{cels} + COD_{hyd}$$

where:
COD_0 = Actual COD after the feed stage.
$COD_{0.45-0.04\mu m}$ = COD of compounds between 0.45 and 0.04 micrometres.
COD_{denit} = COD used in denitrification during the feed stage.
COD_{cels} = COD used in cell growth and maintenance during feed stage.
COD_{hid} = Suspended COD eventually hydrolysed during the feed stage

The difference between COD_0 (actual) and COD_{estim} (estimated) would be:

$$COD_0 - COD_{estim} = 2/3\ COD_{0.45-0.04\mu m} - COD_{denit} - COD_{cels} + COD_{hyd}$$

This difference should be negative because the COD consumed in denitrification and in cell maintenance and growth is higher than the fraction of suspended COD that may be hydrolysed in this stage, because denitrification took place almost completely in the feed stage and the hydrolysis is usually slow compared to other biological processes (Henze *et al.*, 1995). This difference is plotted in Figure 7, where an important accumulation of organic matter from day 15 to day 90 can be observed. This accumulation reached values close to

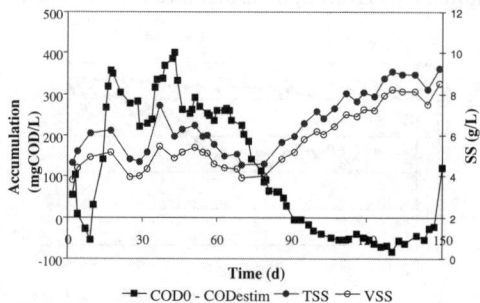

Figure 7 Accumulated COD and SS evolution

400 mg/L, which means that apart from the biological treatment, the MSBR has a complementary retention and a physical removal of organic matter takes place. From day 70 the accumulation starts to decrease, and the difference becomes negative after day 90.

This decrease in organic matter is attributed to an adaptation of biomass to degrade these high molecular weight compounds. Although the origin of these compounds is not yet clear, there are reasons to think that these compounds are subproducts of biomass, maybe exopolymers. Another possibility could be that because of the high SRT, this organic matter could be a product from the cellular lysis of non active cells. In Figure 7 the evolution of accumulated organic matter and suspended solids is shown. It can be seen that the curves are very similar at the beginning. The COD accumulated increases and decreases with biomass, so the COD accumulation is proportional to VSS, it is reasonable to think that this accumulation could come from biomass.

From day 70, the accumulated COD decreases as the suspended solids increase, and also the slopes of these curves are very similar. This contributes to the idea that from day 70, the biomass started to degrade this organic matter, therefore accumulated COD decreases and the biomass increases, approximately at the same rate.

The adaptation of biomass could be due to changes in the microbial population or to the production of enzymes capable of degrading these high molecular weight compounds. The latter possibility is the more feasible, because no important changes were observed in biomass population in this period. Further work is necessary to investigate this behaviour.

Conclusions

The MSBR appears to be a proper system to treat the beamhouse effluent after the oxidation of sulphide compounds. High removal efficiencies were achieved, close to 90% and 100% for organic matter and ammonium respectively. The TN removal efficiency ranged from 60 to 90% but a decrease in the last 30 days was noticed. It seems that the SBR made it possible to achieve a stable population of nitrifying bacteria during the whole period. It was possible to operate during 150 days with no sludge waste, attaining concentrations close to 10 g/L. The observed biomass yield was always under values of 0.12 gSSV/gCOD. The filtration of the effluent through the membrane produced an accumulation of organic matter inside the reactor. This organic matter probably comes from biomass, because its accumulation is directly related to biomass concentration. From day 70, this organic matter decreases possibly because of the production of an enzyme capable of degrading these compounds of high molecular weight.

Acknowledgements

This work was funded by the Xunta de Galicia-Spain (Project PGIDT00MAM20903PR) and the EU by the INCO Project EILT (Ref: ERBIC18–CT98–0286). We also acknowledge the EU TMR Program (Project BIOTOBIO FMRX-CT97-0114) for financing the stay of Andrea Goltara in Santiago.

References

APHA-AWWA-WPCF (1985, 1995). *Standard Methods for Examination of Water and Wastewater* 16th,19th edition, American Public Heath Association, Washington, D.C. USA.

Carucci, A., Majone, M., Ramadori, R. and Rossetti, S. (1994). Dynamics of phosphorus and organic substrates in anaerobic and aerobic phases of a sequencing batch reactor. *Wat. Sci. Tech.*, **30**(6), 237–246.

Carucci, A., Chiavola, A., Majone, M. and Rolle, E. (1999). Treatment of tannery wastewater in a Sequencing Batch Reactor. *Wat. Sci. Tech.*, **40**(1), 253–259.

Choo, K.-H. and Stensel, H. (2000). Sequencing Batch Membrane Reactor treatment: nitrogen removal and membrane fouling evaluation. *Wat. Env. Res.*, **72**(4), 490–498.

Cicek, N., Winnen, H., Suidan, M.T., Wrenn, B.E., Urbain, V. and Manem, J. (1998). Effectiveness of the membrane bioreactor in the biodegradation of high molecular weight compounds. *Wat. Res.*, **32**(5), 1553–1563.

Henze, M., Harremoës, P., LaCour Jansen, J. and Arvin, E. (1995). *Wastewater Treatment: Biological and Chemical Processes*. Springer, Heidelberg.

Cicek, N., Macomber, J., Davel, J., Suidan, M.T., Audic, J. and Genestet, P. (2001). Effect of solids retention time on the performance and biological characteristics of a membrane bioreactor. *Wat. Sci. Tech.*, **43**(11), 43–50.

Lübbecke, S., Vogelpohl, A. and Dewjanin, W. (1995). Wastewater treatment in a biological high-performance system with high biomass concentration. *Wat. Res.*, **29**(3), 793–802.

Muller, E.B., Stouthamer, A.H., van Verseveld, H.W. and Eikelboom, D.H. (1995). Aerobic domestic waste water treatment in a pilot plant with complete sludge retention by cross-flow filtration. *Wat. Res.*, **29**(4), 1179–1189.

Nah, Y.M., Ahn, K.H. and Yeom, I.T. (2000). Nitrogen removal in household wastewater treatment using an intermittently aerated membrane bioreactor. *Env. Tech.*, **21**, 107–114.

Oh, J. and Silverstein, J. (1999). Effect of air on-off cycles on activated-sludge denitrification. *Wat. Env. Res.*, **71**(7), 1276–1282.

Orhon, D., Ates, E. and Ubay Çokgör, E. (1999). Characterization and modelling of activated sludge for tannery wastewater. *Wat. Env. Res.*, **71**(1), 50–63.

Orhon, D., Sözen, S., Ubay Çokgör, E. and Ates Genceli, E. (1998). The effect of chemical settling on the kinetics and design of activated sludge for tannery wastewaters. *Wat. Sci. Tech.*, **38**(4–5), 355–362.

Rivela, B., Vidal, G., Bolhardt, C. and Mendez, R. (2002). Towards a cleaner production: A case study in a Chilean tannery. *Waste Management and Research*. (submitted)

Tünay, O., Kabdasli, I., Orhon, D. and Ates, E. (1995). Characterization and pollution profile of leather tanning industry in Turkey. *Wat. Sci. Tech.*, **32**(12), 1–9.

Ueda, T. and Horan, N.J. (2000). Fate of indigenous bacteriophage in a membrane bioreactor. *Wat. Res.*, **34**(7), 2151–2159.

Van Loosdrecht, M. and Jetten, M. (1998). Microbiological conversions in nitrogen removal. *Wat. Sci. Tech.*, **38**(1), 1–7.

Van Loosdrecht, M. and Henze, M. (1999). Maintenance, endogenous respiration, lysis, decay and predation. *Wat. Sci. Tech.*, **39**(1), 107–117.

Wiesmann, U. (1994). Biological nitrogen removal from wastewater. *Adv. Bioch. Eng.*, **51**, 113–154.

Nitrogen removal from tannery wastewater by protein recovery

I. Kabdaşlı, T. Ölmez and O. Tünay

Istanbul Technical University, Civil Engineering Faculty, Environmental Engineering Department, Ayazağa Kampüsü, 34469, Maslak, Istanbul, Turkey (E-mail: *ikabdasli@ins.itu.edu.tr*; *tolmez@ins.itu.edu.tr*; *otunay@ins.itu.edu.tr*)

Abstract Nitrogen removal from wastewaters has gained importance in recent years. In this paper protein precipitation and recovery potential of leather tanning industry wastewaters were experimentally evaluated. A protein profile for all sources was prepared. Liming was determined to be the most important protein source. Composite samples were made up to assess the protein precipitation applications. Isoelectric pH precipitation yielded around 50% protein removal between the optimum pH interval of 2.1–3.8. $FeCl_3$ proved to be a very effective means of protein removal providing over 60% efficiency. Polyelectrolyte precipitation did not yield satisfactory results. Magnesium ammonium phosphate precipitation followed by acid precipitation of protein provided 85% ammonia removal in addition to 50% protein removal.
Keywords Leather tanning industry; magnesium ammonium phosphate precipitation; protein precipitation; protein recovery

Introduction

Nitrogen is not a common pollutant in industrial wastewaters. However, several industries discharge significant amounts of nitrogen e.g. tanneries, slaughterhouses, fertilizer industries etc. Nitrogen, in the form of ammonia and organic nitrogen, has generally been removed by biological processes i.e. nitrification-denitrification, where applicable. However biological nitrogen removal frequently poses problems in the industrial applications. These problems include; self-inhibition due to high concentration of ammonia, sensitivity to pH and temperature changes and existence of inhibitors to nitrification. Cost of the biological treatment is high and no recovery possibility exists. Increasing demand for nitrogen removal has triggered the efforts to find out alternative means of nitrogen removal. Recently new methods of treatment such as oxidation and magnesium ammonium phosphate (MAP) precipitation have been successfully applied to industrial wastewaters. Attention has been paid to the methods providing recovery. Tannery effluents contain nitrogen species in high concentrations and several toxic substances, such as chromium and sulfide, which adversely affect the biological nitrogen removal. Nitrogen is found in the effluents as ammonia and organic nitrogen. Several studies conducted for tannery wastewaters to remove ammonia nitrogen by MAP precipitation provided promising results especially for concentrated waste flows (Tünay *et al.*, 1997). MAP precipitation also removes particulate organic nitrogen to a significant extent. In leather tanning wastewater organic nitrogen consists, almost totally, of protein and aminoacids. Proteins are precipitated for different purposes such as preparation, concentration etc. Depending on the purpose, the method employed may vary considering factors such as efficiency, purity of protein, ease of application. The methods used mostly aim to modify either the properties of solvent or of the protein. Among the methods commonly used are: addition of miscible solvents such as alcohol or acetone, addition of metal cations such as Zn^{2+}, Cd^{2+}, Cu^{2+}, or bulky anions such as perchlorate, trichloroacetate, salting out using ammonium sulfate or other salts, isoelectric pH precipitation, precipitation by polyelectrolyte aggregation (England

and Seifter, 1990). As these methods are viewed from the standpoint of industrial wastewater treatment some of them seem to be inapplicable, or at least very difficult or costly to apply. Isoelectric pH precipitation and polyelectrolyte addition are considered as the methods having the greatest potential for industrial wastewater treatment applications. Other methods of treatment must be evaluated not only in terms of technical and economical feasibility but also taking into account the impact of the application on wastewater quality as far as both the subsequent treatment steps and final characteristics are concerned. In this context, metal ion addition may prove feasible for the cases, for instance, where metal ions already exist in wastewater or to be effectively removed subsequently by chemical precipitation with sludge recovery, or other methods may be considered as the wastewater characteristics and/or treatment applications justify the use of a method. Numerous studies in the literature dealt with protein precipitation, although many of them are highly specific. Chan et al. (1986) examined the kinetics of soy protein precipitation using ammonium sulfate, calcium, sulfuric acid and ethanol. For 3 g l^{-1} initial protein concentration over 80% removals were obtained with all reactants. Fisher et al. (1986) studied acid precipitation of soy proteins. Shih et al. (1992) studied the solubilities of lysozyme, α-chymotrypsin and bovine serum albumin (BSA) in aqueous solution as a function of solution characteristics. Iyer and Przybycien (1994) studied the effect of mixing on protein solubility. Iyer and Przybycien (1995) also studied metal precipitation of proteins for a zinc/glycine precipitant system using bovine serum albumin and bovine γ globulins. Iyer and Przybycien (1996) in another study proposed a theoretical model for metal protein precipitation considering metal proton competition for the reaction and employing gelatin theory for the aggregation step. Chen et al. (1992) studied the floc formation mechanism for polyelectrolyte precipitation of protein. Only a few attempts have been made for soluble nitrogen removal from wastewaters. Chen et al. (2000) conducted an experimental study for the recovery of proteins from a food processing wastewater using carboxymethyl cellulose and cellulose triacetate fibrets. They obtained high TKN removals as well as BOD_5 reductions up to 90%. Tünay et al. (2001) experimentally investigated the protein removal from leather tanning wastewaters. In this study emphasis was placed upon synthetic wastewaters containing leather proteins and sheepskin processing wastewaters. Metal ion addition and acidification methods were employed. High protein removals varying between 40–90% were obtained. Cattlehide processing is a more common application. On the other hand, the unit amount of wastewater is lower than other tanning processes i.e. sheepskin, goatskin (Tünay et al., 1999). This makes wastewaters more concentrated which, in turn, makes the protein removal more efficient.

In this paper results of an experimental study conducted on cattlehide processing wastewaters for protein removal were presented and discussed. The studies aimed at determining the protein removal efficiencies by different methods and evaluating the operational aspects of the processes.

Protein sources and nitrogen profile in leather processing
Raw leather is made up of three main layers. The middle layer (corium) is the one to react with the tanning agent and to constitute the leather product, while others are removed during processing. The bottom layer is removed by mostly mechanical means. The upper layer (epidermis) has a complex structure and contains hair, glands, muscles, etc. Removal of this layer is realized by chemical means. Removal reactions are carried out through liming and bating processes. A soaking process is employed for the rehydration of skins before the liming. In parallel with rehydration some of the globular proteins are removed out of the skins. Liming is the process where hair is removed from the skin. The reductive process is the most common method of liming. Two important processes take place during

liming: hydrolysis and chemical reduction of keratin. Hydrolysis occurs at alkaline medium and lime being a base of limited solubility provides the most suitable conditions for the process by bringing and buffering the pH around 12.5. Globular proteins can easily be solubilized. Keratin and collagen have comparable resistance to hydrolysis. To prevent hydrolysis of collagen which is the main fibrous protein of corium, reduction of keratin is realized by the use of alkali sulfide salts. Then in the liming step two important groups solubilized are globular proteins and keratin. Elastin, another strong fibrous protein is not soluble under the conditions of liming. Bating is the removal of all unwanted proteins that remain in the liming step. This is carried out by solubilization of proteins using enzymes. In this step solubilized proteins are a mixture of epidermis proteins and their degradation products. Some chemically resistant fibrous proteins such as elastin are also removed. In the pickling process some globular proteins can be removed by the aid of pretanning agents (Thorstensen, 1985). Nitrogen exists in leather tanning wastewaters as ammonia and organic nitrogen. Organic nitrogen is present in both particulate and soluble forms. The amount and composition of total nitrogen in wastewater vary with the type of leather processed, amount of chemicals used in the processing (particularly ammonium sulfate), water use and several other factors such as duration of process. A detailed source characterization study conducted by Zengin *et al.* (2002) has shown nitrogen speciation and distribution for a bovine leather processing plant. This study indicated that all nitrogen species as well as total nitrogen (TKN) load of the processes following tanning are below 1%. Practically all nitrogen is originating from the processes soaking-liming-deliming-bating-pickling-tanning and their washings. Therefore, for the studies aiming at assessment and control of nitrogen, in particular protein, evaluation of these sources is necessary and sufficient.

Experimental study
Experimental planning
The purpose of this study is to assess the applicability of protein removal as a recovery and wastewater treatment method to leather tanning wastewaters. Within this context, several interrelated experimental studies were conducted. In the first step a characterization was made for all relevant sources to evaluate the protein contents as well as for other parameter values. The second step was preparation and analysis of the group samples which represent unit operations and their washings. Characterization covered both protein analysis and determination of protein characteristics. Protein characterization was made by acid precipitation in a pH interval to determine optimum precipitation pH. Acid precipitation is both a potential method of protein removal and a very informative tool to assess the protein characteristics as delineated above. The results of these experiments helped the planning of the protein precipitation applications in the third step. In this step protein precipitation methods were applied to composite samples. In the fourth step the effect of operational parameters on the protein precipitation was assessed. Mixing conditions were considered as the most important process variable. The final experimental phase covered MAP precipitation experiments in combination with protein removal.

Selection of the precipitation methods
As noted above there are many methods, some of which are highly specific and aim at high purity. In the area of waste treatment and material recovery specific and costly methods cannot be employed. Considering the wastewater characteristics and literature information, three methods were selected. These are acidification to near isoelectric point, polyelectrolyte aided aggregation and metal salt addition. The first two methods enable protein recovery. Acidification was carried out using sulfuric acid considering cost, ease of

application as well as the effectiveness of sulfate ion. Polyelectrolyte addition was made using cationic and anionic polyelectrolytes to compare their effectiveness. Metal precipitation was assessed using zinc and ferric ions. Zinc was selected considering several reasons; firstly it was the relatively less toxic metal, secondly it is one of the most effective metals in protein precipitation, thirdly to compare the effectiveness with ferric ions. Ferric ion, one of the least toxic metals, is cheap and a common coagulant. In the experiments ferric ion was used in two modes: to aid acid precipitation and as a metal binder to precipitate protein. MAP precipitation was realized on composite samples in two different sequences. MAP precipitation was first applied to raw composite and protein removal was applied to supernatant of MAP precipitated sample. This application provided a pure protein precipitate. In the second application the sequence was reversed.

Sampling

The leather tanning plant from which the wastewater samples were taken is a bovine leather processing plant. The raw material is green salted leather and the product is upper shoe leather. Process steps and unit wastewater amounts are shown in Figure 1. Finishing processes include retanning, neutralization, dyeing and fatl iquoring.

Samples were taken from the main sources of protein which also contain other nitrogen forms. Presoaking is a process where mostly dirt and suint were removed. Following presoaking, all steps are important up to finishing and samples were collected from each step. Grab samples were taken from the baths as they were dumped. Washing samples were prepared as composites of grabs that were taken at the beginning, half-time and at the end of washing periods.

Materials and methods

Acidification and metal precipitation were carried out using Jar-Test apparatus with 50 × 15 mm paddle size. These experiments were conducted in 2 minutes flash-mixing at 100 rpm, 15 minutes flocculation at 30 rpm and 30 minutes settling sequence in 200 ml sample volume. Acidification experiments were conducted using H_2SO_4 and H_3PO_4 (0.5 mol l^{-1} and 1 mol l^{-1}). $ZnSO_4.6H_2O$ and $FeCl_3.6H_2O$ were used for metal precipitation. Polyelectrolyte aided precipitation experiments were conducted using Praestol® 2440 (medium anionic) and 853BC (strongly cationic) polyelectrolytes. Composite I, Composite II, Group III and Group IV samples were air-oxidized before precipitation experiments to remove sulfide which achieved up to 600 mgl^{-1} in the samples. Sulfide oxidation experiments were realized in 5–l beakers equipped with porous stone diffusers at the bottom to supply air. MAP precipitations were conducted in a 500 ml Erlen meyer flask with stopper. Mixing was provided with a magnetic stirrer. Chemical addition and initial pH adjustment were made under flash-mixing conditions. pH was monitored during the course of slow mixing. $MgCl_2.6H_2O$ and $NaH_2PO_4.2H_2O$ and/or H_3PO_4 were used as magnesium and phosphate sources for MAP precipitation. pH was adjusted using NaOH. pH measurements were made with an Orion 720A model pHmeter. All samples were subjected to filtration

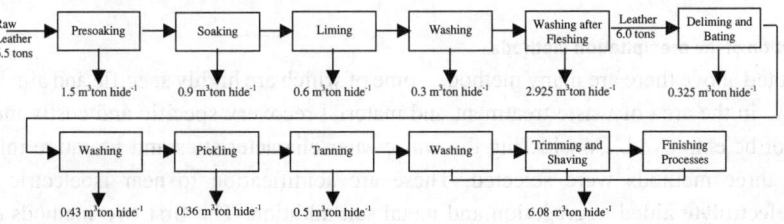

Figure 1 Process scheme

through Millipore AP40 before measurements. All chemicals used were analytical grade. All analyses were performed as defined in the *Standard Methods* (1998) except for protein measurement. Protein was measured by Folin Ciocalteau methods as defined by Gerhardt et al. (1994).

Source-based characterization and selection of composite samples

Source-based characterization of the samples is given in Table 1. As seen from the table, liming and its washing have the highest protein concentrations. Particulate nitrogen load is the highest in washing after fleshing. Ammonia becomes important beginning with washing after fleshing and has comparable concentrations in deliming-bating, its washing and pickling. Table 2 was prepared from Table 1 values to help evaluation and comparisons and to provide a basis for composite sample preparation.

Group samples were prepared for the determination of their characteristics. Four source groups make up the whole protein content of wastewaters. The first group is soaking. The second group is liming-washing-washing after fleshing. The third group is deliming-bating-washing. The fourth group is pickling.

The main protein source is the liming group. However, other groups were also taken into consideration in the analysis program to determine their protein properties and to assess their effect on protein removal when they are combined with other sources. The liming group is important since it contains the majority of the proteins as well as a significant proportion of particulate organic nitrogen. Two additional composite flow weighted samples were prepared to assess protein removal. Composite I was made up of soaking-liming-washing after fleshing and deliming-bating. Composite II was prepared as the mixture of liming-washing, deliming-bating-washing. The purpose of making up these composites was to evaluate the removal of other nitrogen species together with protein removal. Composite I contained as much protein as the liming group but it also had a very high percentage of particulate nitrogen. This composite also contained a high ammonia percentage to assess ammonia removal by MAP precipitation. Composite II had still a high protein content together with a significant ammonia load while its volume was one third of Composite I providing easy and economic operation. Since Composite I contained the

Table 1 Source-based characterization

Source	Stream m³ ton hide⁻¹	pH	Protein mg l⁻¹	NH_3-N mg l⁻¹	TKN mg l⁻¹	Soluble TKN mg l⁻¹
Soaking	0.9	8.93	3,135	95	855	580
Liming	0.6	12.20	135,665	70	7,390	7,200
Washing	0.3	12.60	21,710	10	1,255	970
Washing after fleshing	2.925	9.92	4,040	395	780	630
Deliming and bating	0.325	8.65	7,705	4,930	6,270	5,895
Washing	0.43	8.58	1,260	1,790	1,960	1,940
Pickling	0.36	3.15	2,935	1,560	1,865	1,850
Tanning	0.5	3.48	3,340	900	1,540	1,390

Table 2 Share of loads and volumes as percentage of total wastewater

	Flow	Ammonia	Protein	Soluble N	Org. N	Particulate N
Composite I	34	62	91	77.5	88	84
Composite II	12	52	84	66.4	70	30
Group I	6.5	2	2.6	4.7	9.1	23
Group II	27.5	27	92	58.2	78	58
Group III	5.4	51	2.8	24.5	6.8	12
Group IV	2.6	12	0.9	6	1.4	0.45

highest protein together with the highest particulate nitrogen and approximately two thirds of ammonia load, emphasis was given to this composite in the experimental program. Table 3 shows Composite I and Composite II characterizations. Composite II had a stronger character than Composite I except for ammonia concentration. However the pH of Composite II was very high indicating a higher acid demand for acidification.

Results of experiments
Acidification for protein precipitation

In these experiments acidification of samples was carried out within the selected pH intervals to determine the most suitable pH for the isoelectric point and minimum protein solubility. Group sample acidification results are given in Table 4. For Group I (soaking) optimum pH was around 2.5 providing about 60% protein removal. For Group II (liming) optimum pH was closer to pH 3.0 and protein removal of near 60% was obtained. As the pH increased, a sharp decrease in efficiency was observed. However at pH 3.0 the removal was still over 60%. In Group III a limited removal could be obtained around pH 2.0. At and above pH 2.5 practically no removal was obtainable. In Group IV minimum solubility was at the original pH of the sample. Above and below this pH significant solubility of the existing proteins occurred resulting in several times higher protein concentrations than the initial concentration. Four group sample protein solubilities exhibited different characteristics. Group I and IV optimum pH values of 2.5 and 3.1 were relatively close. This is expected from the structure of proteins, both being mostly globular, but their types are different. Group I did not show a sharp change in solubility while Group IV was extremely sensitive to pH change. The removal percentage in Group I was found to be consistent with the studies with globular proteins (Fisher et al., 1986). Group II and III sample results exhibited closer behavior as expected from the sources, both being keratin originated proteins although deliming-bating proteins were relatively more soluble.

The results of acidification experiments for Composite I and II are outlined in Table 5. Composite I sample exhibited more than one optimum between pH 2.1 and 3.6. This may be due to the mixture of different types of proteins. It is rather difficult to juxtapose the group sample results with composite samples, furthermore the mixture may result in a different

Table 3 Characterization of composite samples

Parameter	Unit	Composite I	Composite II
pH	–	9.60	11.05
TKN	mg l^{-1}	2,010	4,100
Soluble TKN	mg l^{-1}	1,695	3,500
NH_3-N	mgN l^{-1}	820	665
Calcium	mg l^{-1}	1,045	1,350
Protein	mg l^{-1}	18,620	43,490

Table 4 Results of acidification experiments for group samples

Group I		Group II		Group III		Group IV	
pH	Protein mgl^{-1}	pH	Protein mgl^{-1}	pH	Protein mgl^{-1}	pH	Protein mgl^{-1}
7.99*	3,135*	10.8*	15,610*	8.26*	5,930*	3.15*	2,935*
1.92	1,245	2.12	7,015	2.11	3,680	1.80	7,480
2.58	1,220	3.16	6,100	2.56	5,680	3.10	2,930
3.02	1,400	4.13	7,660	2.93	5,785	4.60	9,350
3.45	1,535	4.83	9,705	4.04	5,930	7.00	9,200
3.95	1,800					10.10	8,275

* Original character of sample

composition or proportions. However, the optimum pH range was close to those of groups II and III. The extension of the pH range may be partly due to soaking wastewater. At any rate having such a wide range of pH all with around 50% protein removal is an advantage. The response of Composite II to acidification was different; in that after having a maximum removal at pH 2.2, removal sharply decreased. Then its behavior resembles groups II and III as far as both the optimum pH and sudden change in solubility with varying pH are concerned. However, similar to Composite I, there seems to be another peak, if not maximum, at pH 4.2. This sample requires very close control of pH to ensure high efficiencies.

Precipitation with metals

Composite I was precipitated using $ZnSO_4$ and $FeCl_3$. The results of the experiments are shown in Table 6. In case of zinc at 1,000 mg/l dosage higher efficiencies were obtained around pH 4.0, however protein removals around this pH were at the level of acidification results. The difference began after pH 4.0, where acidification did not work, and it was the pH range preferred for metal precipitation. Around pH 6.0, 32% protein removal was obtained. These removals are consistent with the levels given in the literature for zinc precipitation (Iyer and Przybycien, 1996). Increasing the zinc concentration did not prove useful even at pH 4.0. Excellent results were obtained with ferric chloride as seen from Table 6. The removal, beyond the area in which acidification is effective, at pH 5.15 was 62%. At pH 3.28 where the effect of $FeCl_3$ was presumed via coagulation, around 70% removal was obtained.

Precipitation with polyelectrolytes

Anionic and cationic polyelectrolytes were tried on Composite I at different pH values and with economically acceptable dosages. Anionic polyelectrolyte did not work at the original pH of the sample (8.75) at 50 mgl^{-1} dosage. At pH 6.84 and with 100 mgl^{-1} only a slight removal could be obtained as seen in Table 6. The bridging effect of the anionic polyelectrolyte at pH 3.70 and for 50 mgl^{-1} dosage was also far from being satisfactory. Cationic polyelectrolyte also did not work at the original pH of the sample at 25 mgl^{-1} dosage. Its coagulation effect between pH values 2.94–3.35 and for dosages 50–100 mgl^{-1} did not prove useful.

Table 5 Results of acidification experiments

Composite I				Composite II			
pH	Protein mg l^{-1}	pH	Protein mg l^{-1}	pH	Protein mg l^{-1}	pH	Protein mg l^{-1}
2.10	9,095	3.60	9,175	2.26	13,010	3.45	40,585
2.42	11,330	3.84	10,750	2.69	35,290	3.75	41,490
2.99	9,380	4.31	13,370	3.23	37,820	4.20	34,645
3.16	10,655						

Table 6 Results of precipitation with metals and polyelectrolyte aided precipitation

	Dosage mg l^{-1}	pH	Protein mg l^{-1}		Dosage mg l^{-1}	pH	Protein mg l^{-1}
$ZnSO_4$	1000	3.96	10,490	$FeCl_3$	1,000	3.28	5,730
		4.40	10,690			5.15	7,110
		4.74	11,675	853 BC	50	3.35	12,790
		5.39	12,195		100	2.94	12,370
		5.96	12,670	2440	50	3.70	14,420
		6.33	12,985		100	6.84	17,580
$ZnSO_4$	2000	4.00	15,540				

Effect of mixing

Table 7 shows the effect of mixing conditions and reagent concentration on protein precipitation by acidification for Composite I. For the constant flash mixing of 5 min and 15 min constant flocculation conditions, variation of flash mixing intensity did not affect the precipitation efficiency. For constant flash mixing conditions and constant flocculation duration, flocculation intensity had an effect on the efficiency indicating the optimum intensities being between 20–30 rpm. Increasing the flocculation time from 15 to 30 min did not practically affect the efficiency. Decreasing the flash mixing time to 2 min also had no effect. Addition of more concentrated acid did not seem to affect precipitation efficiency. Effect of mixing was also evaluated for zinc precipitation. For pH values 5.68 and 5.07, flash mixing intensity was reduced to 80 rpm for 2 min mixing and increased to 120 rpm for 5 min mixing, also flocculation time was increased to 30 min at 20 rpm. Protein removal efficiencies with regard to Table 6 values changed only within the interval of 5%.

MAP precipitation

MAP precipitation was applied to Composite I in two different modes. In the first application where MAP precipitation was carried out at pH 9.5 and with stoichiometric doses of chemicals on the supernatant of acid precipitation, it did not prove effective. Ammonia concentration as well as protein concentration was higher than expected. These low efficiencies could be explained by resolubilization of proteins existing in the supernatant in colloidal form. The second application where the order is reversed was found efficient. Results of MAP and following acid precipitation are given in Table 8. MAP precipitation had practically no effect on soluble nitrogen. Ammonia removal was about 85 percent. Residual ammonia concentration around 100 mgl^{-1} was consistent with the literature results given for MAP applications on leather tanning wastewaters (Tünay et al., 1997). Acidification of MAP precipitation affected ammonia concentration slightly. However, the performance of protein removal was lower than for raw wastewater.

Conclusions

In this paper protein precipitation and recovery potential of leather tanning industry wastewaters were experimentally studied. The main conclusions of the study can be summarized as follows. A protein profile was prepared. Liming and its washings were the main source of protein covering over 90% of the whole. Characteristics of the proteins from different sources were found to vary. Acidification proved to be a reliable and efficient method providing 50–70% protein removal in both source group samples and composites. Optimum

Table 7 Effect of mixing on precipitation

H_2SO_4 N	Flash Mixing		Flocculation		Protein mg l^{-1}	H_2SO_4 N	Flash Mixing		Flocculation		Protein mg l^{-1}
	min	rpm	min	rpm			min	Rpm	Min	rpm	
1	5	80	15	30	10,970	1	5	100	15	50	12,440
1	5	100	15	30	10,530	1	5	100	30	30	11,625
1	5	120	15	30	10,790	1	2	100	15	30	10,665
1	5	100	15	20	11,670	1	2	100	30	30	10,885
1	5	100	15	40	11,945	3	2	100	15	30	10,980

Table 8 Results of MAP precipitation and acidification

	N/Mg/P	pH	Protein mg l^{-1}	NH_3 mgN l^{-1}	STKN mg l^{-1}
MAP precipitation	1/1/1.29	9.44	18,420	130	875
Acidification after MAP precipitation	–	2.9	13,610	120	730

pH was generally low, between 2.0–3.0 for all samples. Composite I was considered to be an appropriate combination for nitrogen and protein removal. In addition to acidification, metal precipitation particularly $FeCl_3$ addition yielded very high removals. Polyelectrolyte addition did not provide satisfactory protein precipitation. Effect of mixing on both acidification and metal precipitation was not significant. MAP precipitation preceded acidification proved very effective removing all particulate nitrogen and 85% ammonia. Combining MAP with protein precipitation a very high nitrogen removal was obtained. Another advantage of this application is the recovery potential of almost pure protein. The results and evaluations of the study provided meaningful starting points for further studies for the protein recovery in leather tanning industry wastewaters.

References

Chan, M.Y.Y., Hoare, M. and Dunnill, P. (1986). The kinetics of protein precipitation by different reagents. *Biotech. and Bioeng.*, **28**, 387–393.

Chen, W., Walker, S. and Berg, J.C. (1992). The mechanism of floc formation in protein precipitation by polyelectrolytes. *Chemical Eng. Sci.*, **47**(5), 1039–1045.

Chen, L.A., Carbonell, R.G. and Serad, G.A. (2000). Recovery of proteins and other biological compounds from food processing wastewaters using fibrous materials and polyelectrolytes. *Wat. Res.*, **34**(2), 510–518.

Englard, S. and Seifter, S. (1990). Precipitation techniques. *Methods in Enzymology*, **182**, 285–301.

Fisher, R.R., Glatz, C.E. and Murphy, P.A. (1986). Effects of mixing during acid addition on fractionally precipitated protein. *Biotech. and Bioeng.*, **28**, 1056–1063.

Gerhardt, P., Murray, R.G.E., Wood, W.A. and Krieg, N.R. (1994). *Methods for General and Molecular Bacteriology*. American Society for Microbiology.

Iyer, H.V. and Przybycien, T.M. (1994). Protein precipitation: effects of mixing on protein solubility. *AIChE Journal*, **40**(2), 349–360.

Iyer, H.V. and Przybycien, T.M. (1995). Metal affinity protein precipitation: effects of mixing, protein concentration, and modifiers on protein fractionation. *Biotech. and Bioeng.*, **48**, 324–332.

Iyer, H.V. and Przybycien, T.M. (1996). A model for metal affinity protein precipitation. *J. of Coll. and Inter. Sci.*, **77**, 391–400.

Shih, Y., Prausnitz, J.M. and Blanch, H.W.(1992). Some Characteristics of protein precipitation by salts. *Biotech. and Bioeng.*, **40**,1155–1164.

Standard Methods for the Examination of Water and Wastewater (1998). 20th ed. American Public Health Association/American Water Works Association/Water Environment Federation, Washington DC, USA.

Thorstensen, T.C. (1985). *Practical Leather Technology*. Robert E. Krieger Publishing Company.

Tünay, O., Kabdaşlı, I., Orhon, D. and Kolçak, S. (1997). Ammonia removal by magnesium ammonium phosphate precipitation in industrial wastewaters. *Wat. Sci. Tech.*, **36**(2–3), 225–228.

Tünay, O., Kabdaşlı, I., Orhon, D. and Cansever, G. (1999). Use and minimization of water in leather tanning processes. *Wat. Sci. Tech.*, **40**(1), 237–244.

Tünay, O., Koca, T. and Kabdaşlı, I. (2001). Protein removal and recovery from leather tanning industry wastewaters. *Fresenius Enviro. Bull.*, **10**(2), 170–173.

Zengin, G., Ölmez, T., Doğruel, S., Kabdaşlı, I. and Tünay, O. (2002). Assessment of source-based nitrogen removal alternatives in leather tanning industry wastewater. *Wat. Sci. Tech.*, **45**(12), 205–215.

Keyword Index

accumulation 147
acidic mine waters 199
activated sludge 191
agricultural use 155
algal ponds 67
Anammox 67, 77, 119

BCFS® 77
beamhouse wastewater 207
biofilm carrier 119
black water 27
brown water 111

Canon 77
citizen focus groups 57
COD 131
constructed wetlands 67
consumer attitudes 57
control and automation 95
crystallization 163

denitrification 67
denitrifying dephosphatation 77
dilution 147
draft-tube type reactor 171
duckweed ponds 67

E. coli 131
economic impacts 11
ecosystem impacts 11
energy 1, 103
energy requirements 37
enhanced biological phosphorus removal (EBPR) 87
estuarine and coastal fisheries 11
eutrophication 179
exergy 1
exergy analysis 27

farmers 47
fluidized-bed 163
full-scale control implementation 95

grey water 111

hygiene 19

industrial wastewater 191
in-situ nutrient sensors 95
ion exchange 179, 199

LCA 37
leather tanning industry 215
life-cycle analysis (LCA) 67
lime stabilisation 155

magnesium ammonium phosphate 163
magnesium ammonium phosphate precipitation 215
mail survey 47
membrane bioreactor (MBR) 87, 191, 207
metals recovery 199
methanation 77
microfiltration 87
micropollutants 57
minerals 1

nitrification 67, 119, 131
nitritation 119
nitrogen 27, 37
nitrogen cycle 67
nitrogen removal 171
nutrient management 147
nutrient pollution 11
nutrient recovery and reuse 11
nutrient recycling 19, 155
nutrient removal 95
nutrients 103, 111, 179

participation 47
P-driven Rem Nut® process 179
phosphorus 19, 27, 37, 139, 155
phosphorus content 87
phosphorus recovery 185
phosphorus removal 171
pollution 147
post-treatment 131
potassium fertilisers 199
precipitation 139
protein precipitation 215
protein recovery 215

RBC 131

regulation 19
resource recovery 67
Rottebehaelter 111

sanitation 1
sewage 131
Sharon/Anammox 103
sludge 19, 155, 185
sludge production 207
societal impacts 11
soil 147
source control sanitation 111
struvite 77, 103, 139, 171, 179
supercritical water oxidation 185
supernatant in anaerobic digestion 171
sustainability 1
sustainable nutrient recycling 47, 57

sustainable processes 67
swine 139
synthesis 199

tannery 207
technology transfer 47

UASB 131
urine 103
urine separation 27, 47, 119
urine separation technology 57
urine source separation 37

wastewater 1, 19, 103
wastewater treatment 67

yellow water 111

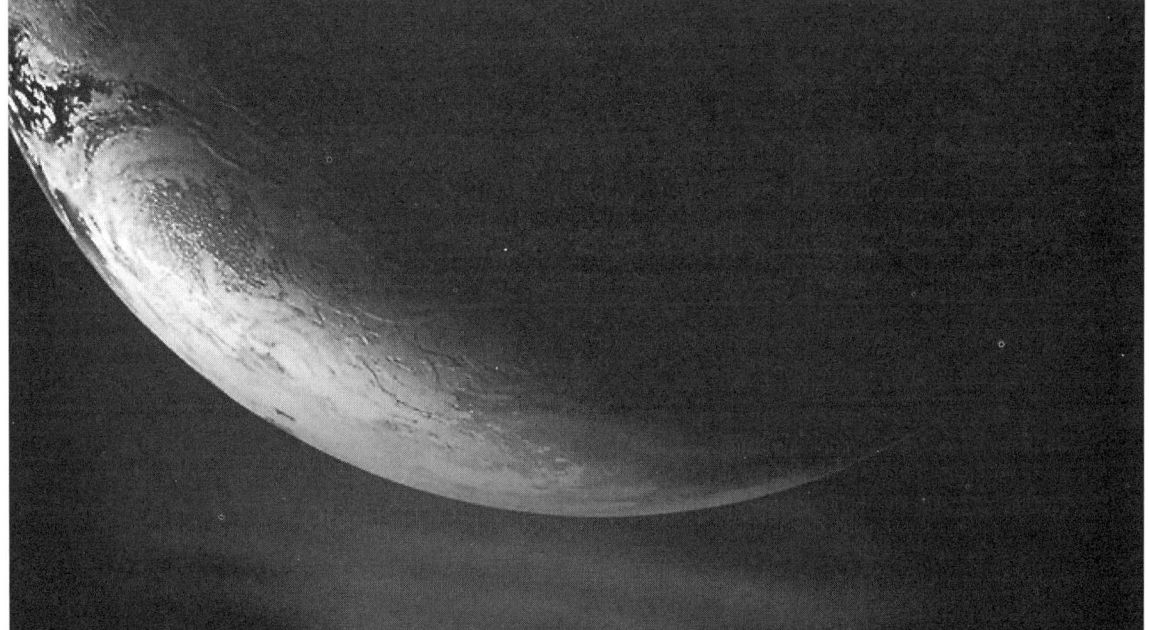

We're all about water

The **International Water Association** (IWA) is a global network of water professionals at the forefront of advancing research and practice, covering all facets of the water cycle

Why you need to be a member...

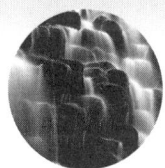

❖ **connect** into IWA's network of professionals and multi-level collaboration

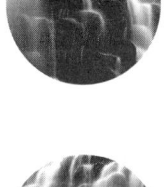

❖ **participate** in biennial congresses, annual Leading Edge conferences on water technology and sustainable urban water management and over 30 speciality conferences every year

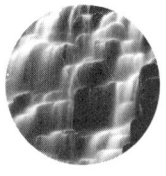

❖ **collaborate** on key issues, develop policy positions and best practice approaches

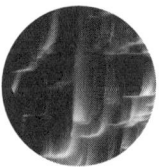

❖ **keep informed** of the latest technical and scientific information with the world's leading journals and information resources on water and wastewater research and practice

JOIN IWA TODAY and become part of a leading-edge network creating innovative, pragmatic and sustainable solutions to global water needs

For information on membership, member benefits or to apply for membership online visit WWW.IWAHQ.ORG.UK, call us on +44 (0)20 7654 5500 or write to us at:

International Water Association
Alliance House
12 Caxton Street
London, SW1H 0QS
United Kingdom

ContentsAlert

A FREE pre-publication alerting service, delivered via email.

This valuable service will enable you to keep up-to-date with articles published in IWA

QUICK

- Journal of Water and Health **NEW**
- Journal of Hydroinformatics

DIRECT

- Journal of Water Supply: Research & Technology-AQUA
- Water Science & Technology
- Water Science & Technology: Water Supply

EASY

Delivers the contents pages of each journal issue approximately 2-4 weeks prior to publication.

Don't miss out on this fast and easy way to keep up-to-date with the latest developments in your field.

Register now at
www.iwapublishing.com

Scientific & Technical Report Series from IWA Publishing

SAMPLING FOR MEASUREMENT OF ODOURS
Authors: P Gostelow, PJ Longhurst, SA Parsons, RM Stuetz
Publication Date: March 2003 · ISBN: 1843390337
IWA Members Price: £37.50 / US$60.00 / €60.00
Non Members Price: £50.00 / US$80.00 / €80.00
Scientific & Technical Report No. 17

ACTIVATED SLUDGE SEPARATION PROBLEMS: THEORY, CONTROL MEASURES, PRACTICAL EXPERIENCES
Editors: V Tandoi, D Jenkins, J Wanner
ISBN: 1900222841 • Pub Date: May 2003
IWA Members Price: £39.75 / US$64.00 / €64.00
Non Members Price: £53.00 / US$85.00 / €85.00
Scientific & Technical Report No. 16

INSTRUMENTATION, CONTROL & AUTOMATION IN WASTEWATER SYSTEMS
Editors : G Olsson, M Nielsen, A Lynggaard Jensen, Z Yuan
ISBN: 1900222833 • Pub Date: July 2003
IWA Members Price: £39.75 / US$64.00 / €64.00
Non Members Price: £53.00 / US$85.00 / €85.00
Scientific & Technical Report No. 15

SOLIDS IN SEWERS
Author: The Sewer Systems and Processes Working Group of the IWA/IAHR Joint Committee on Urban Drainage
ISBN: 1900222914 • Pub Date: July 2003
IWA Members Price: £39.75 / US$64.00 / €64.00
Non Members Price: £53.00 / US$85.00 / €85.00
Scientific & Technical Report No. 14

ANAEROBIC DIGESTION MODEL NO. 1 (ADM1)
Authors: IWA Task Group for Mathematical Modelling of Anaerobic Digestion Processes
ISBN: 1900222787 • Pub Date: Feb 2002
IWA Members Price: £37.50 / US$60.00 / €60.00
Non Member Price: £50.00 / US$80.00 / €80.00
Scientific & Technical Report No. 13

RIVER WATER QUALITY MODEL NO. 1
Editors: P Reichert, D Borchardt, M Henze, W Rauch, P Shanahan, L Somlyody, PA Vanrolleghem (IWA Task Group on River Quality Modelling)
ISBN: 1900222825
IWA Members Price: £42.75 / US$68.00 / €68.00
Non Members Price: £57.00 / US$91.00 / €91.00
Scientific & Technical Report No. 12

RESPIROMETRY IN CONTROL OF THE ACTIVATED SLUDGE PROCESS: BENCHMARKING CONTROL STRATEGIES
Editors: JB Copp, H Spanjers & PA Vanrolleghem
ISBN: 1900222515 • Pub Date: June 2002
IWA Members Price: £44.25 / US$71.00 / €71.00
Non Members Price: £59.00 / US$94.00 / €94.00
Scientific & Technical Report No. 11

SEQUENCING BATCH REACTOR TECHNOLOGY
Editors: PA Wilderer, RL Irvine, MC Goronszy
ISBN: 1900222213
IWA Members Price: £39.00 / US$62.00 / €62.00
Non Members Price: £52.00 / US$83.00 / €83.00
Scientific & Technical Report No. 10

ACTIVATED SLUDGE MODELS ASM1, ASM2, ASM2D & ASM3
By the IWA Task Group on Mathematical Modelling for Design and Operation of Biological Wastewater Treatment
ISBN: 1900222248
IWA Members Price: £43.50 / US$70.00 / €70.00
Non Members Price: £58.00 / US$93.00 / €93.00
Scientific & Technical Report No. 9

CONSTRUCTED WETLANDS FOR POLLUTION CONTROL
By the IWA Specialist Group on Use of Macrophytes in Water Pollution Control
ISBN: 1900222051
IWA Members Price: £45.00 / US$72.00 / €72.00
Non Members Price: £60.00 / US$96.00 / €96.00
Scientific & Technical Report No. 8

RESPIROMETRY IN CONTROL OF THE ACTIVATED SLUDGE PROCESS: PRINCIPLES
Editors: H Spanjers, PA Vanrolleghem, G Olsson, PL Dold
ISBN: 1900222043
IWA Members Price: £26.25 / US$42.00 / €42.00
Non Members Price: £35.00 / US$56.00 / €56.00
Scientific & Technical Report No. 7

SECONDARY SETTLING TANKS
Authors: GA Ekama, JL Barnard, FW Gunthert, P Krebs, JA McCorquodale, DS Parker, EJ Wahlberg
ISBN: 1900222035
IWA Members Price: £41.25 / US$66.00 / €66.00
Non Members Price: £55.00 / US$88.00 / €88.00
Scientific & Technical Report No. 6

MICROBIAL COMMUNITY ANALYSIS
Editors: TE Cloete, NYO Muyima
ISBN: 1900222027
IWA Members Price: £26.25 / US$42.00 / €42.00
Non Members Price: £35.00 / US$56.00 / €56.00
Scientific & Technical Report No. 5

For more information or to order online please visit:
www.iwapublishing.com

ORDER FROM IWA PUBLISHING'S DISTRIBUTOR
Portland Customer Services, Commerce Way, Colchester, CO2 8HP, UK
Tel: 44 (0)1206 796 351   Fax: 44 (0) 1206 799 331   Email: sales@portland-services.com

www.iwapublishing.com

IWA Publishing, Alliance House, 12 Caxton Street, London, SW1H 0QS, UK.
Tel: +44 (0) 20 7654 5500, Tel: +44 (0) 20 7654 5555, Email: publications@iwap.co.uk

New Journal from IWA Publishing for 2003
Journal of Water and Health

Editors:

Charles P. Gerba, *Department of Soil, Water and Environmental Science, University of Arizona, USA*

Paul R. Hunter, *Medical School, University of East Anglia, UK*

Gillian D. Lewis, *School of Biological Sciences, University of Auckland, New Zealand*

Paul Jagals, *Faculty of Health and Environmental Sciences, Technikon Free State, South Africa*

Journal of Water and Health is a new peer-reviewed journal devoted to the dissemination of information on the health implications and control of waterborne microorganisms and chemical substances in the broadest sense. This is to include microbial toxins, chemical quality and the aesthetic qualities of water.

Contributions will be published on the health-related aspects of the following areas:
- epidemiology
- risk assessment
- detection and ecology of pathogens in the environment
- water and wastewater treatment
- disinfection
- disinfection by-products
- indicators of water and waste quality
- regulatory issues and standard development
- water quality surveys
- monitoring
- microbial toxins (including cyanobacteria)
- chemical and physical quality of water as it effects human and animal health
- endocrine disruptors
- taste and odour
- impacts of water quality on food quality
- impact of climate change on water quality

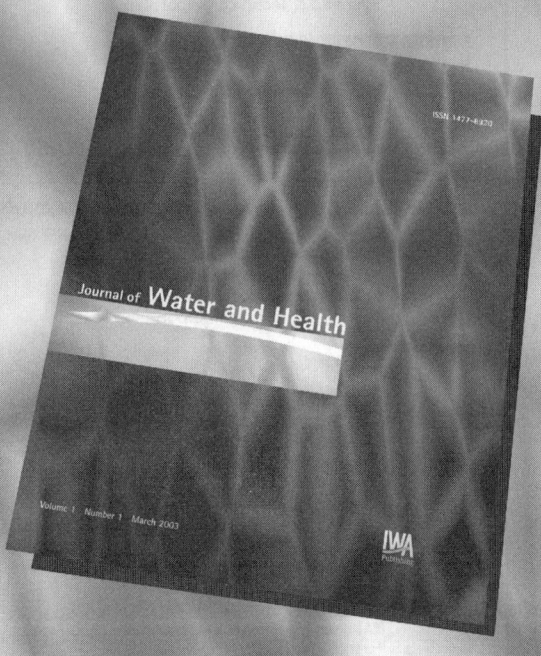

Interested in submitting a paper?
Guidelines for authors are available on our website at www.iwapublishing.com or contact Emma Gulseven, Tel: 44 (0)20 7654 5500, Email: egulseven@iwap.co.uk

To join the mailing list
For further information and a sample copy Email: waterhealth@iwap.co.uk with your full name, address and email.

Sign up for ContentsAlert
ContentsAlert is a free pre-publication alerting service, delivered via email. ContentsAlert will keep you up-to-date with articles published in IWA Publishing journals. ContentsAlert delivers the contents pages of each journal via email approx 2-4 weeks before publication. Sign up via www.iwapublishing.com

Subscription Information:
ISSN: 1477-8920 Volume 1, 4 Issues, 2003
Print plus free online access: £199/US$299/€330

IWA Members
Substantial discounts for IWA members visit www.iwahq.org.uk for more details.

To subscribe contact IWA Publishing's distributor Portland Customer Services, Commerce Way, Colchester, CO2 8HP, UK.
Tel: 44 (0)1206 796351 Fax: 44 (0)1206 799331 Email:sales@portland-services.com

www.iwapublishing.com